Tourism, Hospitality & Event Management

This book series covers all topics relevant to the tourism, hospitality, and event industries. It includes destination management and related aspects of the travel and mobility industries as well as effects from developments in information and communication technologies. "Tourism, Hospitality & Event Management" embraces books both for professionals and scholars, and explicitly includes undergraduate and advanced texts for students. In this setting, the book series reflects the close connection between research, teaching, and practice in tourism research and tourism management and the related fields.

More information about this series at https://link.springer.com/bookseries/15444

João Leitão • Vanessa Ratten • Vitor Braga
Editors

Tourism Entrepreneurship in Portugal and Spain

Competitive Landscapes and Innovative Business Models

Springer

Editors
João Leitão
Department of Management and Economics, NECE - Research Center in Business Sciences
University of Beira Interior
Covilhã, Portugal

Vanessa Ratten
Department of Management, Sport and Tourism, La Trobe Business School
La Trobe Univesity
Melbourne, VIC, Australia

Vitor Braga
School of Technology and Management
Polytechnic Institute of Porto
Felgueiras, Portugal

ISSN 2510-4993 ISSN 2510-5000 (electronic)
Tourism, Hospitality & Event Management
ISBN 978-3-030-89231-9 ISBN 978-3-030-89232-6 (eBook)
https://doi.org/10.1007/978-3-030-89232-6

This Springer imprint is published by the registered company Springer Nature Switzerland AG.
The registered company address is: Gewerbestrasse 11, 6330 Cham, Switzerland

Acknowledgements

We thank Prashanth Mahagaonkar for this helpful feedback and guidance with this volume.

We would also like to thank the following people:

I thank the University of Beira Interior and the NECE Research Center in Business Sciences for their support to this research endeavour. This editorial project is only possible due to my family, friends, and research partners, thanks to their support during my lifetime.

The Portuguese Foundation for Science and Technology (Grants and NECE—UIDB/04630/2020) provided support for this study.

João Leitão

I dedicate this book to my mum Kaye Ratten. I am thankful to have fond memories of our travels together including to the Xian Terracotta Warriors, Beijing, Bangkok, and Angkor Wat, amongst other places. I grew up listening to my mum talk about her time in Kampuchea (now Cambodia) and admiring the temple rubbings at home. It is thanks to my mum that I travelled to these places and had the education that I did. Thank you. I also thank my dad David Ratten, brothers Hamish Ratten and Stuart Ratten, sister-in-law Tomoki Ratten, and niece Sakura Ratten. Thank you.

Vanessa Ratten

I thank the School of Technology and Management, Polytechnic of Porto, for the institutional support, in particular my colleagues in the Business Sciences Department for their enthusiasm and constant willingness to help. I also thank my family and close friends, for their honest support throughout my life.

This work has been supported by national funds through FCT—Fundação para a Ciência e Tecnologia through project UIDB/04728/2020.

Vítor Braga

Contents

List of Contributors

João Almeida GOVCOPP, DCSPT, University of Aveiro, Aveiro, Portugal

João Alves Institute Polytechnic of Porto, School of Hospitality and Tourism, Vila do Conde, Portugal

Xavier Amat-Montesinos Universidad de Alicante, Alicante, Spain

Roberto Antunes NEST, Tourism Innovation Center, Covilhã, Portugal

Francisco Banha Universidadedo Algarve (FEUALG), CEFAGE-UÉvora and (CIEO)-UALG, Faro, Portugal

Francisco Miguel Banha Universidade Católica Portuguesa, Lisbon, Portugal

Fernando Belezas DEGEIT, University of Aveiro, Aveiro, Portugal

José Luís Braga Institute of Higher Studies of Fafe (IESF), Medelo, Portugal

Vitor Braga CIICESI, School of Technology and Management, Institute Polytechnic of Porto, Felgueiras, Portugal

Adília Cabral Polytechnic Institute of Coimbra, Coimbra, Portugal

Ana María Campón-Cerro Universidad de Extremadura, Cáceres, Spain

Alexandra Correia IPCA – Polytechnic Institute of Cávado and Ave, São Martinho, Portugal

Carlos Cortes Universidad de Alicante, Alicante, Spain

Helder Gomes Costa Production Engineering (School of Engineering/UFF), Federal Fluminense University (UFF), Rio de Janeiro, Brazil

Luz María Martín Delgado University of Extremadura, Badajoz, Spain

Paulo Duarte University of Beira Interior, Covilhã, Portugal

Cristina Estevão University of Beira Interior, Covilhã, Portugal

Edgar Tarrés Falcó University of Girona, and Co-Founder at the Institute of Silence, Girona, Spain

Elisabete Gomes Santana Félix Department of Management, University of Évora and CEFAGE-UÉ, Évora, Portugal

Ronnie Figueiredo NECE (Research Center in Business Sciences, University of Beira Interior—UBI), School of Business and Social Sciences, Universidade Europeia, Lisbon, Portugal
Spinner Innovation Centre, Lisbon, Portugal

Manuel Fonseca IPCA – Polytechnic Institute of Cávado and Ave, São Martinho, Portugal

Juan Ignacio Rengifo Gallego Department of Art and Science of the Territory, University of Extremadura, Badajoz, Spain

André Rui Graça Universidade da Beira Interior/Centro de Estudos Interdisciplinares - Universidade de Coimbra, Coimbra, Portugal

Jose Antonio Larrosa Universidad de Alicante, Alicante, Spain

João Leitão NECE—Research Center in Business Sciences, University of Beira Interior, Covilhã, Portugal
ICS, Instituto de Ciências Sociais, University of Lisbon, Lisbon, Portugal
CYFE—Center for Young and Family Enterprise, Università Degli Studi di Bergamo, Bergamo, Italy

Celso Lopes Department of Art and Science of the Territory, University of Extremadura, Badajoz, Spain
NECE - Research Center in Business Sciences, University of Beira Interior, Covilhã, Portugal

Alberto Lorente Universidad de Alicante, Alicante, Spain

Hugo Machado IPCA – Polytechnic Institute of Cávado and Ave, São Martinho, Portugal

Antonio Martínez-Puche Universidad de Alicante, Alicante, Spain

José Manuel Sánchez Martín University of Extremadura, Badajoz, Spain

Raquel Mendes IPCA – Polytechnic Institute of Cávado and Ave, São Martinho, Portugal

Samuel Ortiz Universidad de Alicante, Alicante, Spain

Diana Pereira IPCA – Polytechnic Institute of Cávado and Ave, São Martinho, Portugal

María Cristina Rodríguez Rangel University of Extremadura, Badajoz, Spain

Vanessa Ratten Department of Management, Sport and Tourism, La Trobe Business School, La Trobe University, Melbourne, VIC, Australia

Marcelino Sánchez Rivero University of Extremadura, Badajoz, Spain

Raysa Geaquinto Rocha NECE—Research Center in Business Sciences, University of Beira Interior, Covilhã, Portugal

Isabel Sancho Universidad de Alicante, Alicante, Spain

Cândida Silva Institute Polytechnic of Porto, School of Hospitality and Tourism, CITUR, Vila do Conde, Portugal
Algoritmi Research Center, University of Minho, Braga, Portugal

Oscar Silva Polytechnic Institute of Higher Studies of Gaya (ISPGAYA), Vila Nova de Gaia, Portugal

Susana Silva Institute Polytechnic of Porto, School of Hospitality and Tourism, CEOS.PP and CITUR, Vila do Conde, Portugal

Raquel Reis Soares UNIDCOM-IADE (Design and Communication Research Unit), Cinturs (Research Centre for Tourism, Sustainability and Well-being), Business and Law Research Unit, Universidade Europeia, Lisbon, Portugal

Bruno Sousa IPCA – Polytechnic Institute of Cávado and Ave, São Martinho, Portugal

Carlos Manuel Romão Travanca Department of Management, University of Évora, Évora, Portugal

Laurentina Vareiro IPCA – Polytechnic Institute of Cávado and Ave, São Martinho, Portugal

Carlos Manuel Rodrigues Vieira Department of Economics, University of Évora and CEFAGE-UÉ, Évora, Portugal

Uci Yuliati Universitas Muhammadiyah Malang, Malang, Indonesia

The Role of Tourism Entrepreneurship in the Iberian Peninsula

João Leitão, Vanessa Ratten, Vitor Braga, and Roberto Antunes

Abstract This chapter highlights the importance of tourism for the sustainable development and economic growth of the Iberian Peninsula, both as a destination market and as a destination for investment and technological innovation, particularly through the accelerated digitalization of business models. The importance of fostering opportunity entrepreneurship is also justified, replacing the traditional forms of necessity entrepreneurship, as a lever for growth in the regions of the Iberian Peninsula, including those with low population density, which are opportunities to be explored with regard to the creation of new entrepreneurial units linked to quality of life, wellness, water and air quality, and spirituality. An agenda of business opportunities to be explored is presented, regarding the intensification of the use of new information and communication technologies in the tourism sector, tending to the digitalization of business models. The remaining 13 chapters that make up this volume, which is truly innovative in terms of mapping entrepreneurial initiatives in the tourism sector, as well as analysing the expectations of consumers of tourism

J. Leitão (✉)
NECE—Research Center in Business Sciences, University of Beira Interior, Covilhã, Portugal

ICS, Instituto de Ciências Sociais, University of Lisbon, Lisbon, Portugal

CYFE—Center for Young and Family Enterprise, Università Degli Studi di Bergamo, Bergamo, Italy
e-mail: jleitao@ubi.pt

V. Ratten
Department of Management, Sport and Tourism, La Trobe Business School, La Trobe University, Melbourne, VIC, Australia
e-mail: v.ratten@latrobe.edu.au

V. Braga
CIICESI, School of Technology and Management, Institute Polytechnic of Porto, Felgueiras, Portugal
e-mail: vbraga@estg.ipp.pt

R. Antunes
NEST, Tourism Innovation Center, Covilhã, Portugal

J. Leitão et al. (eds.), *Tourism Entrepreneurship in Portugal and Spain*, Tourism, Hospitality & Event Management, https://doi.org/10.1007/978-3-030-89232-6_1

services against the backdrop of the COVID-19 pandemic crisis, are briefly presented.

The importance of tourism for the Iberian economy is unquestionable, above all through exploitation of the natural resource capital favourable to exporting services and attracting flows of people, expenditure, and investment. The recent visibility of tourism entrepreneurship in the context of the global pandemic crisis, as well as the importance attributed to innovation and reinforcing the export capacity of tourism services makes it an appropriate time to produce a volume devoted to the subject of tourism entrepreneurship in the Iberian Peninsula. This region includes two countries with a common history and different resources, cultures, and landscapes including the Mediterranean Basin, the Algarve coast and the northern and western coasts of the Atlantic.

In terms of the literature considered relevant, Kibedi (1979) was one of the first authors to discuss tourism entrepreneurship, in the scope of a national study reporting the attempts by the Ministry of Industry and Tourism in Ontario (Canada) to train and educate business-people in tourism. However, it should be underlined that this remains a little-explored topic in the literature on entrepreneurship, as the presence of business activity in the sector is generally dealt with in a very wide-ranging way (Koh & Hatten, 2002). Entrepreneurship stresses opportunities to create and manage business initiatives (Zhao et al., 2011), and so is of considerable importance in the formation of tourist destinations. The study by Russel and Faulkner (2004) found a significant relation between innovative individuals (entrepreneurs) and a tourist destination's development. Also noted is the importance given to tourism entrepreneurship, especially in rural areas (Hernández-Maestro & González-Benito, 2014). These authors indicated that here, it is not just one more form of subsistence entrepreneurship, but rather the entrepreneurship of opportunity, pointing towards more evolved forms of qualified entrepreneurship in order to create supplementary income (Getz & Carlsen, 2000). These require articulation with the production and distribution of endogenous products, directed towards environmental transition, as well as intensive incorporation of new information and communication technology, aiming for full digitalization of business.

Added to this is the need to advance knowledge on forms of female entrepreneurship, especially in rural areas, leading genuine family-based businesses (Tinsley & Lynch, 2001). The presence of women in tourism entrepreneurship is also shown by the predominance of micro-firms linked to tourism, namely hotel and catering units (Getz & Carlsen, 2005). Concerning tourism entrepreneurs' motivation, the literature reveals that objectives related to a more rural, healthy, and sustainable lifestyle are particularly important, together with a desire to live in the village, build social networks and be an integral part of a community with its own identity (Bosworth & Farrel, 2011; Getz & Carlsen, 2000), or promoting endogenous products (Lopes et al., 2018) and certified productions with brands linked to places (Lopes et al., 2021).

These are examples of different motivations; it will be interesting to explore in future studies, as entrepreneurs' preferences seem to be changing towards a culture of greater quality of life and a certain emotional and spiritual balance. The aim is to ensure people's internal balance, something in short supply in places with a greater density of population and economic activities.

In tourism entrepreneurship, the place and territorial component cannot be separated from the product/service, being an integral part of both the tourism supply and demand. This is a differentiating element, not only in the Iberian Peninsula explored here, but also in each and every location, where the place, its bio-diversity, natural resources, customs, and cultural heritage affect the interaction between both sides of the market, making them unique, authentic, and unrepeatable.

In this line of thought, previous studies indicate that the identity of local tourism entrepreneurs has a direct and significant influence on business self-efficiency and the support given to the community, which translates into the improved performance of business units (Hallak et al., 2012, 2015). Few studies focus on the tourism entrepreneur ecosystem, despite finding relevant indications with regard to rural tourism (Dana et al., 2014; Kline et al., 2014), forming the above-mentioned but little-explored dimension of the place.

A good starting point for future studies on tourism entrepreneurship should be based on exploring established theories, such as Transaction Cost Theory, the Theory of Resources and Capacities, Network Theory and Stakeholder Theory. However, new research initiatives can be more ambitious and try to explore the adherence of theoretical approaches included in the body of Entrepreneurship Theory, namely the Theory of Knowledge Filter, the Theory of Dynamic Competences and, more recently, the Theory of Entrepreneurial Ecosystems.

In strategic and operational terms, one of the lines of public policy action to explore in the Iberian Peninsula, as regards tourism entrepreneurship, has to include positioning this differentiated area as a centre of international reference in innovation and entrepreneurship, in relation to environmental and digital transition.

According to the important statement of an invited practitioner, Roberto Antunes, the Executive Director of NEST, Tourism Innovation Center, Portugal, we can expect tourism entrepreneurship to contribute to changing the tourism system. As in society in general, so far there has been an excessive focus on quantifiable results and less on people, so we will have to reformulate everything and focus more on other intangible, sustainable, environmental, emotional, and spiritual aspects.

Regarding environmental transition, and incorporating what we are still learning from the current pandemic, travellers will be more demanding in terms of their choice of airlines, cruises, apartments, hotels, and other providers. In addition, tourism will have to concentrate on sustainability and proximity, to recover. It will be necessary to reduce accommodation and improve the infrastructure in many crowded areas of the Iberian Peninsula, as well as devoting renewed efforts to low-density, peaceful destinations.

Inland tourism will benefit greatly from this new form of development. Spain is moving towards transformative tourism where we must specify the extension of the resort model (traditional model) to urban centres and "exclusive" areas, with tourism activity linked to the maintenance of networks of affection and long-term labour, cultural and social activism collaborations. In turn, Portugal has developed interesting projects concerning historical villages and is launching new initiatives devoted to knowledge of rural villages and interconnectable rural hubs.

Looking ahead, tourism will establish itself as future-oriented industry, able to generate positive sustainable impacts rather than being an opportunity for positive economic shifts. It can become a force for good and will become more professional the more it pushes its strategies to deliver qualitative positive impacts. Furthermore, tourism will force a global discussion more than any other, being probably the third biggest export in the world and requiring more connection and alignment rather than sovereignty and economic egoism.

Portugal and Spain are good examples of tourism entrepreneurship with a developed infrastructure in the service sector, ranking among the most visited countries in the world and attracting millions of travellers annually due to their beaches and pleasant weather, the abundant artistic heritage of both countries and the gastronomy. In addition, these two nations are very rich in terms of cultural heritage at a global level, due to the large number of civilizations that left their mark and legacy on the Iberian Peninsula.

Both can represent a paradigm of being developed economies in which tourism is one of the major industries and provides wealth, well-being, and employment. Especially in Portugal, value is generated by thousands of micro and small companies, and entrepreneurs, and this is an opportunity as it builds tourism with thousands of repertoires, curiosities, and unheard narratives, i.e. what tourists now look for. This does not happen in highly concentrated markets, where the offer is extensive and in the hands of a few. Also promising is their predominance when it comes to achievements in sustainability. Portugal and Spain are well positioned to be worldwide references in this matter.

Regarding digital transition, since the pandemic, the adoption of digital tools has been enormous. Solutions have been met for eliminating distance. Chatbots, QR codes, biometrics reading, and others are now a reality, but also for promoting social inclusion and providing an effective answer to special needs and disabilities. A new reading of the changes in demand has opened up great opportunities for entrepreneurs, for example, to succeed in attracting Digital Nomads, when even during the pandemic, demand has been greater than the supply.

Considering the new trends and the rapid changes occurring, the tourism sector in the Iberian Peninsula should be strengthened, by supporting different forms of tourism entrepreneurship, through planned, articulated activities in the public and private sphere, namely by:

(1) Stimulating an entrepreneurial culture through developing entrepreneurship programmes in the hotel and tourism school network and other initiatives targeting micro, small, and medium-sized companies, and business-people in the sector, as well as supporting events to share good entrepreneurial practices;

(2) Promoting access to finance sources, developing and managing lines of financial support for entrepreneurship, and promoting initiatives to join start-ups and investors;
(3) Facilitating innovation in tourism, giving specialized support to start-ups in terms of knowledge transfer and advice, regarding financing instruments and support for the internationalization of start-ups, through participation in international tourism fairs;
(4) Matching the tourism demand for, and supply of, endogenous products, with brand certification and image, linked to the bio-economy and sustainability;
(5) Promoting the adoption of circular economy and eco-efficiency practices, in relation to Sustainable Development Goals (SDGs) in activities of tourism entrepreneurship; and
(6) Stimulating the entrepreneurial ecosystem, promoting agents' coordination, forming partnerships in developing programmes of horizontal and vertical acceleration in tourism, forming a network of incubators, and developing a network of mentors to promote new ideas and projects in tourism.

As proposed by Leitão et al. (Leitão et al., 2021), on the one hand, the relation between entrepreneurship and innovation, and on the other, the relation between innovation and competitiveness, mean that the study of tourism entrepreneurship is of fundamental importance for tourist regions' sustainable development and resilience. It is imperative to consider tourism as a sector of activity that can promote the crossed fertilization of different sectors of economic activity, contributing to greater differentiation of the industrial vertical chains linked to endogenous resources, and to intelligent diversification of production based on horizontal specialization of economic activities where the Iberian Peninsula has comparative advantages, in terms of scale and cultural heritage. This involves policy-makers, scholars, and above all entrepreneurs, who are the drivers of change.

The present volume contains a collection of 15 chapters presenting diverse contributions on the importance of tourism entrepreneurship in the Iberian Peninsula, providing readers with an enriching journey through new trends in tourism and education directed towards tourism entrepreneurship and innovation. It reveals this ancient and multi-faceted territory as an international reference for entrepreneurship, investment, and experimentation in tourism.

After the foreword, in Chap. 2, Figueiredo, Soares, and Costa provide a thoughtful piece of research, categorizing the actors influencing the drivers of access to digital technology in the ecosystem of innovative tourism services. In addition, the influence of alternative factors or drivers is evaluated, with emphasis on situations where there is more than one driver, and more than one evaluator.

In Chap. 3, Machado, Vareiro, Sousa, and Mendes address rural tourism in an innovative way by making use of both physical and emotional experiences. The authors argue that the personalized encounter between places, people/culture, and guests, characteristic of rural accommodation and the village context, can play a central role in the quality of the tourist experience in these areas.

Chapter 4 contains a challenging journey guided by Rocha, Tarrés Falcó, and Leitão, arguing that spirituality and mindful tourism can enhance economic activities in rural towns and villages with no tourism tradition. The authors use a self-narrative method to describe the second author's experiences in mindful travel projects, courses, congresses, and ventures, and discuss mindful tourism, revealing innovative Hispanic cases, as well as providing guidelines for future investigations and innovative entrepreneurial projects in the Iberian Peninsula.

In Chap. 5, Braga and Silva explain the strategies used by the hosts or owners of manor houses to ensure the self-sustainability of Historic Housing Tourism. The research findings, based on conceptual hypotheses derived from the literature, point out that a substantial number of these houses can no longer depend economically on agriculture, since the primary sector has lost relevance in the local economy. Thus, a new logic of multi-functionality for rural areas is required, taking into consideration the value added through cultural and natural heritage.

Chapter 6, by Pereira, Fonseca, Sousa, and Correia, based on the case of the Iberian Euroregion of Galicia and Northern Portugal, analyses stakeholders' perceptions of cooperation as a strategy to enhance new dynamics for a common tourist destination. The authors analyse several alternative marketing strategies and actions, to enhance the region as a tourist destination.

In Chap. 7, Rengifo Gallego, Sánchez Martín, Sánchez Rivero, Martín Delgado, and Rodríguez Rangel present an original approach to the role played by social networks in promoting rural tourism destinations. The authors study rural accommodation in Extremadura, based on the hypothesis that Facebook promotes the visibility of its rural tourism establishments. A Grouping Analysis is applied both with and without spatial restriction, to determine the presence of spatial clusters. The main findings outline the need to incorporate Facebook actively as a basic tool for better performance derived from the positioning of accommodation.

Chapter 8, by Lopes, Leitão, and Rengifo Gallego, assesses to what extent local agri-food products serve as brands of their places of origin and local drivers of creative tourism. Four case studies are presented: Cereja do Fundão (Cherries of Fundão) and Queijo Serra da Estrela (Serra da Estrela cheese) from the Portuguese Central Region, and Cereza del Jerte (Cherries from Jerte) and Torta del Casar (Torta del Casar cheese) from Spanish Extremadura. The empirical findings reveal that the four agri-food products studied are catalysts for tourism activities and events in their places of origin. In addition, these same products are drivers of a range of creative tourism activities.

In Chap. 9, Almeida and Belezas present an interesting study following the effects of the Covid-19 pandemic, positioning remote work and digital nomadism as new entrepreneurial opportunities that can be used by peripheral territories to attract an emerging target group called half-tourists. Two reference projects are assessed, namely Digital Nomads Madeira (Portugal) and Nomad City Gran Canaria (Spain). Through in-depth interviews with the people implementing these strategies, new insights arise concerning the barriers and best practices to innovate in tourism strategies in peripheral places.

Chapter 10, by Travanca, Vieira, and Félix, examines the impact of both internal (e.g. size, age, growth, and debt) and external factors (e.g. real GDP growth rate, sovereign debt crisis, and location) on the profitability of Portuguese tourism firms. Using a panel data analysis, applied to a sample of micro, small, and medium-sized firms, from 2009 to 2017, the empirical findings reveal that profitability is positively affected by firms' size and age and by real economic growth rates. Conversely, the firm's indebtedness ratio and the sovereign debt crisis denote a negative influence. Firms in the main tourism regions are generally more profitable, and relatively less affected by the sovereign debt crisis, which was particularly harmful to micro-firms.

In Chap. 11, Martínez-Puche, Amat, Cortes, Larrosa, Lorente, Ortiz, and Sáncho study a training programme devoted to Smart Tourism Development, illustrating the process, the content of the training programme, the topics and the methodology used, based on the experience of the University of Alicante, targeted to the development of this type of training programme linked to tourism, reinforcing the importance of an integrated rural and territorial approach, with the local population as a fundamental pillar. Technological aspects must be considered, to ensure that "smart rural spaces" are accessible, connectable, and participatory, in order to improve the tourist experience.

Chapter 12 presents a qualitative study by Silva, Silva, and Alves, who assessed the importance of events in promoting a tourist destination, identifying the themes of the events used for that promotion and the role of events in promoting a tourist destination. This study was conducted by local authorities in the north of Portugal that held entertainment, cultural, sports, and educational events. The main findings revealed that those events attracted plenty of visitors, had a good economic impact on the region, and contributed to increased consumption and greater dissemination of their products, facilitating potential new contacts for business and partnerships.

In Chap. 13, Banha, Graça, and Ganha provide an original contribution devoted to the still unexplored topic of tourism entrepreneurship education. The authors present a case study applied to the Tourism Creative Factory programme, in order to identify the methodologies used to foster and promote an entrepreneurial culture throughout the different regions of the country, but also to assess the part this programme played in attaining the strategic goals set by Portugal Tourism. The analysis of this case study may be able to shape the creation of future evidence-based policy-making around entrepreneurship education or encourage further research related to the history and evolution of tourism in Portugal.

Lastly, Chap. 14, by Estevão, Duarte, Cabral, Campón-Cerro, and Yuliati, assesses safety perception and service needs in the Covid-19 situation by exploring seven critical dimensions. The findings indicate differences between Spanish and Portuguese hotel guests in all categories, with special incidence in hygiene-related factors. The differences vary according to gender, and women seem to be more sensitive to safety as they present more differences than men. Level of education and having children also affect the perception of hotel safety and guests' requirements. The need to prove hygiene and having the hotel certificate for Covid-19 emerged as key factors in restoring the client's trust.

References

Bosworth, G., & Farrel, H. (2011). Tourism entrepreneurs in Northumberland. *Annals of Tourism Research, 38*(4), 1474–1494.

Dana, L.-P., Gurau, C., & Lasch, F. (2014). Entrepreneurship, tourism and regional development: A tale of two villages. *Entrepreneurship & Regional Development, 26*(3–4), 357–374.

Getz, D., & Carlsen, J. (2000). Characteristics and goals of family and owner-operated businesses in the rural tourism and hospitality sectors. *Tourism Management, 21*(6), 547–560. https://doi.org/10.1016/S0261-5177(00)00004-2

Getz, D., & Carlsen, J. (2005). Family business in tourism: State of the art. *Annals of Tourism Research, 32*(1), 237–258. https://doi.org/10.1016/j.annals.2004.07.006

Hallak, R., Brown, G., & Lindsay, N. (2012). The place identity—Performance relationship among tourism entrepreneurs: A structural equation modelling analysis. *Tourism Management, 33*(1), 143–154. https://doi.org/10.1016/j.tourman.2011.02.013

Hallak, R., Assaker, G., & Lee, C. (2015). Tourism entrepreneurship performance the effects of place identity, self-efficacy, and gender. *Journal of Travel Research, 54*(1), 36–51. https://doi.org/10.1177/0047287513513170

Hernández-Maestro, R. M., & González-Benito, Ó. (2014). Rural lodging establishments as drivers of rural development. *Journal of Travel Research, 53*(1), 83–95. https://doi.org/10.1177/0047287513481273

Kibedi, G. (1979). Development of tourism entrepreneurs in Canada. *The Tourist Review, 34*(2), 9–11.

Kline, C., Hao, H., Alderman, D., Kleckley, J., & Gray, S. (2014). A spatial analysis of tourism, entrepreneurship and the entrepreneurial ecosystem in North Carolina, USA. *Tourism Planning & Development, 11*(3), 305–316. https://doi.org/10.1080/21568316.2014.890127

Koh, K., & Hatten, T. (2002). The tourism entrepreneur: The overlooked player in tourism development studies. *International Journal of Hospitality & Tourism Administration, 3*(1), 21–48. https://doi.org/10.1300/J149v03n01_02

Leitão, J., Ratten, V., & Braga, V. (2021). *Tourism innovation in Spain and Portugal: New trends and developments*. Tourism, Hospitality & Event Management Series, Springer. https://doi.org/10.1007/978-3-030-80733-7.

Lopes, C., Leitão, J., & Rengifo Gallego, J. (2018). Place branding: Revealing the neglected role of agro food products. *International Review on Public and Nonprofit Marketing, 15*(4), 497–530. https://doi.org/10.1007/s12208-018-0211-9

Lopes, C., Rengifo Gallego, J., & Correia Leitão, J. (2021). La relación de los productos agroalimentarios de calidad diferenciada con el turismo en España y Portugal. *Boletín de la Asociación de Geógrafos Españoles, 89*. https://doi.org/10.21138/bage.3020

Russel, R., & Faulkner, B. (2004). Entrepreneurship, chaos and the tourism area lifecycle. *Annals of Tourism Research, 31*(3), 556–579. https://doi.org/10.1016/j.annals.2004.01.008

Tinsley, R., & Lynch, P. (2001). Small tourism business networks and destination development. *International Journal of Hospitality Management, 20*(4), 367–378.

Zhao, W., Ritchie, J., & Echtner, C. (2011). Social capital and tourism entrepreneurship. *Annals of Tourism Research, 38*(4), 1570–1593. https://doi.org/10.1016/j.annals.2011.02.006

Tourism Services Innovation Ecosystem: A Multicriteria Model Approach Based on Portuguese Higher Education

Ronnie Figueiredo, Raquel Reis Soares, and Helder Gomes Costa

Abstract In recent years, researchers and practitioners shed light on the "platform ecosystem." The platform ecosystem could create an ecosystem that generates unlimited tourism innovation with an open platform, outside complementors, and various consumers [Inoue and Tsujimoto (Technological Forecasting and Social Change 136:235–253, 2018)]. The main objective of this research is to categorize or sort the actors that influence the drivers to access digital technologies tourism services innovation ecosystem. We evaluate the influence of alternatives over factors or drivers, with emphasis on situations where there is more than one driver and even more than one evaluator. The perceptions were mapped by the board of the Portuguese faculty. This challenge arises in subjective situations where perceptions should be mapped as it occurs in multicriteria decision aid/making situations. Roy (Revue Française d'informatique et de Recherche Opérationnelle 2:57–75, 1968) proposed the ELECTRE (ELimination et Choix Traudusiant La REalité) as a non-compensatory algorithm to face choice dilemmas. The assertive "alternative **a** is not worse than another one **b**." We expect that results may encourage practitioners and researchers to influence decisions to access the innovation ecosystem.

R. Figueiredo (✉)
NECE (Research Center in Business Sciences), University of Beira Interior (UBI), Covilhã, Portugal

Centre of Applied Research in Management and Economics (CARME), Polytechnic of Leiria and Spinner Innovation Centre (SIC), Leiria, Portugal

R. R. Soares
UNIDCOM-IADE (Design and Communication Research Unit) and Cinturs (Research Centre for Tourism, Sustainability and Well-being), Lisbon, Portugal
e-mail: raquel.soares@universidadeeuropeia.pt

H. G. Costa
Production Engineering (School of Engineering/UFF), Federal Fluminense University (UFF), Rio de Janeiro, Brazil
e-mail: heldergc@id.uff.br

J. Leitão et al. (eds.), *Tourism Entrepreneurship in Portugal and Spain*, Tourism, Hospitality & Event Management, https://doi.org/10.1007/978-3-030-89232-6_2

1 Introduction

In management research, the use of the term "platform" has been on the increase (Thomas et al., 2014). And recently, researchers and practitioners have agreed on using "platform ecosystem" (Inoue & Tsujimoto, 2018). Thus, some researchers as Thomas et al. (2014) have clarified this concept by assuming it as the platform (system or architecture) and complementary assets or products. Thomas et al. (2014) present this concept as a result of combining the literature of product family and the market intermediary platforms. Besides, an ecosystem that generates unlimited innovation with an open platform, outside complementors, and several consumers could be formed through this platform ecosystem. Likewise, Inoue and Tsujimoto (2018) pointed out that the competition between two platform ecosystems in the same market and the constructed theoretical models of indirect network effects between complementors and consumers are the focus of many studies. Furthermore, becoming a complementor may be attractive, and various platforms attempt to attract them in order to be more valuable to the other participants, since the platform only offers the basic infrastructure, useful boundary resources, and access to its ecosystem (Kenney et al., 2019).

In recent years, there has been increasing attention regarding the service innovation ecosystem platform and an attempt to promote it to cultivate favorable environments systematically and encourage local innovators to create knowledge and capture business value (Xu et al., 2018). The concept of the service innovation ecosystem platform is receiving mounting attention worldwide. Governments and industrialists are keen to foster innovation ecosystems. Since evaluating the influence of alternatives over factors or drivers is a key element, especially when there is more than one driver and even more than one evaluator, the current research aims to categorize or sort the actors that influence the drivers to access digital technologies services innovation ecosystem. This problem rises in subjective situations where perceptions should be mapped as it occurs in multicriteria decision aid/making situations.

The ELECTRE (ELimination et Choix Traudusiant La REalité) method proposed by Roy (1968) as a non-compensatory algorithm to face choice problems is applied to calculate the degree of concordance (C (a,b)) with the predicative "alternative **a** is not worse than another one **b**." The results highlighted that there is a need to align CEO and other members' perspectives, since the CEO favored a disruptive innovation while the other respondents leaned toward an incremental innovation. Nevertheless, results also demonstrated that this unalignment does not cause relevant disagreement around the perception of the influence of the actors that appear in **A** on the drivers of the innovation ecosystem that compose **F**. Moreover, these results demonstrated that "People" were perceived as the most influencer factor on the innovation ecosystems, for both perspectives, as well as Actor Support, Time, Support from Mentors, and Infrastructure were the actors identified as having positive influence on the innovative ecosystem.

The research is structured as follows: Sect. 2 presents a literature review on the digital technologies services innovation ecosystem. Section 3 outlines the methodology of the research, based on the Multicriteria Decision Modeling, and applying the ELECTRE proposed by Roy (1968). Section 4 discusses the results reached by running the steps described in Sect. 3 and presents the implications of the research. Finally, Sect. 5 identifies the limitations of the research that should be considered in future research.

2 Literature Review

Under enhanced global competition and proliferation of information communications technologies (ICT), economic activities have become more knowledge-intensive services, and industrial economies have accelerated their transition to technological knowledge-based systems (Russell & Smorodinskaya, 2018) to create a quality approach. In services, quality can be identified as the outcomes compared with customers' expectations, and service innovation can be regarded as service outcomes beyond their expectations (Yeh, 2020). Service innovation has been considered a major factor in economic growth and ecosystem development (De Silva et al., 2018).

An ecosystem is defined as "A biological system composed of all the organisms found in a physical environment, interacting with it and each other" (Tsujimoto et al., 2018, p. 16). In recent years, researchers and practitioners shed light on the "platform ecosystem." The platform ecosystem could create an ecosystem that generates unlimited innovation with an open platform, outside complementors, and various consumers (Inoue & Tsujimoto, 2018).

In addition, many platforms try to attract complement providers to make the platform more valuable to other participants. Becoming a complementor may be attractive, as the platform offers the basic infrastructure, useful boundary resources, and access to its ecosystem (Kenney et al., 2019).

The concept of the service innovation ecosystem platform is receiving increasing attention worldwide. Governments and industrialists are keen to foster innovation ecosystems to systematically cultivate favorable environments and encourage local innovators to create knowledge and capture business value. Innovation ecosystems require specific attention when faced with fast-developing and emerging industries that closely link science, technology, and business (Xu et al., 2018).

According to Cukier and Kon (2018), startup ecosystems cannot be analyzed as static entities. Like biological ecosystems, they behave like living organisms and change over time. Some changes are planned or somehow controlled, while others result from unexpected forces acting within and outside the ecosystem. In other words, ecosystems acknowledge that firms do not operate in a vacuum and are in fact embedded in the broader social, cultural, and institutional context that shapes their growth and contributes to their chances of success (Jha, 2018).

Furthermore, a digital service ecosystem is a new kind of self-organized environment that addresses openness and dynamicity, enabling collaborative innovation and co-creation among ecosystem members (Immonen et al., 2016).

Digitalization aids services in manufacturing companies, creating new opportunities for different services, platforms, intelligent products, and novel business models (Kohtamäki et al., 2019).

Together, digital technologies enabled the creation of new business ventures and digital startups and incorporated novel technology as a vital component into business models and operations (Elia et al., 2020).

Similar to what happens in incubators, organizations that promote the development of entrepreneurial companies, help them to survive and develop during the initial phase when they are most vulnerable (Tietz et al., 2015).

In today's information society, the concept of an information ecosystem highlights the connection between the properties of an increasingly complex digital informational environment with many types of data available (e.g., accounting information from companies, information from capital markets, statistics on the evolution of consumer and business sentiment) as per Zorio-Grima and Merello (2020).

The concept of disruptive innovation describes how new technology alters a performance improvement trajectory or redefines how performance is understood (Beltagui et al., 2020). According to Beltagui et al. (2020), most radical innovations or new technologies follow an evolutionary pattern of a combination of pre-existing modules, components, and subsystems.

In contrast, incremental innovation is an innovation with a lower degree of novelty, risk, and cost than radical innovation (Martínez-Ros & Orfila-Sintes, 2009), though with considerably less potential for positive impact on a firm's performance.

Parida and Wincent (2019) stated that trends such as digitalization, the circular economy, services, and their associated new business models present a unique value creation opportunity. However, they also entail several challenges associated with greater risk and responsibility related to significant changes in a firm's ecosystem.

Thus, faculties support knowledge spillovers to improve sustainable entrepreneurial ecosystems that benefit economic revitalization or further the development of regions (Wagner et al., 2019). Moreover, strategies should be implemented in order to promote the participation of the main actors involved in the entrepreneurial and innovation tourism ecosystem to ensure the success of technology transfer policies and legislation (Guerrero & Urbano, 2019).

According to Milwood and Maxwell (2020)), innovation and entrepreneurial ecosystem was defined as a community comprising actors and their activities of production and service, have roots in ecological biology, regional development, or closely related fields of systems, networks, and actors. In addition, the ecosystem concept is becoming increasingly important in the field of technology management (Morant-Martínez et al., 2019). Finally, the digital revolution is radically changing the world we live in (Pencarelli, 2020).

3 Methodology

This section is designed to highlight the main concepts used in the proposal, covering the following streams, technology, and tourism innovation ecosystem:

3.1 Multicriteria Decision Modeling

To evaluate the influence of alternatives over factors or drivers is a relevant issue, with emphasis on situations where there is more than one driver and even more than one evaluator. This problem arises in subjective situations where perceptions should be mapped as it occurs in multicriteria decision aid/making situations.

Roy (1968) proposed ELECTRE (ELimination et Choix Traudusiant La REalité) as a non-compensatory algorithm to face choice problems. The fundamental core of ELECTRE is based on building outranking relations, so an alternative **a** outranks another one **b** if its performance is not the worst in most of the criteria considered. Roy (1968, p. 8) proposed a mathematical formulation to calculate the degree of concordance (C (a,b)) with the assertive "alternative **a** is not worse than another one **b**." It follows this formulation as written in Costa (2016)

$$C(a,b) = \frac{1}{\sum_{j=1}^{n} w_j} \sum_{j=1}^{n} c_j(a,b) w_j \tag{1}$$

Where,

- a and b are generic alternatives
- $C(a,b)$ is the whole concordance degree with the assertive a is not worse than b
- w_j is the weight of the jth criterion
- n is the number of criteria
- $c_j(a, b)$ is a "local" concordance with the assertive a is not worse than b.

Thus, as reported by Roy and Skalka (1985), when comparing crisp values (or using true criterion), the local concordance is calculated by applying the Eq. (2).

$$c_j(a,b) = \begin{cases} 1, if\ g_j(a) \geq g_j(b) \\ 0, \text{otherwise} \end{cases} \tag{2}$$

Where $g_j(a)$ and $g_j(b)$ are, respectively, the performance or grade of the alternatives a and b, in the criterion j. For situations with no crisp values or using pseudo criterion as defined in Roy and Skalka (1985), the local concordance is calculated as it appears in Eq. (3).

Table 1 ELECTRE Family of multicriteria methods

Problematic	Method	References
Choice	ELECTRE I	Roy (1968)
	ELECTRE IS	Roy and Skalka (1985)
Ranking	ELECTRE II	Roy and Bertier (1971)
	ELECTRE III	Roy (1978)
	ELECTRE IV	Hugonnard and Roy (1984)
Sorting	ELECTRE TRI	Yu (1992), Mousseau and Slowinski (1998)
	ELECTRE TRI-C	Almeida-Dias et al. (2010)
	ELECTRE TRI-nC	Doumpos and Figueira (2019)

Source: Adapted from Nepomuceno and Costa (2015)

$$c_j(a,b) = \begin{cases} 1, \textit{if } g_j(a) \geq g_j(b) - q_j \\ \dfrac{g_j(a) - g_j(b) + p_j}{p_j - q_j}, \textit{if } g_j(b) - p_j \leq g_j(a) < g_j(b) - q_j \\ 0, \textit{if } g_j(a) < g_j(b) - p_j \end{cases} \tag{3}$$

Where p_j and q_j are, respectively, preference and indifference thresholds. It is interesting to notice that Eq. (2) is a particular case of Eq. (3) when $p_j = q_j = 0$.

As reported by Rodriguez et al. (2013), the ELECTRE algorithm was deployed in several situations, but only to induce other kind of issues beyond the choice problem. This development has generated the ELECTRE family of outranking multicriteria methods, as noted in Table 1.

Most of the methods that appear in Table 1 are designed for a unique decision unity. That is, if there is more than one evaluator, their evaluations should be pre-processed in order to obtain a unique evaluation for each alternative or object in each criterion.

3.2 *Electre Tri Me*

The methods that appear in Table 1 are non-compensatory ones proposed for decision situations with only one evaluator. Recently, it appears that the ME (multiple evaluators) family of non-compensatory methods, ELECTRE ME (Costa et al., 2018), preserve the non-compensatory features of ELECTRE even when using evaluations that come from several evaluators. A variation of such ME proposal is the ELECTRE TRI ME designed for sorting multicriteria and multiple evaluator outranking approach alternatives or objectors into categories. As reported by Costa and Duarte (2019) and Costa et al. (2020), the assumptions and formulas shown in Table 2 should be adopted when applying ELECTRE TRI ME. After these assumptions, the ELECTRE TRI ME compares each alternative in **a** against each profile in

Table 2 ELECTRE family of multicriteria methods

Element	Formula	Description
Alternatives	A = {a1, a2, . . .,ai. . ., am}	a set composed by m alternatives.
Evaluators	E = {e1, e2,. . .,ej,. . ., en}	a set of n evaluators.
Criteria	F = Fe1UFe2U. . .UFejU . . .UFen-1UFen	**Fe**$_1$ = {k_{1e1}, k_{2e1}, . . ., k_{ve1}} is the subset composed by the v criteria adopted by the evaluator e_1;
		Fe$_n$ = {k_{1en}, k_{2en}, . . ., k_{zen}} is the subset composed by the z criteria adopted by the evaluator e_n
Weight of the criteria	**W** = We$_1$U We$_2$ U. . .UWe$_j$U. . .UWe$_{n-1}$UWe$_n$	**We**$_j$ = {w_{1ej}, w_{2ej}, . . ., w_{yej}} is the subset composed by the criteria weights under the perspective of the evaluator e_j.
		We$_n$ = {w_{1en}, w_{2en}, . . ., w_{zen}} is the subset composed by the criteria weights under the perspective of the evaluator e_n.
Performance of the alternartives	**G**(*a*)=Ge$_1$(*a*)UGe$_2$(*a*) U. . .Ge$_{n-1}$(*a*)UGe$_n$	**Ge**$_1$ (a_i) = {g_{1e1}, g_{2e1}, . . ., g_{ve1}}(a_1) is the subset composed by the performance of alternative a_i under the perspective of the evaluator e_1 and the criteria set adopted by this evaluator.
		Ge$_n$ = {g_{1en}, g_{2en}, . . ., g_{zen}} is the subset composed by the z criteria adopted by the evaluator e_n and the criteria set adopted by this evaluator.
Categories	**C**= {C_1, C_2, . . ., C_h, . . ., C_{p+1}}	**C** is the set of *p* categories in which the alternatives will be sorted. These categories are ranked from the worst (C_1) to the best (C_{p+1}). All evaluators should have the same number or categories, in such away h has the same value for all the evaluators.
Boundaires or profiles	$\mathbf{B}_{kej}$ = {b_1, b_2, . . ., b_h, . . ., b_p}	**B** is a set composed by *p* profiles or boundaries that delimit the categories for a criterion k, under the viewpoint of an evaluator e_j. As the categories are adjacent, b_h is both the superior limit of C_h, and the inferior limit of C_{h+1}.In a more general way, considering that there should exist n evaluators:
		B = Be$_1$U Be$_2$ U . . . UBe$_j$U . . . UBe$_{n-1}$UBe$_n$
		OR
		B = {{b_{1e1}, b_{2e1}, . . ., b_{te1}}, {b_{1e2}, b_{2e2}, . . ., b_{te2}}, . . ., {b_{1ej}, b_{2ej}, . . ., b_{tej}}, . . ., {b_{1en}, b_{2en}, . . ., b_{ten}} (8-b)
		Where **B** is the overall vector of profiles, t is the number of profiles, and e_k, is the k evaluator.

Source: Author's creation

b according to Eq. (4) if using crisp or integer value, or Eq. (5) if using pseudo-criteria.

$$c_j(a, b_h) = \begin{cases} 1, if\ g_j(a) \geq g_j(b) \\ 0, \text{otherwise} \end{cases} \tag{4}$$

$$c_j(a, b_h) = \begin{cases} 1, if\ g_j(a) \geq g_j(b) - q_j \\ \dfrac{g_j(a) - g_j(b) + p_j}{p_j - q_j}, if\ g_j(b) - p_j \leq g_j(a) < g_j(b) - q_j \\ 0, if\ g_j(a) < g_j(b) - p_j \end{cases} \tag{5}$$

After calculating the concordance degree $C\ (a_j, b_h)$, the ELECTRE TRI ME compares such values against a cut-level λ using a top-down or descendent direction (from the lower profile of the upper category to the upper profile of the lower category), and classifies the alternative a_j into the first category for which $C\ (a_j, b_h) \geq \lambda$. Check Table 2.

4 Results

In this section, we present all the results reached by running the steps described previously.

4.1 *Setting the Situation in Which the Modeling Was Applied*

The main objective of this paper is to categorize or sort the actors that influence the drivers to access the tourism innovation ecosystem. As a first step, it is necessary to set the problem situation, which comprises the context, the alternatives, the criteria, and the categories in which the actors will be classified.

The situation in which the modeling was applied is adequate to the proposal because the organization undergoes a major strategic change in the business model, migrating from an "analog" system to a "digital" system, blended learning and hybrid in terms of teaching and learning. Complements and highlights relevance with the proposal of a new concept of innovation and entrepreneurship laboratory, aligned with the organization's strategic drivers.

Thus, Innovation Lab (Ilab) will be considered a strategic tool to support the transition of the business model, focused on innovation drivers, digital technological entrepreneurship, and startups. In its quest for speed and innovation, the tech industry has produced a variety of ways of engaging with startups (Prashantham & Kumar, 2019).

This model was developed to better understand the objects or alternatives, maturity levels, and type of tourism innovations, as it appears in Fig. 1.

The set of objects to be evaluated or alternatives (A) was composed as follows:

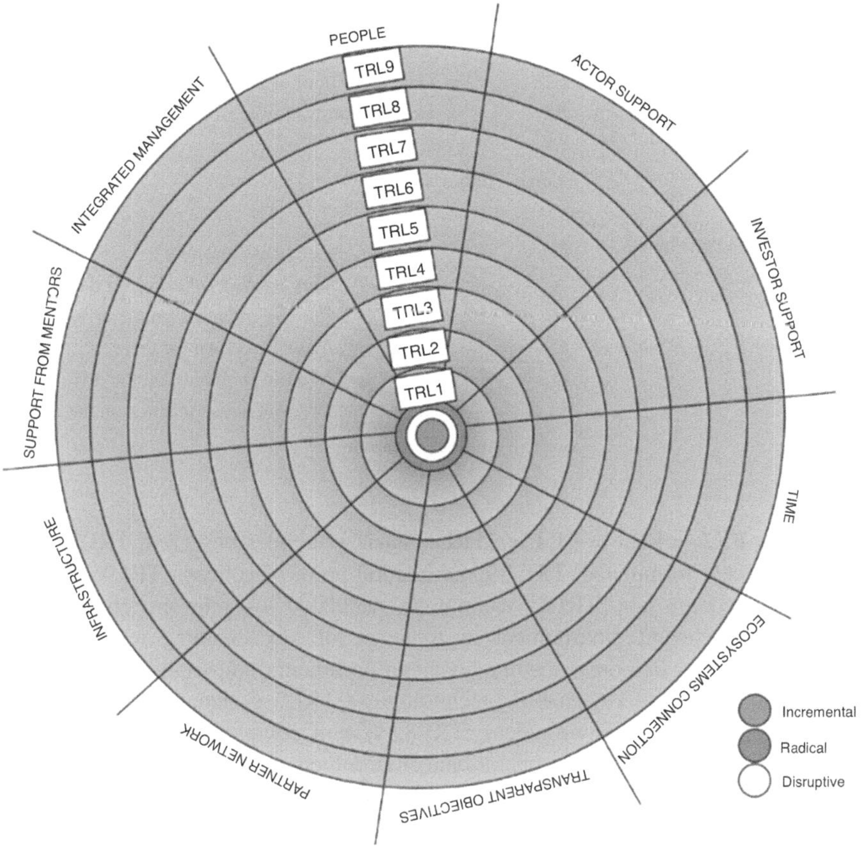

Fig. 1 Drivers to access the digital technologies tourism services innovation ecosystem. Source: Author's creation

A = {People; Integrated management; Actor Support; Investor Support; Mentors Support; Infrastructure; Time; Connection with Ecosystems; Transparent Objectives & Partner Network}.

The choice of the alternatives (A) is due to the fact that these are actors that influence the innovation features of the university, as concluded from the bibliographical review about factors that influence university power to innovate.

The sample of evaluators was composed of three members of the university's decision-making staff and a professor (head of Innovation Lab) closely involved in the business innovation transformation of this university (refer to Table 3).

The innovation maturity levels described in technology readiness levels (TRL's) developed by The National Aeronautics and Space Administration (NASA) were considered as the drivers or the criteria in the multicriteria evaluation mode. As the criteria are set to be the same for all evaluators, we have:

$$\mathbf{F = F_{E1} U F_{E2} U F_{E3} U F_{E4}}$$

Table 3 The sample profiles

Code	Enrollment
E1	Vice-Rector
E2	Dean of School of Business and Social Sciences
E3	Head of Innovation Lab
E4	CEO

Source: Author's creation

Table 4 The Likert-based categories

Code	Linguistic meaning
A	*Very positive influence* on the driver
B	*Positive influence* on the driver
C	*No influence* on the driver
D	*Negative influence* on the driver
E	*Very negative influence* on the driver

Source: Author's creation

Where

FE1 = **FE2** = **FE3** = **FE4** = {TRL1: basic principles observed, TRL2: technology concept formulated, TRL3: experimental proof of concept, TRL4: technology validated in lab, TRL5: technology validated in relevant environment (industrially relevant environment in the case of key enabling technologies), TRL6: technology demonstrated in relevant environment (industrially relevant environment in the case of key enabling technologies), TRL7: system prototype demonstration in operational environment, TRL8: system complete and qualified, and TRL9: actual system proven in operational environment (competitive manufacturing in the case of key enabling technologies, or in space).

Based on the Likert scale, the actors were classified into categories defined as C = {A, B. C, D, E}. The definitions of these categories are presented in Table 4.

4.2 *Collecting Evaluations from the Sample*

After the problem situation was set, the collection of evaluations from the evaluators was performed. A google form (available in https://forms.gle/CFnDNVryZFkBjBAh9) was sent to the sample members. In the first part of the form, we ask the respondents for feedback on what kind of innovation system (incremental, radical, or disruptive) and flow (open or closed) they have in mind when answering the proceeding questions in the form. E1, E2, and E3 answered that they have incremental innovation in mind, while E4 chose the disruptive innovation choice, and all the members of the sample opted for "open" innovation.

In the second part of the form, we ask the members of the sample about their perception regarding the influence of the actors that appears in A over the drivers that compose F. They emitted their answers, taking into account the Likert-based scale that appears in Table 5. In the first column of this table, we have the linguistic value

Table 5 The Likert-based scale adopted for collecting the evaluations

Linguistic meaning	Numerical value
The actor has a very positive influence on the driver	2
The actor has a positive influence on the driver	1
The actor has no influence on the driver	0
The actor has a negative influence on the driver	−2
The actor has a very negative influence on the driver	−1
If the evaluator had not understood or if he/she were not sure how to answer the question, he/she had the choice to leave the answer as" blank"—In such a case, an "NR," a non-numerical value would have been assigned to the answer.	

Source: Author's creation

of the response, while in the second column, we have its numerical value—that was used to run the ELECTRE TRI ME algorithm. Although it was not part of the scale, the evaluators had the choice to leave the answer as "blank" if they did not understand the question or if they were not sure how to answer it—in such a case, an "NR," a non-numerical value would be assigned to the answer (refer to Table 5).

Table 5 shows the data collected which will also be available in the UCI (https://archive.ics.uci.edu/ml/datasets.php) and Mendeley Data Library. We highlight that Evaluators E1, E2, and E3 expressed their perceptions of having an incremental innovation environment in mind, while E4 was thinking of a disruptive innovative scenario.

4.3 *Running Electre Tri Me*

Based on the set of categories and on the scales to evaluate performance that appears in Tables 4 and 5, the set of boundaries was defined as it appears in Tables 6 and 7. The same set of boundaries was adopted for all the evaluators, therefore:

$\mathbf{B} = \mathbf{B}_{E1} \mathrm{U} \mathbf{B}_{E2} \mathrm{U} \mathbf{B}_{E3} \mathrm{U} \mathbf{B}_{E4}$

Where

$\mathbf{B}_{E1} = \mathbf{B}_{E2} = \mathbf{B}_{E3} = \mathbf{B}_{E4}$

As reported by Nepomuceno and Costa (2015) and in Gomes et al. (2017), applying the ELECTRE method to data collected based on a Likert scale, the values of pj and qj are both equal to zero, and one can use the equation to calculate the concordance degree.

By applying Eq. (4) to the data and a cut-level (λ) equal to 0.80, the results for categorization appear in Table 8. In this table, the second column refers to the aggregate results from the evaluators EI, E2, and E3 that have in mind an incremental innovation ecosystem. The third column refers to the data collected from evaluator E4, who has a disruptive ecosystem in mind. The fourth column refers to the

Table 6 The collected answers from evaluators on alternatives

Criteria	Evaluators	Alternatives[a]									
		A1	A2	A3	A4	A5	A6	A7	A8	A9	A10
TRL1	E1	2	1	1	1	1	2	2	2	1	1
	E2	2	NR	2	NR	1	2	1	2	2	NR
	E3	2	0	0	0	1	1	1	0	2	1
	E4	2	1	1	1	1	1	1	1	2	0
TRL2	E1	2	2	2	2	2	2	1	2	1	2
	E2	2	1	2	2	2	2	2	2	2	2
	E3	2	1	1	1	1	1	1	1	2	2
	E4	2	2	1	2	2	1	1	1	2	1
TRL3	E1	0	0	0	0	0	0	0	0	0	0
	E2	2	0	1	1	0	2	1	0	2	0
	E3	2	0	0	0	1	1	1	1	1	1
	E4	2	1	1	1	1	2	1	1	2	1
TRL4	E1	1	1	1	1	1	1	1	1	1	1
	E2	2	1	1	1	1	2	2	2	2	2
	E3	2	1	2	2	2	2	1	1	2	1
	E4	2	2	2	2	2	2	1	1	2	1
TRL5	E1	1	1	1	1	1	1	1	1	1	1
	E2	2	0	1	0	0	2	1	0	2	0
	E3	1	0	1	1	1	1	1	0	1	1
	E4	2	1	1	1	1	1	1	1	2	1
TRL6	E1	2	1	1	1	1	2	1	1	1	1
	E2	2	2	2	2	2	2	2	2	2	2
	E3	2	2	2	2	1	2	1	1	2	1
	E4	2	2	2	2	1	2	2	1	2	1
TRL7	E1	2	2	2	2	2	2	1	2	1	1
	E2	2	0	1	0	1	2	1	0	2	1
	E3	1	0	0	0	1	1	1	0	2	0
	E4	2	1	1	1	1	1	1	1	2	1
TRL8	E1	2	1	1	1	1	2	1	2	1	1
	E2	2	2	2	2	2	2	2	1	2	1
	E3	2	2	2	1	1	1	1	1	2	2
	E4	2	2	2	2	1	1	1	1	2	1
TRL9	E1	2	1	1	2	1	2	1	2	1	2
	E2	2	1	1	0	0	2	0	0	2	0
	E3	2	2	0	2	1	2	1	2	2	2
	E4	2	2	2	1	1	2	2	1	2	1

Source: Author's creation

[a]A1 = People; A2 = Integrated Management; A3 = Support of Actors; A4 = Investor Support; A5 = Mentors Support; A6 = Infrastructure; A7 = Time; A8 = Connection with Ecosystems; A9 = Transparent Objectives; A10 = Partner Network

Table 7 Categories and their boundaries

Category	Code	Boundaries	
		Lower	Upper
Very positive influence on the driver	A	1.5	$+\infty$
Positive influence on the driver	B	0.5	1.5
No influence on the driver	C	−0.5	0.5
Negative influence on the driver	D	−1.5	−0.5
Very negative influence on the driver	E	$-\infty$	−1.5

Source: Author's creation

Table 8 Categorization results

Alternatives (actors)	Incremental	Disruptive	All
People	A	A	A
Integrated management	C	B	C
Actor support	B	B	B
Investor support	C	B	B
Support from mentors	B	B	B
Infrastructure	B	B	B
Time	B	B	B
Connection with ecosystems	C	B	C
Transparent objectives	B	A	B
Partner network	B	B	B

Source: Author's creation

categorization obtained from running together with the evaluations from E1, E2, E3, and E4.

5 Discussion and Conclusion

The research highlighted that there is a need to align the perspectives of the CEO and the other members of the sample since the CEO has disruptive innovation in mind and the other respondents are thinking of incremental innovation. Aligning these perspectives is relevant to reach the organizational goals on innovation drivers.

Despite this unalignment, the results indicate that it does not cause relevant disagreement about the perception of the influence of the actors that appears in **A** on the drivers of the innovation ecosystem that compose **F**.

Obviously, "People" was perceived as the highest influencer factor on the innovation ecosystems for both perspectives: open incremental innovation and open disruptive innovation. From the data collected, and despite the incremental or radical innovation, there is no doubt that the following actors have a positive influence on the innovative ecosystem: Actor Support, Time, Support from Mentors, and

Table 9 Level of influence of the actors on the drivers

Alternatives (actors)	Level of influence
People	Very positive
Transparent objectives	Somewhere between very positive and positive
Actor support	Positive
Support from mentors	
Infrastructure	
Time	
Partner network	
Integrated management	Somewhere between positive and no influence
Investor support	
Connection with ecosystems	

Infrastructure. At the fourth level of influence, the following prevail Investor Support, Integrated Management, Connection with Ecosystems.

Bearing in mind the results from Table 8, it is possible to build Table 9 to show the actors and their level of influence—which appears in the second column of the table.

As evident in Table 9, other points to be highlighted in this research are the following:

- None of the actors in **A** was classified as having a negative influence.
- The number of blank responses "NR" was minimal (only 3 among 280 possibilities), which means that the sample members understood the questions and were willing to answer them.

It is worth mentioning that the sample did not cover the closed innovation system nor the radical, innovative scenario. These situations should be covered in further research.

6 Implications

The current study exhibits some implications. Information and communication technologies (ICTs) have been implemented at different scales to improve tourists' experiences of cultural heritage. In particular, advanced technologies combining augmented reality can offer a deeply immersive visitor's experience aiming both at providing multifaceted information about the site and enhancing the attractiveness of a tourist destination at different scales (Graziano & Privitera, 2020).

Knowledge created over the past two decades can be characterized as two distinct epochs—digitalization, which reflected a common understanding of how technology has changed our society and economy, and innovation, which implies the introduction and use of new technologies to develop the industry (Suyunchaliyeva et al., 2020).

However, digitalization is a new form of communication between producers and consumers of tourist services, becoming a source of competitive advantages for tourist organizations (Natocheeva et al., 2020).

In addition, the development of digital technologies such as Augmented Reality (AR) and Virtual Reality (VR) has evolved rapidly. These technologies are currently in the process of creating driving change in the Cultural and Creative Industries (CCIs), representing innovative means to share information, facilitating access and increasing the value and public awareness of Cultural and Natural Heritage (Bruno et al., 2020).

Finally, digitalization in the manufacturing and industrial sectors has been fast "exported" in the tourism industry as well. The technological revolution has highly affected both the supplier and the consumer of tourism products and services. There is recently an increasing attention from the industry and academia in order to highlight the challenges of ICT use in tourism and the drivers of "smart tourism" in world level (Gjika & Pano, 2020).

7 Limitations and Future Research

The current study exhibits some limitations that should be considered for future research. The non-inclusion of other actors of the institution as decision-makers for access to the innovation ecosystem can represent a limitation, as well as the failure to include more alternatives and expand the sample.

Therefore, one must consider that the limitations do not invalidate the proposal as they were addressed due to the difficulty of initially expanding the study sample. It is also limited to understanding and responding to the institution's internal problem, being considered a latent need in view of the strong changes in the business model imposed by the market.

Moreover, in terms of future research, it is important to analyze whether the technology-based startup market demands a wider volume of alternatives. It is also significant to comparatively analyze the proposed approach with other institutions in Portugal and present a profile of access to innovation tourism ecosystems according to each business size and whether it is a public or private institution.

Funding This study received financing from national funding through the FCT—the Foundation for Science and Technology under the auspices of project UID/GES/04630/2020.

References

Almeida-Dias, J., Figueira, J. R., & Roy, B. (2010). Electre Tri-C: A multiple criteria sorting method based on characteristic reference actions. *European Journal of Operational Research, 204*(3), 565–580. https://doi.org/10.1016/j.ejor.2009.10.018

Bruno, F., Ricca, M., Lagudi, A., Kalamara, P., Manglis, A., Fourkiotou, A., . . . Veneti, A. (2020). Digital technologies for the sustainable development of the accessible underwater cultural heritage sites. *Journal of Marine Science and Engineering, 8*(11), 1–25. https://doi.org/10.3390/jmse8110955

Beltagui, A., Rosli, A., & Candi, M. (2020). Exaptation in a digital innovation ecosystem: The disruptive impacts of 3D printing. *Research Policy, 49*(1), 103833. https://doi.org/10.1016/j.respol.2019.103833

Costa, H. G. (2016). Graphical interpretation of outranking principles: Avoiding misinterpretation results from ELECTRE I. *Journal of Modelling in Management, 11*(1), 26–42. https://doi.org/10.1108/JM2-08-2013-0037

Costa, H. G., de Oliveira Nepomuceno, L. D., & Pereira, V. (2020). A proposal for avoiding compensatory effects while using ELECTRE TRI with multiple evaluators. *Advances in Intelligent Systems and Computing, 923*, 470–480. https://doi.org/10.1007/978-3-030-14347-3_46

Costa, H. G., & Duarte, M. B. T. (2019, October 28–31). Applying ELECTRE TRI ME for evaluating the quality of services provided by a library. In P. Kommers & F. Lou (Eds.), *{ICETC} 2019, 11th International Conference on Education Technology and Computers, Amsterdam, The Netherlands* (pp. 278–281). ACM. https://doi.org/10.1145/3369255.3369313.

Costa, H. G., Nepomuceno, L. D. D. O., & Pereira, V. (2018). ELECTRE ME: A proposal of an outranking modeling in situations with several evaluators. *Brazilian Journal of Operations & Production Management, 15*(4), 566–575. https://doi.org/10.14488/bjopm.2018.v15.n4.a10

Cukier, D., & Kon, F. (2018). A maturity model for software startup ecosystems. *Journal of Innovation and Entrepreneurship, 7*(1), 1–32.

De Silva, M., Howells, J., & Meyer, M. (2018). Innovation intermediaries and collaboration: Knowledge–based practices and internal value creation. *Research Policy, 47*(1), 70–87. https://doi.org/10.1016/j.respol.2017.09.011

Doumpos, M., & Figueira, J. R. (2019). A multicriteria outranking approach for modeling corporate credit ratings: An application of the ELECTRE TRI-NC method. *Omega (United Kingdom), 82*, 166–180. https://doi.org/10.1016/j.omega.2018.01.003

Elia, G., Margherita, A., & Passiante, G. (2020). Digital entrepreneurship ecosystem: How digital technologies and collective intelligence are reshaping the entrepreneurial process. *Technological Forecasting and Social Change, 150*, 119791. https://doi.org/10.1016/j.techfore.2019.119791

Gjika, I., & Pano, N. (2020). Effects of ICT in Albanian tourism business. *Academic Journal of Interdisciplinary Studies, 9*(6), 252. https://doi.org/10.36941/ajis-2020-0129

Gomes, C. F. S., Costa, H. G., & de Barros, A. P. (2017). Sensibility analysis of MCDA using prospective in Brazilian energy sector. *Journal of Modelling in Management, 12*(3), 475–497. https://doi.org/10.1108/JM2-01-2016-0005

Graziano, T., & Privitera, D. (2020). Cultural heritage, tourist attractiveness and augmented reality: Insights from Italy. *Journal of Heritage Tourism, 15*(6), 666–679. https://doi.org/10.1080/1743873X.2020.1719116

Guerrero, M., & Urbano, D. (2019). Effectiveness of technology transfer policies and legislation in fostering entrepreneurial innovations across continents: An overview. *Journal of Technology Transfer, 44*(5), 1347–1366. https://doi.org/10.1007/s10961-019-09736-x

Hugonnard, J., & Roy, B. (1984). Ranking of suburban line extension projects for the Paris metro system by multicriteria method. *Transportation Research, 16*(4), 301–331.

Immonen, A., Ovaska, E., Kalaoja, J., & Pakkala, D. (2016). A service requirement engineering method for a digital service ecosystem. *Service Oriented Computing and Applications, 10*(2), 151–172. https://doi.org/10.1007/s11761-015-0175-0

Inoue, Y., & Tsujimoto, M. (2018). New market development of platform ecosystems: A case study of the Nintendo Wii. *Technological Forecasting and Social Change, 136*, 235–253. https://doi.org/10.1016/j.techfore.2017.01.017

Jha, S. K. (2018). Entrepreneurial ecosystem in India: Taking stock and looking ahead. *IIMB Management Review, 30*(2), 179–188. https://doi.org/10.1016/j.iimb.2018.04.002

Kenney, M., Rouvinen, P., Seppälä, T., & Zysman, J. (2019). Platforms and industrial change. *Industry and Innovation, 26*(8), 871–879. https://doi.org/10.1080/13662716.2019.1602514

Kohtamäki, M., Parida, V., Oghazi, P., Gebauer, H., & Baines, T. (2019). Digital servitization business models in ecosystems: A theory of the firm. *Journal of Business Research, 104*, 380–392. https://doi.org/10.1016/j.jbusres.2019.06.027

Martínez-Ros, E., & Orfila-Sintes, F. (2009). Innovation activity in the hotel industry. *Technovation, 29*(9), 632–641.

Milwood, P. A., & Maxwell, A. (2020). A boundary objects view of entrepreneurial ecosystems in tourism. *Journal of Hospitality and Tourism Management, 44*, 243–252. https://doi.org/10.1016/j.jhtm.2020.06.008

Morant-Martínez, O., Santandreu-Mascarell, C., Canós-Darós, L., & Roig, J. M. (2019). Ecosystem model proposal in the tourism sector to enhance sustainable competitiveness. *Sustainability (Switzerland), 11*(23). https://doi.org/10.3390/su11236652

Mousseau, V., & Slowinski, R. (1998). Inferring an ELECTRE TRI model from assignment examples. *Journal of Global Optimization, 12*(2), 157–174. https://doi.org/10.1023/A:1008210427517

Natocheeva, N., Shayakhmetova, L., Bekkhozhaeva, A., Khamikhan, N., & Pshembayeva, D. (2020). Digital technologies as a driver for the development of the tourism industry. *E3S Web of Conferences, 159*. https://doi.org/10.1051/e3sconf/202015904002.

de Nepomuceno, L. D. O., & Costa, H. G. (2015). Analyzing perceptions about the influence of a master course over the professional skills of its alumni: A multicriteria approach. *Pesquisa Operacional, 35*(1), 187–211. https://doi.org/10.1590/0101-7438.2015.035.01.0187

Parida, V., & Wincent, J. (2019). Why and how to compete through sustainability: A review and outline of trends influencing firm and network-level transformation. *International Entrepreneurship and Management Journal, 15*(1), 1–19. https://doi.org/10.1007/s11365-019-00558-9

Pencarelli, T. (2020). The digital revolution in the travel and tourism industry. *Information Technology and Tourism, 22*(3), 455–476. https://doi.org/10.1007/s40558-019-00160-3

Prashantham, S., & Kumar, K. (2019). Engaging with startups: MNC perspectives. *IIMB Management Review, 31*(4), 407–417. https://doi.org/10.1016/j.iimb.2019.01.003

Rodriguez, D. S. S., Costa, H. G., & Do Carmo, L. F. R. R. S. (2013). Métodos de auxílio multicritério à decisão aplicados a problemas de PCP: Mapeamento da produção em periódicos publicados no Brasil. *Gestao e Producao, 20*(1), 134–146. https://doi.org/10.1590/S0104-530X2013000100010

Roy, B. (1968). Classement et choix en présence de points de vue multiples. *Revue Française d'informatique et de Recherche Opérationnelle, 2*(8), 57–75. https://doi.org/10.1051/ro/196802v100571

Roy, B. (1978). ELECTRE III: Un alghoritme de méthode de classements fonde sur une representation floue des préfé-rences em présence de critéres multiples. *Cahiers Du CERO, 20*(1), 3–24.

Roy, B., & Bertier, P. M. (1971). *La méthode ELECTRE II: Une méthode de classement en présence de critéres multiples* (No. 142). SEMA (Metra International).

Roy, B., & Skalka, J. M. (1985). *ELECTRE IS: Aspects methodologiques et guide dutilisation.* Université Paris-Danphine. Document du LAMSADE, 30.

Russell, M. G., & Smorodinskaya, N. V. (2018). Leveraging complexity for ecosystemic innovation. *Technological Forecasting and Social Change, 136*, 114–131. https://doi.org/10.1016/j.techfore.2017.11.024

Suyunchaliyeva, M., Shedenova, N., Kazbekov, B., & Akhmetkaliyeva, S. (2020). Digital economy: Information technology and trends in tourism. *E3S Web of Conferences, 159*. https://doi.org/10.1051/e3sconf/202015904029.

Thomas, L., Autio, E., & Gann, D. M. (2014). Architectural leverage: Putting platforms in context. *Academy of Management Perspectives, 28*(2), 198–219.

Tietz, G., Anholon, R., Ordoñez, R. E. C., & Quelhas, O. L. (2015). Business incubators in Brazil: Main gaps to be explored by academic researchers. *Journal of Technology Management and Innovation, 10*(4), 18–27. https://doi.org/10.4067/S0718-27242015000400003

Tsujimoto, M., Kajikawa, Y., Tomita, J., & Matsumoto, Y. (2018). A review of the ecosystem concept—Towards coherent ecosystem design. *Technological Forecasting and Social Change, 136*, 49–58. https://doi.org/10.1016/j.techfore.2017.06.032

Wagner, M., Schaltegger, S., Hansen, E. G., & Fichter, K. (2019). *University-linked programmes for sustainable entrepreneurship and regional development: how and with what impact?* Small Business Economics (Sedlacek 2013). https://doi.org/10.1007/s11187-019-00280-4.

Xu, G., Wu, Y., Minshall, T., & Zhou, Y. (2018). Exploring innovation ecosystems across science, technology, and business: A case of 3D printing in China. *Technological Forecasting and Social Change, 136*, 208–221. https://doi.org/10.1016/j.techfore.2017.06.030

Yeh, H. (2020). Factors in the ecosystem of Mobile payment affecting its use: From the customers' perspective in Taiwan. *Journal of Theoretical and Applied Electronic Commerce Research, 15*(1), 13–29. https://doi.org/10.4067/S0718-18762020000100103

Yu, W. (1992). *Aide multicritère à la décision dans le cadre de la problématique du tri: Concepts, méthodes et applications* (Doctoral dissertation, Paris 9).

Zorio-Grima, A., & Merello, P. (2020). Consumer confidence: Causality links with subjective and objective information sources. *Technological Forecasting and Social Change, 150*, 119760. https://doi.org/10.1016/j.techfore.2019.119760

Tourist Experience in Rural Areas in Portugal: The Case of the "Quadrilátero Do Minho"

Hugo Machado, Laurentina Vareiro, Bruno Sousa, and Raquel Mendes

Abstract Rural tourism (both physical and emotional experiences) has been increasingly sought after for a multitude of reasons. The personalised encounter between places, people/culture and guests, characteristic of rural accommodation and the village context, can play a central role in the quality of the experience of tourism in these areas.

If so, understanding the profile of the tourist in Portuguese rural areas becomes increasingly relevant from a research perspective. In this study, using an online questionnaire with a sample size of 133, we look at the case of the rural tourism in the "Quadrilátero do Minho" Region.

Findings show that the profile of the tourist in rural areas in Portugal is mostly composed of higher educated (81.2%) women (72.2%), aged 35 through 44 years (32.3%), who are not self-employed (59.4%) and that live in the Northern region (45.1%). With respect to their accommodation profile, the most representative sample of respondents are those who stay with relatives (61.7%) in rural tourism units once a year (64.7%), from 1 to 3 days (68.4%) and spend anywhere from between 50€ and 100€ per day (64.7%). The tourist in rural areas in Portugal appreciates calmness and tranquillity in a low-density populated area and contact with nature, seeking out what is genuine and endogenous to the region. Rural tourism units with a swimming pool, cultural activities and the possibility of participating in experiences stand out as the favourite attributes among the respondents/tourists.

1 Introduction

In this study we focus on the territory known as the "Quadrilátero do Minho." It consists of a competitive and innovative urban network that seeks to promote and strengthen an already highly industrialised and exports driven region. It comprises

H. Machado · L. Vareiro · B. Sousa (✉) · R. Mendes
IPCA – Polytechnic Institute of Cávado and Ave, São Martinho, Portugal
e-mail: bsousa@ipca.pt

J. Leitão et al. (eds.), *Tourism Entrepreneurship in Portugal and Spain*, Tourism, Hospitality & Event Management, https://doi.org/10.1007/978-3-030-89232-6_3

four municipalities of the Braga district: Barcelos, Braga, Guimarães and Vila Nova de Famalicão. The first two municipalities belong to the NUT III Cávado region while the latter two to the NUT III Ave region.

Considering the growing importance of rural areas in the national, as well as in the international tourism sectors (Kastenholz, 2010; Pato, 2016; Saxena & Ilbery, 2008), rural tourism has become an area of interest. The fact that the tourist phenomenon has offered a wide variety of accommodation and complementary service options only enhance the demand for a holistic experience (Machado et al., 2021).

Tourists consider rural tourism as an escape from the stress of everyday life, a refuge from a massified and popularised environment and an opportunity to witness and feel new, very different emotions and realities (Kastenholz et al., 2012). Kastenholz (2010) states that rural tourism can also contribute to sustainable development of rural areas given that it increases the quality of life of residents and contributes to conserving natural, social and cultural resources. In addition, rural tourism is considered to be a way of culturally strengthening a given territory and preserving and fostering the development of its community, in this way, contributing to an increased civic awareness in environmental and sociocultural treatment (Komppula et al., 2007). Finally, another important factor of rural tourism is the continuance of a connection to the land and the property by those who have invested the requalification of the buildings themselves and by those who have begun to reside in rural areas in the meantime, something highly unlikely if there were no rural tourism (Silva, 2005/2006).

Using an online questionnaire survey applied in 2020, this study seeks to generally profile the typical tourist in Portuguese rural tourism, specifically those in the "Quadrilátero do Minho" region. We aim to identify the contributing factors to the quality of the tourist experience in rural areas. The theoretical foundations of this study focus on tourism in rural areas and on the multidimensional experience of joining the physical tourist experience with its emotional aspect.

2 Literature Review

2.1 Rural Tourism

There is no singular definition of rural tourism because it is a multifaceted concept. It is a type of tourism with various singularities of the rural environment (Nunes, 2012). According to Rodrigues (2013, p. 24), "the rural tourism product is a complex reality where there are numerous demographic and functional variants that contribute to its definition and that result in various types of accommodation and experiences".

Lane (1994, p. 20) states that "there are 14 factors that promote the growth of rural tourism, these being: (1) increased levels of education, (2) growing interest in heritage, (3) increased leisure time, (4) improved transport and communications,

(5) awareness of health care, (6) improved outdoor clothing, which helps in the activities practised in rural tourism establishments (because most of them are outdoors and require adequate clothing necessary to practice them), (7) the growing interest in specialised menus, (8) environmental issues, (9) authenticity, (10) peace and tranquillity, (11) an ageing albeit more active population, (12) "real" tourism, (13) individualism and (14) rural agencies".

Of all definitions of rural tourism that the literature provides, which comes closest to the attributes mentioned above is provided by the European Federation of Rural Tourism—EuroGites (2005). This federation defines rural tourism as a sustainable and multifunctional activity that is related to local resources: traditional agriculture, culture, or natural values in open areas or small populations whose tourist activity does not represent the main source of income (EuroGites, 2005). Because of this, rural accommodation establishments should offer a small-scale service with harmonised personalised care, calm and peacefulness and quality standards in accordance with good environmental conservation, human and cultural authenticity.

Widely accepted is the notion that rural tourism is a composite of agricultural products, ecological products, cultural resources and environmental charms. These include multiple functions, namely: economic, social, educational, environmental, recreational and therapeutic activities (Hwang & Lee, 2015). Regional authenticity and endogeneity have become much sought-after aspects in tourism (Figueiredo, 2003; Pereiro & Conde, 2005) creating a great demand for new tourist destinations and novel experiences. Silva (2014) shares the same opinion due to the wide range of consumption opportunities, activities and rewarding experiences that are offered to urban visitors.

Considering rural tourism in Portugal, the dynamics, specificities and origins are similar to other examples of this type of tourism. Rural tourism used to be limited to families of higher income and greater socio-economic privileges during the holidays, but its relevance died down, giving way to more fashionable coastal tourism linked to the sun and the sea. The generalised increase in leisure time, the democratisation of tourism (Fernandes, 2002), and the subsequent technological development of transport and communications were decisive to a greater mobility that gave rise to new flows to the inland (DGADR, 2019). This propensity was exasperated by the consolidation of national road accessibility and environmental concerns (Gomes & Renda, 2016; Villanueva-Álvaro et al., 2017).

Family hospitality, the location in rural regions, natural and/or protected spaces are inherent characteristics of rural tourism that foster a more direct coexistence with nature, people, culture, monuments and traditions. It is a way to diversify national tourism and promote culture and local development (Carson & Carson, 2018; Dinis et al., 2019; Dubois et al., 2017).

Rural tourism can alleviate some economic and social dilemmas associated with the scarcity of economic opportunities and the decline in population that has evolved with the decline of agricultural activity (Doh et al., 2017; Eusébio et al., 2017; van der Ploeg, 2018) and can also be an alternative to family farms' increasing income (Riva & Bertolini, 2017; Villanueva-Álvaro et al., 2017).

Rural tourism is the result of the reconstruction and a favourable revaluation of existing properties and heritage reconverted into accommodation units. It has also been a result of underused land or labour resources that allow for the preservation of the land and links to ownership (Dubois et al., 2017; Lane & Kastenholz, 2015). Small rural units are generally not oriented towards growth and other motivations other than the economic benefits that come with it.

2.2 *Multidimensionality in the Perception of the Rural Tourism Experience*

A tourism experience refers to events that lead to emotions, which arise from the stay in a given place. These emotions, in turn, contribute to the creation of memories (Dolcos & Cabeza, 2002; Oh et al., 2007). Therefore, full and pleasant sensory stimulation and the creation of memories can result in a positive memorable experience in tourism (Oh et al., 2007). Stimulation or an arousal of sorts is a state of heightened activation (Finn, 2005; Oliver et al., 1997), in regard to the extent an individual feels excited, uplifted and active during the consumption of the experience (Kaltcheva & Weitz, 2006; Rufín et al., 2012). Oliver et al. (1997) and Finn (2005) define stimulation/arousal as a function of surprising levels of consumption. In the context of rural tourism, Loureiro and Kastenholz (2011) highlight rural tourists' need for an experience that allows for the possibility of arousal that can promote positive effects. Memorable experiences are associated with strong emotions, that, in turn, are correlated with the moment in which they occur (Anderson & Shimizu, 2007; Ballantyne et al., 2011). These memories influence the attachment to a given place, the intention of returning, and the willingness to recommend it to family and friends (Martin, 2010). Vada et al. (2019), using a survey of 430 travellers, found that memorable experiences in tourism significantly influence place attachment; it also found that well-being fully mediates this relationship. Their findings showed that the benefits of well-being in tourism favour its place attachment. Therefore, well-being, as a resource of tourism product, can promote the sustainability of the tourism industry via the development of place attachment and the promotion of loyalty to certain tourism destinations (Ferreira & Ratten, 2017; Lubowiecki-Vikuk et al., 2021).

Rural tourism refers to a number of niche activities integrated into segments such as ecotourism, nature tourism, agriculture, adventure, sports, food and wine and cultural tourism. As such, it results in a complex and multifaceted activities, marked by a continuously increasing diversity (Lane, 2009; Lopes et al., 2018). Rural tourism experiences are in demand for a multitude of reasons (Lane & Kastenholz, 2015). The personalised encounter between place, people/culture and guests, as envisaged in the rural accommodation and village context, can play a central role in the quality of the tourism experience in these areas (Kastenholz et al., 2013). Many times, tourists refer to the rural tourism experience as a positive contrast to the stress and other negative conditions of everyday urban life, reflected in opposite affective images of home and the rural destination visited (Kastenholz et al., 2012). However, even if the main benefits found to be associated with tourism in rural areas are considered complex and diverse, relatively little is still known as to how the experience of rural tourism is concretely lived and which dimensions can be distinguished and how these relate to emotions derived from their experience, recall and overall satisfaction (Kastenholz et al., 2018).

The rural tourist experience, in particular, should be understood as the overall experience of a large number and diversity of resources, attractions, services, people and environments provided by a destination. Many times, aspects of this package are not always designed for tourism use, nonetheless, its impact on the experience is valued by tourists (Kastenholz et al., 2012).

The tourism experience is lived by several agents: tourists, local people and tourism service providers. Tourists assume an active role in defining their experience, initiating the process by searching for information well in advance and activating a network of services available at the destination place. Furthermore, the nature of their involvement during the experience (e.g. active or passive, characterised by absorption or immersion) is fundamental to the way they live and remember that experience (Knutson & Beck, 2004). Participation and interaction are relevant because preferences for tourist activities in destinations have evolved towards a more participatory behaviour. It is important to recognize active participation and interaction in co-creation experiences, considering that on-site tourism experiences involve parties interconnected in different ways (emotional, cognitive, physical and social) (Campos et al., 2018).

The possible consequences of a multidimensional experience range from arousal, memorization, quality (perception) to satisfaction (Oh et al., 2007). Marques (2007) highlights the role of arousal as a motivational force in the tourism process. This is because a positive expectation of arousal is associated with the motivation to seek experiences that imply a certain degree of sensory, intellectual or social stimulation. Gnoth et al. (2000) associate this type of effect from the tourism experience with the individual characteristics, namely curiosity and leisurely and challenging experiences. Results suggest that experience is likely to have a positive impact on arousal (Oh et al., 2007; Song et al., 2015). Multiple dimensions of experience contribute to a differential impact on arousal in different contexts.

Another outcome of the tourist experience is memory, which plays an important role in the economic conceptualisation of the experience (Pine & Gilmore, 1998). Associating memory to experience translates the dynamic and holistic nature of the tourist experience itself felt prior to, during, and after visiting a specific destination or attraction. Memories related to past tourist experiences are important factors that promote pleasant memory, as well as the expectation of affectionate expectations which, in turn, condition future evaluation and memory (Goossens, 2000; Klaaren et al., 1994; Tung & Ritchie, 2011).

3 Methodology

In seeking to contribute to the profile of the tourist in rural areas in Portugal, this chapter aims to identify the factors that contribute to the quality of the tourist experience in these areas. We focus on the tourist experience in the "Minho Quadrilateral" region.

We use an online questionnaire survey with 18 items. According to authors, such as Evans and Mathur (2018), online questionnaires have the advantage over the traditional paper and pencil ones. The ease of analysis and the diversity of questions are some of the benefits of opting for online questionnaires over the more traditional ones. They are also more convenient and flexible, quicker to obtain, and they reach a greater number of individuals.

Our survey is organised in three distinct parts:

- Part I—Characterisation of rural tourism.
- Part II—Diversification of tourism supply.
- Part III—Characterisation of the socio-demographics.

This organisation promotes the understanding of the profile of the tourist in rural tourism in Portugal, particularly with regard to aspects such as the importance attributed to the factors. These factors can complement the accommodation units, as well as the importance of the emotional experience in combination with the physical experience (multidimensional experience).

The survey was carried out between April 19th and June 12th of 2020. To foster reliability, we made the online survey available in two groups of the social network Facebook, providing links to the rural territory entitled "Rural Tourism in Portugal". This territory with a sample of 4500 members, the group "Tourism in rural areas" with 1700 members, the group "Higher Technicians of Tourism" with 1600 members and the group "Tourism Researchers" with 5000 members.

At the end of the data collection, it was possible to obtain a total of 133 survey responses for further processing. The Statistical Package for the Social Sciences (SPSS) was used for data analysis. In addition to data analysis using descriptive statistics, statistical tests were also performed, namely the Chi-square test and the t-test (Maroco, 2010; Pestana & Gageiro, 2005). In this research, the main purpose of these tests was to understand whether there are differences between individuals who have already stayed in rural tourism units in the "Quadrilátero do Minho", designated as "Quadrilátero" and those who have never stayed in rural tourism units in the "Quadrilátero do Minho", designated as "Others".

4 Results

Table 1 shows us the characteristics of our sample. We can see that, with the exception of the place of residence ($X^2 = 11.663$; Sig = 0.040; $p < 0.05$), there are no statistically significant differences in the respondents' socio-demographic characteristics. In both groups, there are more female respondents, aged 25–44 years. Academically speaking, an overwhelming majority (81.2%) is university graduates and 59.4% were employed at the time.

In his study, Gonçalves (2020) draws very similar conclusions when evaluating the profile of the tourists of rural tourism units in the Centre region. He finds that almost three-fourths of the sample (73.8%) is female, and that the most representative age group is that of individuals aged 18–30 years (32.9%), followed by

Table 1 Socio-demographic profile of respondents

	Quadrilátero	Others	X^2	Sig.
	n (50)	n (83)		
Gender			1.524	0.217
Female	33	63		
Male	17	20		
Age			5.377	0.251
Up to 24	5	10		
25–34	12	18		
35–44	11	32		
45–54	18	18		
55–64	4	5		
Profession			2.082	0.721
Unemployed	5	11		
Retired	2	1		
Self-employed	12	17		
Employee	28	51		
Student	3	3		
Education			3.036	0.219
Primary education	2	1		
Secondary education	11	11		
Higher education	37	71		
Place of residence			11.663	**0.040**
Norte	30	30		
Centro	10	20		
Lisboa e Vale do Tejo	8	24		
Alentejo	1	4		
Algarve	0	5		
Madeira	1	0		

A p-value less than 0.05 (typically ≤ 0.05) is statistically significant denoted in bold
Source: Elaborated by the authors

individuals in the age bracket of 31–40 years (26%). Roughly 71% of the sample has a university degree and about half (50.3%) was employed at the time the survey was carried out. Other studies present similar conclusions, namely Almeida and Pinto-Correia (2012), Fernandes (2016) and Gonçalves (2020).

5 Characterisation of the Experience in Rural Tourism

This part of the questionnaire was designed to characterise the experience of the tourist in rural tourism in Portugal, namely: places of accommodation, frequency of accommodation, average length of stay, those with whom they usually stay and the average amount spent per day. The main characteristics most valued by tourists in rural tourism are also analysed, along with the degree of satisfaction and the probability of recommending and/or returning. Table 2 presents the accommodation profile in rural tourism.

Table 2 Accommodation profile in rural tourism

	Quadrilátero	Others	X^2	Sig.
	n (50)	n (83)		
Frequency of accommodation			12.853	**0.005**
Once a year	29	57		
Twice a year	13	17		
3 or more times a year	8	2		
Other	0	7		
Average length of stay			0.725	0.696
1–3 days	32	59		
4–6 days	15	20		
7 or more days	3	4		
With whom do you usually stay			8.696	0.122
Alone	4	3		
Family	31	51		
Friends	6	17		
Alone, family	0	5		
Alone, friends	1	0		
Family, friends	8	7		
Average daily amount spent			2.655	0.448
Between 50€ and 100€	34	52		
Between 150€ and 200€	12	27		
Between 250€ and 300€	3	4		
Between 350€ and 400€	1	0		

A *p*-value less than 0.05 (typically $\leq$0.05) is statistically significant denoted in bold
Source: Elaborated by the authors

The frequency of accommodation is the only characteristic that reveals a statistically significant difference between the two groups under analysis ($X^2 = 12.853$; Sig $= 0.005$; $p < 0.05$). A study carried out by Eusébio and Kastenholz (2010) show that the average length of stay in rural tourism units in the Centre region is that of 5 days and that tourists around 72€ per day. We can also see that 63% of respondents enjoy these units with family members. This result is corroborated by studies, namely Milheiro et al. (2017, p.84), when they assert that "the region, and the units in particular, seem to be appealing for family travel". With respect to the frequency of accommodation, Fernandes (2016), corroborates the results of this study, since he also concluded that the most representative sample concerns individuals who visit rural tourism units only once a year.

6 Most Valued Characteristics in Rural Tourism

Tourist attraction to certain destinations depends on the physical factors, environmental or sociocultural attributes (Jafari, 1982). Physical and environmental attributes include climatic conditions, landscape and ecology; and sociocultural attributes include history, politics, art, economic activities, ways of life, monuments, individual buildings and man-built environments (Benur & Bramwell, 2015). Table 3 shows the characteristics of rural tourism valued by the respondents.

Calmness and tranquillity stand out from the other characteristics, reaching a total of almost 80% of the answers. These are followed by "hospitality", with 55.6%, "escape from daily stress" and "local gastronomy", each with 45.9% of total answers. A distant fourth factor is "authenticity of heritage" with a total of 28.6% and "novel experiences" with 27.1%. Only one respondent mentioned other characteristics in addition to those mentioned above.

Table 3 Characteristics of rural tourism

	n (133)	%
Hospitality	74	55.6
Calmness and tranquillity	105	78.9
Heritage authenticity	38	28.6
Escape from daily stress	61	45.9
Local gastronomy	61	45.9
Relationship with the local community	23	17.3
Novel experiences	36	27.1
Other	1	0.8

Source: Elaborated by the authors

Table 4 Satisfaction, recommendation and likelihood of revisiting rural tourism

	Quadrilátero *n* (50)		Others *n* (83)			
	Satisfied/ probably (%)[a]	Mean (SD)	Satisfied/ probably (%)[a]	Mean (SD)	*T*	Sig.
How satisfied are you with your experience of rural tourism?[b]	98.0	4.58 (0.538)	95.2	4.51 (0.592)	0.722	0.472
How likely are you to recommend your stay at a RT unit to friends and family?[c]	98.0	4.66 (0.519)	94.0	4.65 (0.593)	0.093	0.926
How likely are you to stay at a RT unity again?[c]	96.0	4.66 (0.557)	96.4	4.66 (0.590)	−0.026	0.980

Source: Elaborated by the authors
[a]Percentage of respondents answering 4 or 5 on the respective Likert scales
[b]Scale from 1 = not at all satisfied to 5 = very satisfied
[c]Scale from 1 = Very unlikely to 5 = Very likely

7 Satisfaction, Recommendation and Likelihood of Revisiting Rural Tourism

As for the degree of satisfaction regarding experiences of rural tourism, the percentage of respondents who replied that they were at least satisfied was 96.2%. This value corresponds to an average of 4.53, based on the Likert scale, where 1 is "not at all satisfied" and 5 is "very satisfied". Martínez-Roget et al. (2015) found that the degree of satisfaction of the respondents was 66%. This corresponds to a mean value of 3.8, using the same scale.

The percentage of respondents who say they are likely or very likely to recommend rural tourism (95.5%) is extremely high. This value corresponds to an average of 4.65, based on the Likert scale, where 1 is "not at all likely" and 5 "very likely". Fernandes (2016) found that the likelihood of recommending a rural tourism unit was also quite high, at 87.7%. Silva (2007, p. 157) asserted that "It is . . . symptomatic that some tourists write in their honour books that they will not recommend the rural tourism places and units in which they stayed to many people", coming to the conclusion that this happens, "so that these people do not destroy with their presence the tranquillity and charm of these places".

With respect to the probability of returning to a rural tourism unit, the results are in line with previous studies with about 97% responding that it is likely or very likely that they would return. This value corresponds to an average of 4.66, based on the Likert scale, where 1 is "not at all likely" and 5 is "very likely". According to Martínez-Roget et al. (2015, p. 78), "the success of rural tourism destinations will depend largely on the potential loyalty of their tourists, and the recommendation to third parties plays a decisive role in the election of a tourist destination" (Table 4).

8 Diversification of Supply in Tourism

In this section, we seek to understand which factors influence the experiences of tourists in rural tourism. We are particularly also interested in the most valued characteristics, apart from accommodation. Again, using a Likert scale of five points, tourists rated their stance in regard to a given number of statements, where 1 represents "strongly disagree" and 5 "strongly agree".

The data show that respondents agree with all of the statements presented when asked about a more general perspective of rural tourism (Table 5). With regard to the first statement that the typical tourist is drawn to a calm and peaceful environment, contact with nature and the possibility to enjoy a holiday in a non-massified environment, it is very highly rated in the Quadrilátero, as well as in rural tourism in general (4.72 and 4.47 out of 5, respectively), thereby making these factors greatly associated with an increase the demand for tourism in rural areas.

Although having failed the difference in means test, the second and third statements, highly rate both genuine endogenous experiences in the region (4.6 and 4.48, respectively, in the Quadrilátero and rural tourism in general) and novel destinations and experiences, albeit a little less so (4.36 in the "Quadrilátero" group and 4.25 in the "Others" group).

No statistically significant results pertaining to the multidimensional experience were found (Table 6). Despite the slight difference found between the two groups concerning all of the statements, it is worth noticing that the "Quadrilátero" group always scores higher than the "Others" group. The first statement, a high satisfaction of the rural tourism experience resulting from the recollection and emotion of the moments lived, shows the highest average response, both in the "Quadrilátero" group and in the "Others" group, with values of 4.58 and 4.46, respectively. Social interaction with the community enhances the meaningfulness and lasting impression of the rural tourism experience obtained an average of 4.36 in the first group and 4.19 in the second group.

Table 5 Introductory questions

	Quadrilátero *n* (50)	Others *n* (83)		
Statements	Mean (SD)	Mean (SD)	*T*	Sig.
A calm and peaceful environment, contact with nature and the possibility to enjoy holidays in a non-massified environment boost the demand for rural tourism	4.72 (0.454)	4.47 (0.721)	2.454	**0.015**
The rural tourist wishes to experience genuine experiences endogenous to the region	4.60 (0.639)	4.48 (0.669)	1.002	0.318
The rural tourist continuously seeks new destinations and new experiences	4.36 (0.663)	4.25 (0.730)	0.847	0.398

A p-value less than 0.05 (typically ≤ 0.05) is statistically significant denoted in bold
Source: Elaborated by the authors

Table 6 Multidimensional experience in RT

Statements	Quadrilátero *n* (50) Mean (SD)	Others *n* (83) Mean (SD)	*T*	Sig.
The high satisfaction of the rural tourist experience results from the memory and emotion of the moments lived	4.58 (0.642)	4.46 (0.611)	1.096	0.275
Social interaction with the community enhances the meaning and memorisation of the rural tourism experience	4.36 (0.722)	4.19 (0.706)	1.312	0.192
The memories of tourist experiences are important elements for a pleasant recollection and condition the evaluation of future memories	4.54 (0.613)	4.42 (0.627)	1.063	0.290
The local community plays an active role in creating the tourist experience	4.30 (0.647)	4.17 (0.746)	1.032	0.304
Diversified offer promotes destination loyalty.	4.48 (0.614)	4.31 (0.825)	1.236	0.218
Loyalty to the destination promotes environmental sustainability, due to the preservation of the authenticity of the rural area	4.36 (0.722)	4.13 (1.021)	1.380	0.170

Source: Elaborated by the authors

The third statement regarding memorable tourism experiences and their potential effects on future memories reached an average of 4.54 in the "Quadrilátero" group and 4.42 in the "Others" group. The extent to which the local community plays an active role in shaping the tourist experience scores an average value of 4.30 in the "Quadrilátero" group and 4.17 in the "Others" group. Roughly the same difference in average scores, diversity provided by rural tourism, the "Quadrilátero" group averaged 4.48 and the "Others" group 4.31. Finally, the idea that client loyalty to the destination promotes environmental sustainability because of the preservation of the authenticity of the rural area presents as average value 4.36 in the "Quadrilátero" group and 4.13 in the "Others" group.

Of all the services/activities complementary to the type of accommodation, only the existence of a spa shows a statistically significant difference between two groups ($M = 3.26$; $t = 2.231$; Sig $= 0.027$; $p < 0.05$). Despite it is worth a look at the data. Table 7 shows the types of services/activities that can be provided as complements to a simple accommodation. Respondents from the "Quadrilátero" group rated the availability of a swimming pool the highest, with an average score of 4.08, followed by the provision of cultural activities (4.02) and participation in experiences (3.96). Services/activities towards which respondents of the "Quadrilátero" group seem to appreciate to a lesser degree are the existence of a golf course (1.74), a meeting room (2.08) and the possibility of fishing (2.22).

As for the "Others" group, respondents valued the existence of cultural activities the most, scoring an average of 3.96. This is closely followed by the availability of a swimming pool (3.95) and participation in experiences (3.90). Tennis and golf courses (2.27 and 1.67, respectively) and the existence of a meeting room come in last (1.94).

Table 7 Complementary offer in rural tourism

Complementary offer	Likert scale (1–5) %					Quadrilátero *n* (50)	Others *n* (83)		
	1	2	3	4	5	Mean (SD)	Mean (SD)	*T*	Sig
Golf course	62.4	12.0	19.5	5.3	0.8	1.74 (0.965)	1.67 (1.037)	0.361	0.719
Swimming pool	2.3	12.8	8.3	36.1	40.6	4.08 (1.085)	3.95 (1.114)	0.649	0.517
Spa	9.8	14.3	25.6	40.6	9.8	3.54 (0.994)	3.10 (1.175)	2.231	**0.027**
Gym	17.3	17.3	33.8	24.1	7.5	3.00 (1.195)	2.80 (1.177)	0.967	0.336
Tennis court	32.3	18.8	38.3	8.3	2.3	2.34 (1.099)	2.27 (1.072)	0.387	0.699
Meeting room	48.1	12.0	33.1	6.0	0.8	2.08 (1.140)	1.94 (1.016)	0.736	0.463
Horseback riding	33.8	11.3	33.1	18.0	3.8	2.50 (1.165)	2.45 (1.281)	0.244	0.807
Birdwatching	20.3	10.5	32.3	29.3	7.5	2.92 (1.291)	2.94 (1.203)	−0.089	0.929
Fishing	40.6	10.5	35.3	10.5	3.0	2.22 (1.148)	2.27 (1.211)	−0.212	0.832
Trails	6.8	9.0	16.5	41.4	26.3	3.62 (1.244)	3.77 (1.097)	−0.731	0.466
Cultural activities	3.8	5.3	15.8	39.1	36.1	4.02 (0.892)	3.96 (1.120)	0.301	0.764
Experiences	6.8	3.0	17.3	36.8	36.1	3.96 (1.124)	3.90 (1.133)	0.279	0.781

A p-value less than 0.05 (typically ≤ 0.05) is statistically significant denoted in bold
Source: Elaborated by the authors

9 Conclusions

This study sought to understand the profile of the tourist in rural areas in Portugal, specifically that of the tourist in the "Quadrilátero do Minho" region. It also aimed to identify the factors that contribute to the quality of the tourist experience in these areas. Findings show that the profile of the tourist in the rural area in Portugal is mostly composed of higher educated (81.2%) women (72.2%), aged 35 through 44 years (32.3%), who are not self-employed (59.4%) and live in the Northern region (45.1%). These results support other studies which present similar conclusions, namely, that of Almeida and Pinto-Correia (2012), Fernandes (2016) and Gonçalves (2020).

In analysing the accommodation or housing profile, we found that typically respondents are individuals who stay in rural tourism units once a year (64.7%), with family members (61.7%), for 1–3 days (68.4%) and spend anywhere from €50 to €100 per day (64.7%); similar findings can be found in Eusébio and Kastenholz (2010), Fernandes (2016), and Milheiro et al. (2017).

Of the 133 respondents, only 50 had already stayed in a rural tourism unit in the "Quadrilátero do Minho": 57.4% in Braga, followed by 29.8% in Guimarães, 12.8% in Vila Nova de Famalicão (12.8%) and, lastly, 10.6% in Barcelos. With regard to the characteristics considered most relevant in the RT, "calmness and tranquillity" stands out among the others, reaching a total of 78.9% of responses. This is followed by "hospitality" with 55.6%, "escape from daily stress" and "local gastronomy", each with a total of 45.9% of the responses. Factors such as "authenticity of heritage" and "new experiences", come in last with 28.6% and 27.1%, respectively.

With respect to the degree of satisfaction concerning the experiences in rural tourism, an overwhelming 96.2% of individuals replied that they were satisfied or very satisfied. Roughly 96% stated that they would stay at a rural tourism unit with friends and family. When asked if they would again return, 96.6% of the respondents answered yes. Taking into account the respondents' level of agreement with the statements on rural tourism, it is important to highlight the three with the highest level of agreement. Highly in line with Martínez-Roget et al. (2015) and Milheiro et al. (2017), and using the Likert scale, where 1 corresponds to "totally disagree" and 5 to "totally agree", "maintaining the authenticity of the rural destination fosters tourist attraction" was the statement most agreed with, with an average value of 4.57, followed by the idea that "a calm and peaceful environment, contact with nature and the possibility of enjoying a holiday, in a non-massified environment, boosts the demand for tourism in rural areas" with an average value of 4.56. Finally, "the tourist in rural areas looks for what is genuine and endogenous to the region" reached an average value of 4.53.

The limitation of this study is the sample size of tourists staying in the "Quadrilátero do Minho". Future research could be understanding the profile of the tourist in rural areas in the "Quadrilátero do Minho" region, using in-person surveys distributed by the rural tourism units of the region.

References

Almeida, M., & Pinto-Correia, T. (2012). Novas procuras na paisagem rural do Alentejo: as visões e preferências dos urbanos. *Geografia: Revista da Faculdade de Letras da Universidade do Porto, 1*, 101–121.

Anderson, D., & Shimizu, H. (2007). Factors shaping vividness of memory episodes: Visitors' long-term memories of the 1970 Japan world exposition. *Memory, 15*(2), 177–191.

Ballantyne, R., Packer, J., & Sutherland, L. A. (2011). Visitors' memories of wildlife tourism: Implications for the design of powerful interpretive experiences. *Tourism Management, 32*(4), 770–779.

Benur, A. M., & Bramwell, B. (2015). Tourism product development and product diversification in destinations. *Tourism Management, 50*, 213–224.

Campos, A. C., Mendes, J., Valle, P. O. D., & Scott, N. (2018). Co-creation of tourist experiences: A literature review. *Current Issues in Tourism, 21*(4), 369–400.

Carson, D. A., & Carson, D. B. (2018). International lifestyle immigrants and their contributions to rural tourism innovation: Experiences from Sweden's far north. *Journal of Rural Studies, 64*, 230–240.

DGADR. (2019). Direção Geral de Agricultura e Desenvolvimento Rural. O interesse pelo turismo no espaço rural. Acedido dia 26 de Dezembro 2019. Disponível em https://www.dgadr.gov.pt/diversificacao/turismo-rural/o-interesse-pelo-turismo-no-espaco-rural

Dinis, I., Simões, O., Cruz, C., & Teodoro, A. (2019). Understanding the impact of intentions in the adoption of local development practices by rural tourism hosts in Portugal. *Journal of Rural Studies, 72*, 92–103.

Doh, K., Park, S., & Kim, D. Y. (2017). Antecedents and consequences of managerial behavior in agritourism. *Tourism Management, 61*, 511–522.

Dolcos, F., & Cabeza, R. (2002). Event-related potentials of emotional memory: Encoding pleasant, unpleasant, and neutral pictures. *Cognitive, Affective, & Behavioral Neuroscience, 2*(3), 252–263.

Dubois, C., Cawley, M., & Schmitz, S. (2017). The tourist on the farm: A 'muddled' image. *Tourism Management, 59*, 298–311.

EuroGites. (2005). EuroGites Public Documentation. Product Criteria Rural Accommodation. http://eurogites.org/documents/.

Eusébio, C., Carneiro, M. J., Kastenholz, E., Figueiredo, E., & da Silva, D. S. (2017). Who is consuming the countryside? An activity-based segmentation analysis of the domestic rural tourism market in Portugal. *Journal of Hospitality and Tourism Management, 31*, 197–210.

Eusébio, C., & Kastenholz, E. (2010). Quem procura o turismo em espaço rural? Especificidades do mercado da Região Centro de Portugal. In E. Figueiredo, E. Kastenholz, M. C. Eusébio, M. C. Gomes, M. J. Carneiro, P. Batista, & S. Valente (Eds.), *Mundos rurais em Portugal: Múltiplos olhares, múltiplos futuros* (pp. 334–346). Aveiro: Universidade de Aveiro.

Evans, J. R., & Mathur, A. (2018). The value of online surveys: A look back and a look ahead. *Internet Research, 28*(4), 854–887.

Fernandes, A. (2016). *Turismo em espaço rural – As preferências dos turistas* [Master's thesis]. Universidade Europeia.

Fernandes, A. T. (2002). Poder Local e Turismo Social. *Revista da Faculdade de Letras: Sociologia, 12*, 9–26.

Ferreira, J., & Ratten, V. (2017). Competitiveness of locations: The effects of regional innovation and entrepreneurial practices. *Competitiveness Review: An International Business Journal incorporating Journal of Global Competitiveness, 28*(1), 2–5.

Figueiredo, E. (2003). Quantas mais aldeias típicas conseguimos suportar? Algumas reflexões a propósito do turismo como instrumento de desenvolvimento em meio rural. In O. Simões & A. Cristóvão (Org.) *Turismo em Espaços Rurais e Naturais* (pp. 65–81). IPC.

Finn, A. (2005). Reassessing the foundations of customer delight. *Journal of Service Research, 8* (2), 103–116.

Gnoth, J., Zins, A. H., Lengmueller, R., & Boshoff, C. (2000). Emotions, mood, flow and motivations to travel. *Journal of Travel & Tourism Marketing, 9*(3), 23–34.

Gomes, A., & Renda, A. (2016). Turismo rural—A região Centro de Portugal na perspetiva dos promotores turísticos. *Omnia, 5*, 129–136.

Gonçalves, P. (2020). *Estudo das (Des) motivações para a prática de Turismo em Espaço Rural: o caso da Região Centro* [Master's thesis]. Instituto Politécnico de Viseu.

Goossens, C. (2000). Tourism information and pleasure motivation. *Annals of Tourism Research, 27*(2), 301–321.

Hwang, J., & Lee, S. (2015). The effect of the rural tourism policy on non-farm income in South Korea. *Tourism Management, 46*, 501–513.

Jafari, J. (1982). The tourism market basket of goods and services. The components and nature of tourism. In T. Singh, J. Kaur, D. Singh, & S. Misra (Eds.), *Studies in tourism, wildlife, parks, conservation* (pp. 1–12). Metropolitan.

Kaltcheva, V. D., & Weitz, B. A. (2006). When should a retailer create an exciting store environment? *Journal of Marketing, 70*(1), 107–118.
Kastenholz, E. (2010). Experiência global em turismo rural e desenvolvimento sustentável das comunidades locais. In E. Figueiredo, E. Kastenholz & Outros (Eds.), IV *Congresso de Estudos Rurais Mundos Rurais em Portugal, Múltiplos Olhares Múltiplos Futuros* (pp. 420–435). Universidade de Aveiro.
Kastenholz, E., Carneiro, M. J., Marques, C. P., & Lima, J. (2012). Understanding and managing the rural tourism experience—*The case of a historical village in Portugal. Tourism Management Perspectives, 4*, 207–214.
Kastenholz, E., Carneiro, M. J., Marques, C. P., & Loureiro, S. M. C. (2018). The dimensions of rural tourism experience: Impacts on arousal, memory, and satisfaction. *Journal of Travel & Tourism Marketing, 35*(2), 189–201.
Kastenholz, E., Eusébio, C., Carneiro, M., & Figueiredo, E. (2013). Host-guest relationships in rural tourism—Evidence from two Portuguese villages. *Anatolia: An International Journal of Tourism and Hospitality Research, 24*(3), 367–380.
Klaaren, K. J., Hodges, S. D., & Wilson, T. D. (1994). The role of affective expectations in subjective experience and decision-making. *Social Cognition, 12*(2), 77–101.
Knutson, B. J., & Beck, J. (2004). Identifying the dimensions of the experience construct—Development of the model. *Journal of Quality Assurance in Hospitality & Tourism, 4*(3), 23–35.
Komppula, R., Rhodri, T. & Marcjanna, A. (2007). Developing rural tourism in finland through entrepreneurship. In *Tourism in the new Europe* (pp. 123–133). Elsevier.
Lane, B. (1994). What is rural tourism? *Journal of Sustainable Tourism, 2*(1/2), 7–21.
Lane, B. (2009). Rural tourism: An overview. In T. Jamal & M. Robinson (Eds.), *The SAGE handbook of tourism studies* (pp. 354–370). Sage.
Lane, B., & Kastenholz, E. (2015). Rural tourism: The evolution of practice and research approaches–towards a new generation concept? *Journal of Sustainable Tourism, 23*(8–9), 1133–1156.
Lopes, C., Leitão, J., & Rengifo-Gallego, J. (2018). Place branding: Revealing the neglected role of agro food products. *International Review on Public and Nonprofit Marketing, 15*(4), 497–530.
Loureiro, S. M. C., & Kastenholz, E. (2011). Corporate reputation, satisfaction, delight, and loyalty towards rural lodging units in Portugal. *International Journal of Hospitality Management, 30* (3), 575–583.
Lubowiecki-Vikuk, A., Djercan, B., & Sousa, B. (2021). Sustainable development and leisure services: Changes and trends. In A. Lubowiecki-Vikuk et al. (Eds.), *Handbook of sustainable development and leisure services, World Sustainability Series*. https://doi.org/10.1007/978-3-030-59820-4_1.
Machado, H., Vareiro, L., Caldas, I., & Sousa, B. (2021). Supply diversification and coopetition in rural tourism. In J. V. de Carvalho, Á. Rocha, P. Liberato, & A. Peña (Eds.), *Advances in Tourism, Technology and Systems. ICOTTS 2020. Smart Innovation, Systems and Technologies* (Vol. 208). Springer. https://doi.org/10.1007/978-981-33-4256-9_18.
Maroco, J. (2010). *Análise estatística: Com utilização do SPSS* (3a ed.). Edições Sílabo.
Marques, C. P. (2007). Seeking to escape: Sights over approach-avoidance dialectics. In *Progress in tourism marketing* (pp. 221–236). Routledge.
Martin, D. (2010). Uncovering unconscious memories and myths for understanding international tourism behavior. *Journal of Business Research, 63*(4), 372–383.
Martínez-Roget, F., Moutela, J. A. T., & Núñez, J. C. E. (2015). Chaves do êxito do Turismo em Espaço Rural (TER): evidências a partir da lealdade turística na Rede das Aldeias do Xisto. *Revista Portuguesa de Estudos Regionais, 40*, 65–81.
Milheiro, E., Kastenholz, E., & Carneiro, M. J. (2017). O perfil do turista e a experiência vivida em espaço rural no Alto Alentejo. *Revista Turismo & Desenvolvimento, 27*(28), 81–84.

Nunes, S. (2012). *Turismo no Espaço Rural na Região Autónoma da Madeira – Um estudo para o desenvolvimento de um Sistema de Gestão de Qualidade no setor. Madeira* [Master's thesis]. Universidade da Madeira.

Oh, H., Fiore, A. M., & Jeoung, M. (2007). Measuring experience economy concepts: Tourism applications. *Journal of Travel Research, 46*(2), 119–132.

Oliver, R. L., Rust, R. T., & Varki, S. (1997). Customer delight: Foundations, findings, and managerial insight. *Journal of Retailing, 73*(3), 311–336.

Pato, L. (2016). O Modesto Contributo do Turismo Rural no Douro, Portugal: um estudo baseado nos promotores e na oferta turística. *Revista Turismo em Análise, 27*(3), 624–643.

Pereiro, X., & Conde, S. P. (2005). Turismo e oferta gastronómica na Comarca de Ulloa (Galiza): Análise de uma experiência de desenvolvimento local. *Revista de Turismo y Patrimonio Cultural, 3*(1), 109–123.

Pestana, M., & Gageiro, J. (2005). *Análise de dados para ciências sociais: A complementaridade do SPSS*. Edições Sílabo.

Pine, B. J., & Gilmore, J. H. (1998). Welcome to the experience economy. *Harvard Business Review, 76*, 97–105.

Riva, G., & Bertolini, G. R. F. (2017). Perspectiva do Turismo Rural como Alternativa de Renda para Agricultura Familiar: Análise de Trabalhos Científicos. *Desenvolvimento em Questão, 15*(38), 197–227.

Rodrigues, A (2013). *Projeto urbanístico de turismo rural—Casas da Fajã de Baixo (Calheta)* [Doctoral dissertation]. Universidade da Beira Interior.

Rufín, R., Medina, C., & Rey, M. (2012). Adjusted expectations, satisfaction and loyalty development. *The Service Industries Journal, 32*(14), 2185–2202.

Saxena, G., & Ilbery, B. (2008). Integrated rural tourism a border case study. *Annals of Tourism Research, 35*(1), 233–254.

Silva, L. (2005/2006). Os impactos do turismo em espaço rural. *Antropologia Portuguesa, 22*, 295–317.

Silva, L. (2007). A procura do turismo em espaço rural. *Etnográfica, 11*, 141–163.

Silva, L. (2014). The two opposing impacts of heritage making on local communities: Residents' perceptions: A Portuguese case. *International Journal of Heritage Studies, 20*(6), 616–633.

Song, H. J., Lee, C. K., Park, J. A., Hwang, Y. H., & Reisinger, Y. (2015). The influence of tourist experience on perceived value and satisfaction with temple stays: The experience economy theory. *Journal of Travel & Tourism Marketing, 32*(4), 401–415.

Tung, V. W. S., & Ritchie, J. B. (2011). Exploring the essence of memorable tourism experiences. *Annals of Tourism Research, 38*(4), 1367–1386.

Vada, S., Prentice, C., & Hsiao, A. (2019). The influence of tourism experience and Well-being on place attachment. *Journal of Retailing and Consumer Services, 47*, 322–330.

van der Ploeg, J. D. (2018). From de-to repeasantization: The modernization of agriculture revisited. *Journal of Rural Studies, 61*, 236–243.

Villanueva-Álvaro, J. J., Mondéjar-Jiménez, J., & Sáez-Martínez, F. J. (2017). Rural tourism: Development, management and sustainability in rural establishments. *Sustainability, 9*(5), 818–825.

Spirituality and Mindful Tourism: Experiencing Innovative Hispanic Cases

Raysa Geaquinto Rocha, Edgar Tarrés Falcó, and João Leitão

Abstract Organizational contexts demand superhuman productivity, and work-related illnesses are becoming increasingly common. While there is no change in companies" strategies from savage capitalism to a humanized management, alternatives are required for employees" well-being. Spirituality and mindful tourism can be one of the options for fostering it, simultaneously enhancing economic activities in the countryside cities and villages with no tourism legacy. It requires unique places that are uncrowded, allowing tourists to connect with others and with nature. We apply a self-narrative method of the second author's experiences in mindful travel projects, courses, congresses, and ventures to discuss mindful tourism unveiling innovative Hispanic cases. We analyze those cases through the lenses of secular spirituality. In this chapter, we go through this tendency in tourism and see the practical benefits of spirituality, mindful travel, and holidays. We also provide suggestions for future investigations and innovative entrepreneurial projects in the Iberian Peninsula.

R. G. Rocha (✉)
NECE—Research Center in Business Sciences, University of Beira Interior, Covilhã, Portugal
e-mail: geaquinto.rocha@ubi.pt

E. T. Falcó
University of Girona, and Co-Founder at the Institute of Silence, Girona, Spain
e-mail: go@edgartarres.com

J. Leitão
NECE—Research Center in Business Sciences, University of Beira Interior, Covilhã, Portugal

ICS, Instituto de Ciências Sociais, University of Lisbon, Lisbon, Portugal

CYFE—Center for Young and Family Enterprise, Università Degli Studi di Bergamo, Bergamo, Italy
e-mail: jleitao@ubi.pt

J. Leitão et al. (eds.), *Tourism Entrepreneurship in Portugal and Spain*, Tourism, Hospitality & Event Management, https://doi.org/10.1007/978-3-030-89232-6_4

1 Introduction

The number of people affected by burnout syndrome (Maslach et al., 2001), anxiety, depression (Sanne et al., 2005), stress (Parker & DeCotiis, 1983), and other pathologies and diseases is increasing in organizational contexts (Ul-Haq, 2020; Leitão et al., 2021). Our capitalistic, volatile, uncertain, complex, and ambiguous society demands companies to be highly adaptable and innovative (Schick et al., 2017). This pressure falls on the shoulders of the employees, who are getting sick from excessive demands, bad habits in lifestyle, and dehumanizing performance related to the workplace (Lang et al., 2007; Ul-Haq, 2020).

While there are no visible changes in organizations toward a humanized strategy (Bratianu, 2015; Nonaka & Takeuchi, 2021; Ul-Haq, 2020), other means are needed to reduce the harmful effects on employees" health. One could suggest spiritually-informed management (Steingard, 2005). However, until there is not a systemic change in society and governance, other palliative solutions arise.

In this chapter, we address mindful tourism through self-reported experiences of spiritual travel. The latter has several formats; for example, the silent landscape experience provides a context for discovering other forms of bonding and being (Da Paz et al., 2009), providing well-being (Buzinde, 2020). Mindful travel is an innovative tendency to increase it for people, destinations, and organizations.

This chapter aims to present, in innovative terms, a pioneering approach, which addresses a still unexplored relationship between spirituality and mindful tourism. In addition, it aims to present innovative cases of mindful tourism in the Iberian Peninsula through the self-reporting of experiences.

The exploratory study now presented contributes to the literature on tourism, evidencing the importance of promoting well-being through travel tourism linked to spirituality, self-discovery, and self-knowledge. This pioneering contribution maps several reference cases in the Iberian space, where it is possible to learn how to better deal with stress, anxiety, burnout, and emotional imbalances through innovative tourism experiences. More than rest, it is essential to know how to stop, experience, and listen to the deepening of our soul so that we can continue stronger, balanced, and leaders of our emotions.

We apply a self-narrative method of the second author's experiences to discuss mindful travel. We analyze Spanish projects through the lenses of secular spirituality. In this chapter, we go through this tendency in tourism and see the practical benefits of mindful travel and holidays. We also make suggestions for future investigations and innovative entrepreneurial projects in the Iberian Peninsula.

The remainder of the chapter is structured as follows. Firstly, a literature review is presented, linking spirituality and mindful tourism. Secondly, a set of illustrative Spanish cases is self-reported. Thirdly, a future action agenda and project suggestions are presented. Lastly, the concluding remarks are provided.

2 Literature Review

2.1 Spirituality

Secular spirituality is a way of being, living, and experiencing that comes about through an awareness of a transcendent dimension. Specific identifiable values characterize that regarding self, nature, life, others, and what one believes to be the Ultimate (Elkins et al., 1988, p. 10). One understands that all life is valuable and sacred because everything is part of the whole; nothing is isolated or separated. Everything is linked, and every action matters. It is a construct related to religion, but they are not synonyms (Bratianu, 2015; Elkins et al., 1988; Rocha & Pinheiro, 2020). "A religion is a solidarity system of beliefs and practices concerning sacred things" (Durkheim, 1960, p. 65).

Spiritual knowledge concerns understanding the meaning of our existence, our values in society and organizations, and how they influence decision-making (Bratianu, 2015). The application of spiritual knowledge is related to spiritual intelligence; it is a framework for recognizing and arrange competencies and capabilities for the use of spirituality (Bratianu, 2015; Emmons, 1999). Spiritual intelligence can be a powerful transformational force, broadening our understanding of reality (Bratianu, 2015).

The process of spiritual growth starts with questions, "what is my purpose?" "why am I alive?" "how can I be better?" "what I need to do differently?" and so on (Cavanagh & Bandsuch, 2002; Mitroff & Denton, 1999). To address these questions, one needs to look inside to find the answers because each one of us has a different spiritual path to walk. Moments of peace, silence, pray, meditation, and artistic expressions (singing, painting, sculpting, and handicraft) are means that help people an insight into their spirituality. Nowadays, spiritual values, philosophies, and practices are in organizations, travels, retreats, best-sellers, mass media, social networks, and magazines. We cannot deny that they are even more present in our everyday life.

Everyone has their path to walk. So, not everybody is ready to understand spirituality. One should be aware of their spirituality, but it is not necessary (Nogués, 2016). One can continue living with no purpose, unconsciously avoiding whatever comes from our deep inside. Believing only in the material realm, in what we see and touch.

We highlight that all necessities of the body ought to be satisfied in order to be able to search for spiritual growth (Tischler, 1999). Once that spirituality goes beyond the material realm, a person needs to have their body fully healthy, safe, and well cared for to begin their spiritual evolution (Tischler, 1999). It invites us to relish every moment of our existence, forget about what is unessential, and learn to develop all aspects of our lives. To create significant values and take all the actions to accomplish them. A person cannot do that with an empty stomach or worrying about their small payment.

In the last decades, spirituality has been included in the management literature (Brown, 2003; Friedman et al., 2005; Karakas, 2010; Mitroff & Denton, 1999; Poole, 2009; Rocha & Pinheiro, 2020). There are some warnings about it. First, the rhetorical use of spiritual discourse is targeted to increase productivity and serve as a tool for capitalistic goals (Ul-Haq, 2020). Second, the mystic around it can inhibit organizational learning and drive out organizational development (Friedman et al., 2005). Spirituality, thus, ought to be an end in itself. The path of awareness and transcendence with practical outcomes in one daily life (Elkins et al., 1988). It is not just a discourse neither a managerial tool. It is the metaphysical part of each of us, and we need to take care of it, as we do with our body and mind.

Following Rocha and Pinheiro (2020), organizational spirituality is a construct that addresses spirituality in management in the holistic view of a company, integrating spiritual leadership, members" spirituality, workplace spirituality, and organizational spirituality, combining all others. The former is also an organizational identity, resulting from its values, discourse, and practices. It is visible in its image, mission, vision, and values. In this line of action, the leader guides individual and collective spirituality toward organizational spirituality.

Members" spiritual experience in the workplace is called workplace spirituality (Pawar, 2017). Giacalone and Jurkiewicz (2003) defend that "workplace spirituality is a framework of organizational values evidenced in the culture that promotes employees experience of transcendence through the work process, facilitating their sense of being connected to others in a way that provided feelings of completeness and joy." It can be addressed according to three main perspectives: "(a) Human resources: Spirituality enhances employee well-being and quality of life; (b) Philosophical: Spirituality provides employees a sense of purpose and meaning at work; and (c) Interpersonal: Spirituality provides employees a sense of interconnectedness and community" (Karakas, 2010, p. 92).

In this vein, people also can find meaning and life purpose in their jobs (Driver, 2007; Thakur & Singh, 2016). It is a crucial and valuable aspect of everyday life. We need to consider the professional field regarding meaning and spirituality because we spend countless hours in our workplace, enriching society, and moving the world. Creativity, productivity, and performance are the quality of a person's inner work life, the mix of perceptions, emotions, and motivations over a working day. That is their happiness, motivation, and interest in the work they do, the company's appreciation, and their relationships with their boss, their team, or themselves. All combined will lead to success or drag one to failure (Amabile & Kramer, 2011). This success or failure is a vital aspect of our well-being. Science and spirituality tend to converge again (Kapra, 1983). Spirituality can drive quality of life (Jeste et al., 2021; Negi, 2020), job satisfaction, commitment (Rego & Pina e Cunha, 2008), well-being, and resilience (Gogo et al., 2019).

We must distinguish between wellness and well-being. Both refer to the sensation of feeling good in our lives, but with different connotations. Wellness is superficial and refers to hedonistic happiness. Hedonism promotes positive experiences and pleasure sensations, trying to avoid painful and negative experiences. For its turn, well-being refers to Eudemonia (good spirit), something more than just objective feels good, but to identify and cultivate the strengths, positive, solid, and transcendental values, and significances (Pallarés, 2018).

Furthermore, happiness is not a state but a process. An eternal path to keep on becoming our better version of ourselves. Hence, feeling good goes beyond avoiding painful and negative experiences and failures. We understand them as ways to learn and grow up. It is only possible when we learn to use our inner capabilities, tools and manage our emotions (Kabat-zinn, 2012).

2.2 *Mindful Tourism*

The typical image of tourism in Hispanic countries is just the sun and the beach. Fortunately, this is far more than this for the country and the tourists in Spain's case, although we bear in mind the broad picture of basic tourism. Society is evolving and changing, so does tourism, and how and why we travel. The COVID-19 pandemic has boosted these changes. After the lockdown and the measures to fight the virus, we realize how important is our well-being; we see how the professional trends are changing; how markets became more globalized, complex, and, especially, uncertain (Noah Harari, 2018).

Employees ought to have other capabilities besides rational knowledge, such as soft skills, creativity, emotional intelligence, leadership, entrepreneurship, decision-making, among others (Noah Harari, 2018). New pathologies are emerging because of the changes we live as a society (Maslach et al., 2001; Parker & DeCotiis, 1983; Sanne et al., 2005; Ul-Haq, 2020). Stress and anxiety provoke a significant number of burnout syndromes, depressions are spreading faster, and that is because we cannot manage the intricate lifestyle we create with new technologies, responsibilities, and work–life balance. If we do not learn to cope with it, the results will worsen exponentially (Noah Harari, 2018). Society progress (mostly measured through economic and technological indicators) shows a decrease in society's well-being, with a high level of depression, stress, and even suicide rates (Wiking et al., 2020).

The World Tourism Organization defines tourism as "a social, cultural and economic phenomenon which entails the movement of people to countries or places outside their usual environment for personal or business/professional purposes. These people are called visitors (which may be either tourists or excursionists; residents or non-residents), and tourism has to do with their activities, some of which involve tourism expenditure" (UNWTO, glossary terms). These purposes are related to people's interests. For example, those who like literature and want to visit where their favorite writers lived, or those who like cinema and wish to visit movie locations. The success of movies, such as the Crocodile Dundee increased travels to Australia; the books and movies of The Lord of the Rings influenced tourists flying to New Zealand and Paulo Coelho and the pilgrimage to Santiago (Chias, 2005). The best-seller Eat Pray Love address its author's spiritual journey popularizing it (Putcha, 2020).

According to the Global Wellness Institute, in 2018, wellness tourism represented a market of $639 B (Sánchez, 2020) and increased more than 40% between 2014 and 2017. Studies reported that wellness tourism would rise to nearly $919 B by 2022. This growth in the demand for well-being in the last 5 years is reflected in several segments (Tarrés, 2020): best-sellers books, social media content, specialized

Fig. 1 A police department in Toronto, Canada (Omar, 2016). Source: See specific information on: https://www.huffingtonpost.ca/2016/04/13/peel-police-meditation_n_9684274.html

magazines, and practices like Mindfulness, Yoga, Tai-chi are present in programs for companies (Google, Apple, and others). For example, the Army and police departments (see Fig. 1) also develop mindfulness among the troops, professional sports leagues like the NBA (Esgueva, 2016), and everyday life. People are more interested in such matters; it is expected that they want to practice it during their holidays.

The pandemic has truncated everything relating to tourism. However, this crisis has also aroused a sense of importance about well-being among people, the tourism which is embracing this field will grow up faster than others. Not only because that was the tendency before COVID-19, but also because it helps people to transform themselves into a better lifestyle. "In the spa and wellness world, the transformational benefits of treatments, services, or programs can honestly be considered as meaningful. Respecting local, natural resources, cultural traditions, and heritage rituals can generate transformative experiences" (Puczkó et al., 2021, p. 11).

In the holiday season in 2020, people were looking for places with no agglomerations, in the countryside and nature, in small and rural hotels with known people (friends or relatives) or little groups, and with authentic experiences useful for their well-being and health. A great number of people will keep this way of travel (Sánchez, 2020). This kind of tourism is the Mindful Travel or Wellness Tourism, although both terms have different connotations.

Mindful travel is a kind of tourism that focuses on the individual's well-being, destinations, and organizations (Tarrés, 2020). It is not just a relaxing experience but also a transformative one (Reisinger, 2013; Puczkó et al., 2021). As presented in Table 1, mindful travel congregates activities and practices with touristic experiences,

Table 1 Mindful travel activities and practices: examples and description

Activities	Description
Yoga (Iyengar, 2017)	It intends to balance body and mind through different movements called asanas. It has been practiced in India for more than 5000 years, and exists in different kinds, like Iyengar, kundalini, Vinyasa, Ashtanga, prenatal, and Aereal. It was introduced on occident during the late 1960s.
Tai-Chi (Bunnag, 2006)	It is a millenary practice born in China. It consists of a sequence of slow movements symbolizing different aspects of nature and animals, harmonizing body and mind. It is a way of living and philosophy to understand life and the cosmos through movement and consciousness.
Mindfulness (Kabat-Zinn, 2012)	It is based on breathing focused to bring attention to the here and now. A way to sense the present time consciously with no irrelevant thoughts and emotions. Mindfulness is the name given to the meditation practiced by Buddhism for thousands of years when occidentals brought it to the occident.
Sound and music therapy (Goldman, 2002)	It is a range of therapies using sounds to treat physical and mental conditions. It is based on the idea that our body is vibration, and the practitioner also uses vibrations of different frequencies belonging to special instruments like Tibetan or crystal singing bowls, tuning forks, gongs, drums, and the voice. It restores a healthy balance to the body, reduces stress, high blood pressure, and depression.
Aroma therapy (Buckle & Buckle, 2003)	It is based on aromatic materials, including essential oils and other aromatic compounds, to improve psychological or physical Well-being. It uses mixtures of essential oils and other products derived from vegetables. It can be used by inhalation, immersion in flavored water, or topical application.
Silence (Nht & Woren, 2015)	Noise is a source of stress, and as our society becomes noisier, silence provides many benefits to our mental health. It is crucial to experience silence around us, but also inside us. Our thoughts with no control become a high level of inner noise. Through mindfulness practices, we learn to manage our thoughts, manage our emotions, and increase a sense of silence inside us. Listen and feel nature sounds like birds singing, calmy waters of rivers or beaches, and tree leaves" movement also produce the sense of silence and harmony within.
Shinrin Yoku or Forest Bathing (Li, 2018)	It consists of taking about 2 h of mindful wandering through a forest, focusing the attention on the sensation of sounds, silences, colors, lights, and textures, producing high doses of Well-being. Old trees also produce phytoncides, oil essentials we can breathe with a proven benefit for our immune system. This practice which started in Japan in the 8 has expanded in the last years. It is also helpful for digital detox experiences, reduces stress and anxiety, even blood sugar.
Qi-Gong (Cohen, 1997)	It is a Chinese medicinal therapy based on breathing control. It helps eliminate tensions and stress, being a trustworthy source of peace that provides health and vitality. It is similar to tai-chi.
Cold Bath (Hof, 2020)	It is a well-extended practice in the northern countries. People immerse the body into cold water and stay for some minutes

(continued)

Table 1 (continued)

Activities	Description
	while breathing deeply and slowly. It is also practiced in some forest bathing in Japan, where the participants finish it under a waterfall at five or six degrees. This uncomfortable activity has many health benefits, such as developing the vagus nerve and the central nervous system, easing aching muscles, improving circulation, relieves depression, keeping skin healthy, strengthening immunity, and increasing energy and Well-being.

most of them belonging to a very ancient spiritual tradition or philosophy adapted to nowadays. Depending on the person and the destination, we can use different techniques, and they even can be mixed with themselves or with some other kind of activity, for example, gastronomy, enology, sports, arts, culture, hiking, and trekking.

We face new and complex challenges, which we must solve in a new level of awareness. We have a lot of information and knowledge, but we lack wisdom; we do not know how to use that knowledge to expand the whole of our life: individual, professional, and social (Nonaka & Takeuchi, 2019). A still limited number of institutions teach such subjects concerning well-being, creativity, emotional intelligence, personnel and effective communication, entrepreneurship, citizenship, sustainability, social responsibility, and social leadership, especially, to children and young students. Primary and Secondary schools keep on teaching the same technical and academic classes as it has been doing for centuries. We also have much more new knowledge, but there is a lack of practical wisdom to use it properly and significantly (Noah Harari, 2018; Nonaka & Takeuchi, 2019).

If one thinks that just 2 or 4 weeks of relaxing doing nothing in a swimming pool with an all-included bracelet is enough to feel better, it is wrong. To successfully face the current challenges and problems, we ought to change our minds and transform our lifestyles (Tarrés, 2020). If not, after 2 weeks after the holidays, we will be feeling stressed again; and this gets worse year after year (Tarrés, 2020). Therefore, by using our holidays to learn how to improve our lives and our world, we can relax and learn some tips, techniques, and beneficial habits to know what to do when we feel overwhelmed and stressed. Mindfulness is a great tool to do it; so are Yoga and other techniques and practices (Esgueva, 2016).

Positioning Mindful Travel as a way of tourism helps transform mentalities because it uses well-being and transcendent awareness to change our lives. People who like that way of travel require that destinations take care of the natural environment, develop environmental, social responsibility, and sustainability policies. Everything keeps as natural and authentic as possible. It is also of great value that organizations and companies develop programs where their employees and teams feel better (sciences of well-being and happiness) (Johnson & Park, 2020). Mindful tourists are usually capable of feeling and see the details of a destination. If employees feel happy and sound, the atmosphere will be positive for the experience and everything involved (Johnson & Park, 2020). There are benefits for the employees, destinations, companies, and all indirectly involved.

3 Spanish Cases: Reporting Self-Experiences

Mindful Travel tendencies have grown all over the world (Fraiz & Rodríguez, 2017). It has been developing because of people's natural interests and attitudes that enjoy such experiences or think that they are helpful to improve their lifestyle (Tarrés, 2020). In this section, the second author reports his experience with mindful travel.

In Spain, two destinations acknowledged the importance of implementing this new tendency. In the north, both in the east, Girona Costa Brava—Pirineu, in the north of Catalunya (Fig. 2), and the other in the west, and Ribeira Sacra, Galicia (Fig. 3).

The first destination that started to develop was the Patronat de Turisme Girona Costa Brava—Pirineus,[1] a tourism organization aiming to promote Girona's beautiful region. In Girona and the Bay of Roses, one can ski in the morning and swim on the beach in the afternoon. They changed the marketing club of Beauty and Health

Fig. 2 Map of Spain with the Girona destination in red. For more details, please consult: https://en.wikipedia.org/wiki/Province_of_Girona#/media/File:Girona_in_Spain.svg

[1] Additional information available here: https://en.costabrava.org/

Fig. 3 Map of Ribeira Sacra, in Galicia, northwest of Spain. Additional information available here: https://en.wikipedia.org/wiki/Ribeira_Sacra_(DO)

into Health and Wellness, and new projects relating to wellness, well-being, awareness, health, and medical joined the project. I was one of them. Before that, I was performing as a sound therapist, using sound as a therapy. It is handy to calm down, develop inner peace, reduce stress, and anxiety. When they invited me to join the club, I realized that the possibilities were unique by mixing the fields of awareness and travel: develop something new and bring my vocation to a very different area, full of possibilities.

Initially, I saw no information or academic courses, so I had to learn from myself, and it was something that I wanted to dive into. Moreover, the more I went through it, the more I enjoyed it. So, I ended up experimenting and writing about the experiences I was designing and organizing, which was the beginning of my book, Mindful Travel XP. After some time creating experiences, I contacted professionals with great knowledge concerning tourism, such as professors, professionals, and researchers. That was crucial to realizing that well-being concerning tourism was a great field to develop. Therefore, I started to organize two crucial things relating to

Wellness Tourism in 2019: Tourism and Spirituality, a postgraduate in the University of Girona,[2] and the Mindful Travel Meet, a congress in Andorra.

Both projects were successful in the first edition. Unfortunately, due to the COVID-19 pandemic, in 2020, we delayed the continuation of both projects. However, now in 2021, we are starting to start them over.

Mindful travel Destinations[3] is an organization led by Jimmy Pons. We work together to spread the Mindful Travel philosophy through experiences, training, and events. They have a long experience in the tourism field and realized the importance of business development and improving society's lifestyle. They were part of the organization in the first Mindful Travel Meet Congress,[4] and now we are planning the second edition. In the first one, there came projects from all over Spain. We realize that few projects are working with the same objective, but they go on their way, with no leading association. Some of them did not know about this kind of tourism, although they were doing something similar. After the congress, all the participants and the organization signed a commitment agreement to work in the same direction,[5] values, and ideas.

Concerning the postgraduate of Tourism and Awareness, the main purpose is to contribute with an academic background to a new tendency which is growing fast. For one side, to give tools and knowledge to professionals of tourism. For the other, to show to professionals of well-being, therapists, and coaches that there is another exciting field to develop, and so they can spread their practices. It aims to support their entrance to a new niche expanding their business by giving them tools to innovate and adapt their practices and techniques toward mindful tourism. The first edition had students from different backgrounds from all Catalunya.

At the same time, I started a new project as a training consultant in the Ribeira Sacra. The Ribeira Sacra is closer to becoming a World Heritage Site. His candidacy for World Heritage has managed to be among the 24 international nominations to be held in 2021.[6]

In this context, they realized that adding a new dimension, of well-being, to such an incredible and fantastic ambiance would be of great value. In October 2019, I imparted a course about Forest Bathing, Nature and well-being applied to tourism experiences. That was in Ribeira Sacra's core, for a group of more than 25 people, and I taught how to use nature to create well-being experiences for tourism. The course was sold out in 3 days. There came people from all over Galicia and other communities of Spain. Another significant sign of the importance of this tendency.

---

[2] Additional information available here: https://www.fundacioudg.org/es/curs-postgrau-turisme-espiritualitat-eines-innovacio-benestar-consciencia.html

[3] See specific information on: http://mindfultraveldestinations.com

[4] Additional information available here: https://www.facebook.com/Mindfultravel.meet

[5] For more details, please consult: https://www.edgartarres.com/2021/03/08/mindful-travel-meet/

[6] See specific information on: https://turismo.ribeirasacra.org/en/news/oficialmente-candidata-a-patrimonio-de-la-humanidad-2021

Fig. 4 Bay of Roses, Costa Brava, Spain (Sara Lladó personal archive)

In the north of the Costa Brava, just 20 km from the French border, the Bay of Roses is in the club of The Most Beautiful Bays of the World (Fig. 4). In 2019, Roses, Castelló d'Empúries, Sant Pere Pescador, and L'Escala municipalities, which belong to the bay, started to work together to promote them as well-being destinations. We designed the program "Feel the Bay" to experience the destination through well-being related to Nature, Silence, and History Heritage, in November offseason. The outcomes were better than we expected because we were sold out in all the activities. We had people coming from different parts of Catalunya, and although it was a morning or an afternoon session (depending on the municipality), some of them decided to stay the whole weekend. In 2020, though COVID-19, we could celebrate the second edition. Because of the pandemic regulations, we had to move the experiences from November to Christmas vacations. We must say that we were afraid of the change because we could not go beyond 2020 and have some days left during these winter festivities. Furthermore, again, we had most of the activities sold out. The first one was empty because of the Tramontana, the strong wind (about 100 km/h), which blows in that area of Alt Empordà (Girona), and it was impossible to stay outside and meditate.

In the "Feel the Bay" program, we use Yoga, tai-chi, mindfulness, forest bathing to design it. We mix experiences with cultural heritage and nature. People love to take care of themselves in such beautiful places.

After all these experiences, we can ensure that we will continue developing and spreading this travel way. In Figs. 5 and 6, two samples of experiences. It is an enjoyable experience, and participants feel very good when they live it. They feel they are doing something extraordinary. Figure 5 is a concert in the Astronomy Observatory of Albanyà (Girona, Spain). They were sitting in hot wooden banks,

Fig. 5 A concert in the Astronomical Observatory of Albanyà, Girona (Edgar Tarrés personal archive)

Fig. 6 A concert in the sea at Cap de Creus's Natural Park, Girona (Edgar Tarrés personal archive)

tiny lights, and aromatic herbs. An astronomer explained shortly about the Moon, Saturn, Galaxy of Andromeda, and the Deep Space, as we watched it live. At the same time, I performed a concert with Tibetan bowls, gongs, overtone singing chants, and drums. The sound drives the participants to understand and feel the Universe.

Figure 6 is a concert out in the sea, under some cliffs of Cap de Creus's Natural Park (Girona). The participants and I saw the sunset, having a bath in an old crater underwater. The dinner was typical Catalan food. We finish it with a concert of Tibetan bowls and gongs under the full moon about midnight.

Both experiences are connecting the participants with the Universe. They felt special to be able to connect with nature, feeling new experiences. It also makes the participants have a good time enjoying nature, having fun, and producing positive results in their health and well-being.

Mindful Travel is an open field to tourism innovation. Part of my training is about creativity. I do not teach "a specific way" to create and design an experience. Every designer must use what they are passionate about, what they like, what they know to create singular experiences.

Hence, there are no specific activities to develop mindful travel experiences. We can use different techniques, such as Mindfulness, Yoga, Tai-chi, Forest Bathing, Sound Therapy, and others in spiritual and well-being. Such practices aim to increase a person's health and well-being. In brief, Mindfulness is about focus and breathing through inner silence (Kabat-Zinn, 2016). Yoga and Tai-Chi empower our vitality, strengthen our muscles, and bring inner harmony through physical movement (Bunnag, 2006; Iyengar, 2017). Forest Bathing is a mindful wandering through a forest, going slowly, in silence, feeling the sensations, and the beauty (Li, 2018). Sound therapy uses the frequencies and harmonics of some special instruments like Tibetan bowls, gongs, drums, didgeridoos, the voice, and others handmade belonging to different spiritual traditions to calm and bring inner peace for body and mind (Jauset, 2010). All those techniques are helpful to strengthen our immune system, reduce stress and anxiety, reduce heart rate and sugar blood, and have lots of more health benefits for our body and mind. It depends on the designer's knowledge, creativity, and wisdom; they will offer something valuable and singular in this field. The experience enriches when we can mix it with art, gastronomy, cultural heritage, and sports.

When I teach forest bathing in my training program, I teach the basics about the forest. However, I ask participants to use what they like, which can be art, gastronomy, sports, or anything. They include this as a particular value for the experience. If not, we will find that everybody will be performing the same forest bathing. I believe that every person's uniqueness can create something great, remarkable, and extraordinary, significantly impacting people visiting a destination. If such experience leads to improving and empowering people's well-being, we can help participants decrease their stress and work-related illnesses.

4 Future Action Agenda and Project Suggestions

We present insights about mindful travel in this topic, both to a future research agenda and project suggestions. First, every crisis always shows a new path to walk, a new change to apply. It will not be easy. However, those with a mindful mentality will have more significant opportunities in the new context we are starting to live in from now on. All those ancient practices have implications on well-being and creativity, two aspects very useful in the new times we are starting to live: very important for an entrepreneurship attitude.

Destinations with a long tourism tradition, especially those of sand and beach, have more difficulties in becoming mindful travel destinations. Furthermore, this is because the mindful traveler requires a place with no agglomerations, authentic, and singular natural ambiance. Although it is a topic, most sun and beach destinations are all quite similar, with big hotels in the sandy beachfront row, various shops selling the same kind of souvenirs with many restaurants. However, becoming a mindful travel destination is possible; they would have to fight against something hardly embedded in tourists minds. Consequently, municipalities with no tourism tradition have an opportunity to work together if they know how to, and public administration, private sector (accommodation, restaurants, services, and stakeholders). Even among different municipalities, when a whole group of them work together, they create a significant destination with riches and different resources.

Along with the pandemic, some of them have realized that they own an opportunity. They have places that nobody has been to before. Therefore, if they can design great experiences, they will be willing to gather people looking for something new. Such destinations are not in the tourist's mind, so as a mindful travel experience is new and trendy, the media quickly talks about it, which is beneficial for promotion. It is a value that we cannot obviate.

From the Institute of Silence,[7] they have started to work on wellness tourism to enrich a region. It is trendy a region embracing this way of travel, and it might bring prosperity. With specific training, a community of hotels, restaurants, and people interested in well-being can transform their projects into mindful experiences. Employees and entrepreneurs can develop functional capabilities and tools for well-being and creativity, so a region enriches in very different ways.

In Portugal, places like the natural park of Serra da Estrela, or the Peneda-Gerês and the transboundary biosphere reserve of Gerês-Xurés, are projects and professionals" spiritual well-being can be easily transformed into mindful travel destinations including the use of thermal water can be added to the mindful activities. In both places, we can practice mindfulness and forest bathing along the different routes. There are countless resources, like caves and rivers, in which one can enjoy the silence, develop senses, and rest.

For example, in Serra da Estrela, Ananda Marga or Ananda Valley can bring much value to the zone if they converge with other ventures and create a big project for the region. There is also the GeoparkEstrela,[8] recognized by UNESCO, in which we can find places to stay in silence, trees, caves, little rivers, trekking routes, viewpoints, and protected landscapes.

---

[7] Additional information available here: http://institutdelsilenci.com

[8] See specific information on: https://www.geoparkestrela.pt/

Also, in this geographical context, the recent initiative promoted by the Dark Sky Aldeias do Xisto and Observation of Stars[9] deserve to be outlined since it has the potential to promote cross-fertilization between the reported mindful tourism and astrotourism, targeted to the promotion of well-being and personnel balance.

It is not about organizing Yoga or mindfulness classes, for instance. One ought to adapt these activities into a tourism experience. Both professionals of well-being and tourism should combine ventures to develop such a project. Training is of the utmost importance when we want to develop a project of such characteristics.

Spirituality is increasingly present in management-related research (Steingard, 2005). However, research is still sparse and in need of orientation. Concerning academics" proposals, mindful tourism can be a new field. The need to use empirical methods is latent, mainly qualitative methods that can grasp spirituality's metaphysical essence (Rocha & Pinheiro, 2020). Concerning spiritual tourism and entrepreneurship in the Iberian Peninsula, we present the following proposals.

Empirical research with experiments is the best way to understand the results of mindful travel projects. Analyzing the feelings and emotions before and after the employees participate in the projects. Case studies with successful and failed cases. It will be valuable to design other ventures. Longitudinal studies are also of great value, as they can identify the medium- and long-term effects of participation in these projects, especially concerning employees" well-being and economic and social changes in the projects" host cities. Inclusive, these qualitative researches can build grounded theories and provide implications for designing a new generation of public policies targeted to sustainable development and quality of life.

Investigation about these kinds of tourists" expectations and characteristics, such as their nationality, religion, languages, age, gender, number of persons traveling together, traveling with pets or not, and budget. It will be an essential decision-making basis for future endeavors, investments, and policies centered on the end-users, that is, the persons.

5 Concluding Remarks

This chapter introduced in the literature on tourism a still unexplored relationship, that is, the cross-fertilization between spirituality and mindful tourism. Despite being an exploratory study, this is a small step to trigger the development of future research and planned policy actions on new modalities of end-user-focused tourism, i.e., the person.

The innovation here lies in matching the identified and poorly satisfied need to improve the person, as a complex and spiritual being, who continuously seeks the individual balance in the context of a highly demanding, pressing, and uncertainty-laden society.

[9] For more details, please consult: https://aldeiasdoxisto.pt/noticia/5633

This is also an entrepreneurial business opportunity to explore, not only in the reported context of the Iberian Peninsula but, above all, in the landscapes and mountains of opportunities to explore in low-density territories, where the ancestral nature of culture, traditions, uses, and customs, allows to ensure the ideal conditions to silence and refined the essence of our soul.

An important insight of this chapter originates from the experiences reported by the second author, which draw attention to the importance of transforming and innovating tourism, transporting it to a new transcendental and metaphysical level, where the goal is not simply to rest but to raise the person to an upper level of self-knowledge and improvement, which allows her/him to face, with greater intelligence, effectiveness and balance, the pressures of society, increasingly unequal and unbalanced.

At the level of policy decision-making, it is not a question of lifting tourism policies but of truly innovating the usefulness afforded to people through the enjoyment of renewed experiences, which empower them and make them healthier and more resilient. In addition, the importance of designing a new generation of public policies that use mindful tourism to add value to the main elements of nature through the valorization of spirituality should be underlined. In conjunction with new maps of protected landscapes, the future creation of spirituality destinations may push positive spillovers to the interrelations between the five tangible elements, which coexist in nature in the purest form.

Finally, making use of the lessons coming from the traditional Chinese philosophy, which highlights the interrelations between earth, wood, fire, water, and metal, symbolizing the changing and complementary character that is present in everything that surrounds living beings, it is necessary to harmonize the cycle between the five elements, adding spirituality to it, as a catalyst for new forms of mindful tourism, balanced and sustainable, where each element gives rise to another, without compromising the chain of unity and the sustainability of ecosystems.

References

Amabile & Kramer. (2011). *The progress principle: Using small wins to ignite joy, engagement and creativity at work*. Harvard Business Review Press.

Bratianu, C. (2015). Spiritual knowledge. In *Organizational knowledge dynamics: Managing knowledge creation, acquisition, sharing, and transformation* (pp. 72–102). IGI Global. https://doi.org/10.4018/978-1-4666-8318-1.ch004.

Brown, R. B. (2003). Organizational spirituality: The sceptic's version. *Organization, 10*(2), 393–400. https://doi.org/10.1177/1350508403010002013

Buckle, J., & Buckle, J. (2003). *Clinical aromatherapy: Essential oils in practice*. Churchill Livingstone.

Bunnag, T. (2006). *La esencia del T'ai Chi Ch'uan*. La liebre de marzo.

Buzinde, C. N. (2020). Theoretical linkages between well-being and tourism: The case of self-determination theory and spiritual tourism. *Annals of Tourism Research, 83*(September 2019), 102920. https://doi.org/10.1016/j.annals.2020.102920

Cavanagh, G. F., & Bandsuch, M. R. (2002). Virtue as a benchmark for spirituality in business. *Journal of Business Ethics, 38*, 109–117. https://doi.org/10.1023/A:1015721029457
Chias, J. (2005). *El Negocio de la Felicidad*. Pearson Educación S.A.
Cohen, K. (1997). *The way of Qigong: The art and science of Chinese energy healing*. Bantam.
Driver, M. (2007). Meaning and suffering in organizations. *Journal of Organizational Change Management, 20*(5), 611–632. https://doi.org/10.1108/09534810710779063
Durkheim, E. (1960). *Les formes élémentaires de la vie religieuse: Le système totémique en Australie (4a)*. Quadrige PUF.
Elkins, D. N., Hedstrom, L. J., Hughes, L. L., Leaf, J. A., & Saunders, C. (1988). Toward a humanistic-phenomenological spirituality. *Journal of Humanistic Psychology, 28*(4), 5–18. https://doi.org/10.1177/0022167888284002
Emmons, R. A. (1999). Religion in the psychology of personality: An introduction. *Journal of Personality, 67*(6), 874–888. https://doi.org/10.1111/1467-6494.00076
Esgueva, M. (2016). *Mindfulness, una guía práctica*. Zenith.
Fraiz, J. A. (dir.), & Rodríguez, L. (2017). *Turismo Médico y Turismo de Salud. Tendencias actuales del turismo en el mundo*. Ed. Síntesis.
Friedman, V. J., Lipshitz, R., & Popper, M. (2005). The mystification of organizational learning. *Journal of Management Inquiry, 14*(1), 19–30. https://doi.org/10.1177/1056492604273758
Giacalone, R. A., & Jurkiewicz, C. L. (2003). Toward a science of workplace spirituality. In C. L. Jurkiewicz & R. A. Giacalone (Ed.), *Handbook of workplace spirituality and organizational performance* (pp. 3–28). M.E. Sharpe.
Gogo, A., Osta, A., McClafferty, H., & Rana, D. T. (2019). Cultivating a way of being and doing: Individual strategies for physician well-being and resilience. *Current Problems in Pediatric and Adolescent Health Care, 49*(12), 100663. https://doi.org/10.1016/j.cppeds.2019.100663
Goldman, J. (2002). *Healing sounds: The power of harmonics*. Element.
Hof, W. (2020). *The Wim Hof method: Activate your potential, transcend your limits*. Random House.
Iyengar, B. (2017). *Light on yoga*. Harper Collins Publishing India.
Jauset, J. A. (2010). *Sonido, Música y Espiritualidad*. Gaia Ediciones.
Jeste, D. V., Thomas, M. L., Liu, J., Daly, R. E., Tu, X. M., Treichler, E. B. H., Palmer, B. W., & Lee, E. E. (2021). Is spirituality a component of wisdom? Study of 1,786 adults using expanded San Diego wisdom scale (Jeste-Thomas wisdom index). *Journal of Psychiatric Research, 132*(September 2020), 174–181. https://doi.org/10.1016/j.jpsychires.2020.09.033
Johnson, K., & Park, S. (2020). *Mindfulness Training for Tourism and Hospitality frontline employees*. Emerald Group.
Kabat-Zinn, J. (2012). *Mindfulness for beginners: Reclaiming the present moment—And your life*. Sounds True.
Kabat-Zinn, J. (2016). *Vivir con Plenitud las Crisis*. Editorial Kairós.
Kapra, F. (1983). *The Tao of physics: An exploration of the parallels between modern physics and Eastern mysticism*. Shambhala.
Karakas, F. (2010). Spirituality and performance in organizations: A literature review. *Journal of Business Ethics, 94*(1), 89–106. https://doi.org/10.1007/s10551-009-0251-5
Lang, J., Thomas, J. L., Bliese, P. D., & Adler, A. B. (2007). Job demands and job performance: The mediating effect of psychological and physical strain and the moderating effect of role clarity. *Journal of Occupational Health Psychology, 12*(2), 116–124. https://doi.org/10.1037/1076-8998.12.2.116
Leitão, J., Pereira, D., & Gonçalves, Â. (2021). Quality of work life and contribution to productivity: Assessing the moderator effects of burnout syndrome. *International Journal of Environmental Research and Public Health, 18*(5), 1–20.
Li, Q. (2018). *Shinrin Yoku. The art and science of forest bathing*. Roca Editorial.
Maslach, C., Schaufeli, W. B., & Leiter, M. P. (2001). Job Burnout. *Annual Review of Psychology, 52*(1), 397–422. https://doi.org/10.1146/annurev.psych.52.1.397

Mitroff, I. I., & Denton, E. A. (1999). A study of spirituality in the workplace. *Sloan Management Review*, 83–92.
Negi, A. S. (2020). *Spirituality Leads to Quality of Life: A Conceptual Study., 12*(11), 40–50.
Nht, H., & Woren, D. (2015). *Silence: The power of quiet in a world full of noise. Unabridged.* Blackstone Audio.
Noah Harari, Y. (2018). *21 lessons for the 21st century*. Edicions 62.
Nogués, R. (2016). *la salut espiritual*. Fragmenta editorial.
Nonaka, I., & Takeuchi, H. (2019). *The wise company: How companies create continuous innovation*. Oxford University Press.
Nonaka, I., & Takeuchi, H. (2021). Humanizing strategy. *Long Range Planning*. https://doi.org/10.1016/j.lrp.2021.102070
Omar. (2016). *Meanwhile in Canada, peel regional officers meditate in a temple*. Retrieved March 21, 2021, from https://www.huffingtonpost.ca/2016/04/13/peel-police-meditation_n_9684274.html.
Pallarés, E. (2018). *la felicidad asequible y sostenible*. Ediciones mensajero.
Parker, D. F., & DeCotiis, T. A. (1983). Organizational determinants of job stress. *Organizational Behavior and Human Performance, 32*(2), 160–177. https://doi.org/10.1016/0030-5073(83)90145-9
Pawar, B. S. (2017). The relationship of individual spirituality and organizational spirituality with meaning and community at work. *Leadership & Organization Development Journal, 38*(7), 986–1003. https://doi.org/10.1108/LODJ-01-2016-0014
Da Paz, M. D. G. T., Gosendo, E. E. M., Dessen, M. C., & Mourão, R. G. V. (2009). Justiça organizacional e bem-estar pessoal nas organizações. *Estudos, Goiânia, 36*(1/2), 95–115.
Poole, E. (2009). Organisational spirituality—A literature review. *Journal of Business Ethics, 84*(4), 577–588. https://doi.org/10.1007/s10551-008-9726-z
Puczkó, L., Jacka, A., & others. (2021). *From wellness to transformation. Wellness during and post Covid-19 – Managing experiences*. Hall of Wellness Awards.
Putcha, R. S. (2020). After eat, pray, love: Tourism, orientalism, and cartographies of salvation. *Tourist Studies, 20*(4), 450–466. https://doi.org/10.1177/1468797620946808
Rego, A., & Pina e Cunha, M. (2008). Workplace spirituality and organizational commitment: An empirical study. *Journal of Organizational Change Management, 21*(1), 53–75. https://doi.org/10.1108/09534810810847039
Reisinger, Y. (2013). *Transformation and transformational learning theory. Transformational tourism*.
Rocha, R. G., & Pinheiro, P. G. (2020). Organizational spirituality: Concept and perspectives. *Journal of Business Ethics*. https://doi.org/10.1007/s10551-020-04463-y.
Sánchez, S. (2020). *Mindful(ness) y Turismo*. Ostelea School of Management.
Sanne, B., Mykletun, A., Dahl, A. A., Moen, B. E., & Tell, G. S. (2005). Testing the job demand–control–support model with anxiety and depression as outcomes: The Hordaland health study. *Occupational Medicine, 55*(6), 463–473. https://doi.org/10.1093/occmed/kqi071
Schick, A., Hobson, P. R., & Ibisch, P. L. (2017). Conservation and sustainable development in a VUCA world: The need for a systemic and ecosystem-based approach. *Ecosystem Health and Sustainability, 3*(4), e01267. https://doi.org/10.1002/ehs2.1267
Steingard, D. S. (2005). Spiritually-informed management theory: Toward profound possibilities for inquiry and transformation. *Journal of Management Inquiry, 14*(3), 227–241. https://doi.org/10.1177/1056492605276841
Tarrés, E. (2020). *Mindful travel XP*. A Mindful Hacker Edition.

Thakur, K., & Singh, J. (2016). Spirituality at workplace: A conceptual framework. *International Journal of Applied Business and Economic Research, 14*(7), 5181–5189. https://www.scopus.com/inward/record.uri?eid=2-s2.0-84986199865&partnerID=40&md5=706ac4511a33320e8f8ac89bfe51e586.

Tischler, L. (1999). The growing interest in spirituality in business. *Journal of Organizational Change Management, 12*(4), 273–280. https://doi.org/10.1108/09534819910282117

Ul-Haq, S. (2020). Spiritual development and meaningful work: A Habermasian critique. *Human Resource Development International, 23*(2), 125–145. https://doi.org/10.1080/13678868.2019.1679570

Wiking, M., et al. (2020). *Wellbeing adjusted life years. A universal metric to quantify the happiness return on investment*. Leaps by Bayer & Happiness Research Institute.

Considerations About the Economic Sustainability of Historic Housing Tourism

José Luís Braga and Oscar Silva

Abstract The methodology of data collection and analysis used in this study was Grounded Theory (GT). Its methods allowed the generation of a theory that emerged from the action that took place in the substantive area of HHT. GT has the advantage of not being restricted to one unit of analysis, since the categories that emanate from the data are generalizable. In this way, GT allows for the development of theory grounded in the data collected. Our sample for this study consisted of 53 interviews with owners and/or hosts of HHT and five participant observations held in the substantive area of HHT.

This chapter aims to explain the strategies used by the hosts or owners of manor houses to ensure the self-sustainability of the HHT. The research findings presented here should be regarded as an integrated set of conceptual hypotheses (i.e., probabilistic assertions) that intend to explain a substantial part of the behavioral patterns apprehended in the manor houses.

1 Introduction

Tourism has been the engine of economic growth in Portugal in recent years. In 2019, tourism in Portugal provided 1.047 million jobs, which is equivalent to 21.8% of total employment (PRESSTUR, 2019). Furthermore, it represents about 14.6% of the national GDP (TravelBI, 2019). Portugal is the fourth country according to the World Travel & Tourism Council (WTCC), with the highest rate of wealth generation by tourism (WEF, 2018). The increase of tourism activity in Portugal in recent years was due, among other factors, to the power of initiative of entrepreneurs and

J. L. Braga (✉)
Research, Development and Innovation Center, Institute of Higher Studies of Fafe (CIDI-IESF), Medelo, Portugal
e-mail: josebraga@iesfafe.pt

O. Silva
Polytechnic Institute of Higher Studies of Gaya (ISPGAYA), Vila Nova de Gaia, Portugal
e-mail: osilva@ispgaya.pt

J. Leitão et al. (eds.), *Tourism Entrepreneurship in Portugal and Spain*, Tourism, Hospitality & Event Management, https://doi.org/10.1007/978-3-030-89232-6_5

the natural conditions offered by the territory (Moreira, 2018). Lately, there has been international recognition of Portugal's attractiveness as a tourist destination. Hence this destination was distinguished in the World Travel Awards 2019, for the third consecutive year, and in the 11th edition of the Marketeer Awards. Moreover, in the World Golf Awards 2018, Portugal earned recognition as the best World and European golf destination. In Portugal, there are three regions where tourism has experienced significant development: Lisbon, Algarve, and Madeira.

The success of a tourist destination depends on its ability to meet the needs and preferences of tourists. Tourist needs are globally identified in the literature and substantiated in the look for new experiences and knowledge, rest and relaxation, and in the search for leisure activities, among others (Simková & Holzner, 2014; Wong et al., 2013). According to Mahika (2011), tourists are also moved by the desire to meet new places, people, cultivate new friendships, or strengthen personal relationships. The destination Porto and North of Portugal is characterized by a multiproduct offering, targeted at different market segments, in line with the National Strategic Plan for Tourism (Ministério da Economia e da Inovação, 2007) and the Action Plan for Tourism Development in Northern Portugal (Comissão de Coordenação e Desenvolvimento Regional do Norte-CCDRN, 2011). As regards the strategic guidelines relating to tourism products and differentiating attributes, the abovementioned destination decided to develop the following segments: (i) urban tourism; (ii) city breaks; (iii) business tourism; (iv) cultural tourism, events, scientific tourism, food and wine, touring and cultural landscape; (v) natural tourism, health, and wellness; (vi) nature tourism; and (vii) religious tourism.

The Strategic Plan of Tourism of Portugal, "Estratégia 2027," states as one of its strategic challenges to ensure the preservation and sustainable economic enhancement of cultural and natural heritage and of local identity.

Historic Housing Tourism (HHT) fits perfectly with the strategic vision of Turismo de Portugal since it is a type of tourism that aims at the recovery of manor houses (Pereiro, 2018). Moreover, this tourism product also includes the immersion in the history and culture of the house's surroundings through the family that owns it (Pereiro, 2009).

The methodology of data collection and analysis used in this study was Grounded Theory (GT). Its methods allowed the generation of a theory that emerged from the action that took place in the substantive area of HHT. GT has the advantage of not being restricted to one unit of analysis, since the categories that emanate from the data are generalizable. In this way, GT allows for the development of theory grounded in the data collected. Our sample consisted of 53 interviews with owners and/or hosts of HHT and five participant observations held in the substantive area of HHT.

This chapter aims to explain the strategies used by the hosts or owners of manor houses to ensure the self-sustainability of the HHT. Likewise, the research results presented here should be regarded as an integrated set of conceptual hypotheses (i.e., probabilistic assertions) that intend to explain a substantial part of the behavioral patterns apprehended in the manor houses (Braga, 2016).

This chapter is divided into six sections. Initially, the pandemic crisis and its impact on tourism are addressed. The opportunities that COVID-19 can bring and the need to focus on more sustainable, more informal forms of tourism, in which HHT is included, and that can contribute to the integrated development of tourism are also addressed. Next, a characterization of heritage tourism is made, in which HHT is inserted as a distinctive tourist product. In turn, the methods section and the research findings are explained in detail in Sects. 4 and 5. Finally, a model is devised to explain the problems faced and the solutions found by the hosts and owners of manor houses when trying to ensure the self-sustainability of the HHT.

2 Tourism and Sustainability in Times of Uncertainty

The pandemic effects caused by COVID-19 are having impact on the development of tourism activities worldwide (Gössling et al., 2021). Although many travel and tourism companies are used to include risk management and assessment models in their business planning, like it is stated in Ural (2015) and Ritchie and Jiang (2019), the nature of this pandemic has put these models under strain as their impacts manifest themselves on a global scale. Measures to contain and restrict the mobility of citizens have resulted in a significant reduction in tourism demand, especially in international travel, with massive flight cancellations by the world's leading airlines. According to Horta (2020), Portugal is one of the European countries where a strong reduction in tourism activity is expected, with a forecast of more than 40% decrease in the number of visitors. These values of economic contraction in tourism are only surpassed by Italy and Spain.

This strong impact on international tourism, whose real economic and social effects are still far from being accounted for, materialized in major crashes in commercial aviation, closed hotels, and businesses. These have resulted in heavy losses and large-scale unemployment (ILO, 2020). This panorama is worrying and has an impact on all economies. However, its effect will be potentially greater in countries with a small internal market and where foreign dependence is higher. Tourism is extremely important for Portugal as it is indicated in the studies conducted by Bento (2016) and Oliveira (2014).

Currently, there is a worldwide consensus that tourism is a key strategic axis for the sustained development of the territory. Portugal has been nominated in recent weeks as a destination of choice for living in the post COVID-19 period (Responsible Tourism, 2020).

Tourists are expected to bring new hygiene and sanitization demands, which causes tourism agents to have to implement clean and safety measures. This is intended to strengthen Portugal's image as a safe destination that has been able to adapt quickly to these new times. It is expected that tourists' priorities may change and that there will be greater demand for alternative destinations, where they can have unique experiences and where quality prevails over quantity. Destinations where there is a demand for social and environmental sustainability, with more nature and fewer crowds. With this in mind, small hotel units and resorts with a familiar approach, may become the most sought-after products by tourists in the post COVID-19 era. This opens a window of opportunity for the emergence of underexploited tourist sites, in areas like the countryside and peripheral regions. This could also be beneficial in combating the desertification certain regions of the country endure and reducing the social asymmetries between the coast and the interior of Portugal. In recent years, Portugal has attracted an increasing number of elderly people from Northern and Central Europe, who seek to live in this country for the remaining years of their lives. According to Lopes et al. (2020), key reasons for this choice are weather conditions and the existence of a good private health system that offers good conditions for a person to retire. Considering that this segment of the population is one of the most affected by COVID-19, the incentives to move to a safe and comfortable country in the last years of life are even greater. Another identified opportunity is the digitalization of tourist operations. The strong role of information technologies and alternative payment methods in the growth of tourism activity and in attracting new audiences is recognized (Almeida et al., 2019). Tour operators should accelerate these digitization processes and promote the establishment of distance businesses and the adoption of new digital platforms (e.g., social networks and virtual congresses) that allow the formalization of distance businesses and greater diversification of tourism promotion.

3 Heritage and Tourism

Heritage tourism dates back to ancient times. It is the oldest mode of tourism. There are documentary records alluding to the travels of merchants, sailors, and adventurers to visit the Great Pyramids and the Nile River. According to Timothy and Boyd (2003), these early explorers were among the first tourists. Centuries later, in the mid of the Middle Ages, there was a displacement of people who went to places where ancient remnants of culture lay. There they encountered great buildings, cathedrals, and works of art.

In the early modern period, the Grand Tour found adherents among the privileged classes. In the sixteenth and seventeenth centuries, this experience—which included visits to historic cities such as Paris, Milan, Venice, Florence, Rome, and Naples—was considered a cultural and educational enhancement (Goethe, 2018). The pioneers of this type of travel were the English, generally belonging to the landed gentry. However, by the late 1800s, this circuit was popular among lawyers, doctors, bankers, and merchants (Boyer, 2003; Roberts & Hall, 2001; Timothy, 2021). Even today, many of the towns that were part of this itinerary are still highly sought after for cultural and heritage tourism.

The terms "cultural tourism," "heritage tourism," "ethnic tourism," and "art tourism"—according to Timothy and Boyd (2003)—are used almost interchangeably, with no agreement as to their meaning.

In fact, cultural tourism goes beyond the mere visit to monuments or places to encompass the consumption of the way of life of the places visited. Thus, this activity brings together cultural products and contemporary culture, incorporating both heritage tourism (related to artifacts of the past) and art tourism (associated with contemporary cultural production).

Heritage tourism can also be regarded as an activity focused on what has been inherited, whether historic buildings, works of art, or stunning landscapes. According to Timothy (2021) heritage is "what we inherit from the past and use in the present day" (p. 3). This market segment can also be defined as the economic activity that uses sociocultural heritage resources to attract visitors. Thus, heritage tourism includes folk traditions, arts and crafts, ethnic history, social customs, and cultural celebrations (Chhabra, Healy & Sills, Chhabra et al., 2003).

Poria et al. (2003) distinguished two main approaches to this phenomenon:

1. View it as tourism in places considered as heritage or historical. This is the most common approach and establishes as a minimum condition the presence of tourists.
2. Another point of view emphasizes the connection between the elements of the place and the phenomenon. In this case, the role of history is emphasized as being an integral part of the experience and this area of knowledge is associated, in part, with the motivations for travel.

The same authors emphasize the importance of demand in defining this type of tourism, rather than the nature of the artifacts displayed. Therefore, according to them, this market segment should not include tourists who find themselves in certain places because they have been declared heritage resources by experts or stakeholders. Therefore, heritage tourism is a market niche in which the main motivation to visit lies in the uniqueness of the place, given the tourists' perception of their own heritage. This type of tourism, then, stems from the relationship between supply and demand. It is not only about attributes, but, above all, about the perceptions related to them.

On the other hand, Zeppal and Hall (as cited in Chhabra et al., 2003) also emphasize motivation and see heritage tourism as being based on nostalgia and the desire to experience different cultural forms and landscapes. Thus, on the demand side, this activity embodies the desire of many contemporary tourists to experience and consume in situ various cultural landscapes—past or present—performances, delicacies, crafts, and performed activities. In effect, heritage tourists are well-educated, since education can spur people's interest in witnessing historical sites and attending cultural events (Timothy, 2021). On the supply side, this type of tourism is widely regarded as a tool for economic development and is often actively promoted by local governments and private activities.

For Waitt (2000), the demand for heritage tourism is due to several factors:

> An increasing awareness of heritage, an ability to express individuality through recognition of historical environment or staged history, greater affluence, increased leisure time, mobility, access to the arts, the need to transcend contemporary experiences to compensate for

Table 1 Differences between the tourism sector and the cultural sector

	Cultural sector	Tourist sector
Actors	Public sector and third sector	Private sector
Primary end	Search for social benefits	Search for economic benefits
Basic management activity	They try to protect it and spread it	They try to market it
Values assigned	Symbolic value	Use value (increases the attractiveness of a destination)
Recipients	End consumer	Tourist
Motivation	Intellectual enjoyment, learning entertainment	Leisure
Main attractions of the assets	Cultural value, exclusivity, beauty, uniqueness	Convenience, simplicity, distance, time

Source: Pulido Fernández et al. (2013)

> their deficiencies and demands, and/or to fulfill psychological needs for continuity through an appreciation of personal family history (p. 838).

In addition, heritage tourism has the capability to increase appreciation of the past and to galvanize connections between past, present, and future—an increasingly challenging endeavor in a rapidly changing world.

Heritage assets are generally nonrenewable and irreplaceable, requiring maintenance and conservation. In this context, tourism has been seen both as an opportunity and as a threat to heritage. In this respect, Table 1 shows the differences between the tourism sector and the cultural sector.

If there is an excessive valorization of the economic aspects linked to the development of the activity, this may result in a devaluation and demoralization of the social and cultural values of heritage (Pulido Fernández et al., 2013). Nevertheless, heritage is not an immutable resource, due to its limited, nonrenewable, or non-replicable character. Both tourism and heritage need to be considered as a process and performance, as they are subject to incessant negotiation and renegotiation in the local and global contexts (Park, 2014).

Timothy and Boyd (2003) warn about the current trend of having an oversupply of heritage resources, without, however, questioning the role of heritage as an indispensable tool to bring tourism to urban and rural areas in rapid decline, regenerating them.

Indeed, heritage tourism is closely linked to the experience of cultural assets, both material (tangible) and socio-psychological (intangible) of the national or local past (Timothy, 2021). This segment of tourism is above all unrelated to the valorization of a hegemonic mode of reconstruction and reinterpretation of collective memory. Rather, it favors ways for individuals to position themselves in the broader context of cultural construction and symbolic embodiment of national identity (Timothy & Boyd, 2003).

3.1 Historic Housing Tourism

In Portugal, Historic Housing Tourism (HHT) is defined by Decree-Law No. 80/2017, Article 17, as "family-run businesses set up in old private properties that are representative of a particular era due to their architectural, historical, or artistic value, namely palaces and manor houses, and can be located in rural or urban areas."

According to Pina (1988), HHT was introduced in Portugal long after it had been established abroad, namely in Austria, Germany, France, England, or Switzerland. From the beginning, HHT had a strong implantation in the north of the country (Pina, 1988; Bote Gómez, 2001[1988]). This new tourist product, as Pina (1988) calls it, brought together the allure of a humanized and refined lodging and the taste of the privileged and informal conviviality of the outsider with a traditional Portuguese family nucleus, inserted in its genuine sociocultural atmosphere (p. 190).

According to Edwards and Fernandes (1999), HHT was initially promoted in Germany, Scandinavia, and the United Kingdom, with room prices being equal to or higher than four-star hotels. At a conceptual level, this accommodation was sought to be distinguished from British bed and breakfast, as HHT guests were often privileged to stay in stately homes, along with family members who often were the descendants of the mansion's founders.

In Portugal, HHT started as an experience in 1978 under the vernacular designation of *Turismo de Habitação*. Its creation was aimed at: (1) Reducing the disparity between Portuguese inland and coastal regions, and (2) Offering differentiated hotel capacity in areas where it was scarce (Braga & Dionísio, 2021a).

In an extremely simplified way, we can argue that the HHT provides a stay in a manor house, as well as the cohabitation and socializing with the owners of the house. The latter is also assumed as a repository of national and regional identity. However, HHT is distinguished from Rural Tourism (RT), since it is a manifestation of cultural tourism that occurs mainly in rural areas. At HHT, the cultural product is essentially the accommodation in a house of classical architecture. In other words, this type of offer values the eminently rural geographical situation of the house, its historical heritage, and its small size, which influences the way in which the reception takes place. The service at HHT is "always personalized, unprofessional, domestic and familiar" (Cavaco, 1999, p. 294).

HHT can contemplate recreation complementary activities such as hunting and fishing, sailing, riding, visits, and tours or even activities like those that usually happen in swimming pools or tennis courts attached to the house (Cavaco, 1999; Silva, 2010). Additionally, the owners can also complement their income with the lodging modality by organizing parallel events: renting of spaces for parties, weddings, christenings, conferences, among others.

HHT has been contributing to restoring countless houses of erudite heritage, contributing to the "return to the origins" of countless families once installed in urban areas and for the emergence of new types of tourism in rural areas. In addition, together with RT, HHT has contributed to the diversification of the tourist offer; the reduction of seasonality; cooperation between local actors; the improvement of tourism animation projects; and the provision of services that support tourism activities.

However, HHT has proved disappointing relative to the expectations of farmers and with little impact on rural areas, except perhaps for viticulture. Rural employment has hardly benefited from tourism. In addition, the business has been accused of being elitist and of spending public funds inappropriately—particularly, financial nonrepayable support received by the hosts to preserve their properties and start HHT businesses (Silva, 2005–2006).

Silva (2009), in turn, confirms the scarce relevance of the number of jobs directly generated in the country by the sector. Much of the rural tourism units employ only one or two permanent workers, and some do not hire any permanent workers. This is also the case with seasonal and part-time workers. As regards the latter, it is observed that most establishments hire only one worker. Furthermore, a significant portion of HHT staff tends to be multitasking, being primarily assigned to the family household and secondarily to tourist business. Most staff are female, undertaking to perform gardening cores, cleaning and maintaining rooms, serving breakfasts, and responding to customer queries by telephone, among other duties.

HHT also contributes to the preservation of traditional gastronomy as hospitality units offering catering services are required by law to include national and local dishes in their menu. Therefore, the advantages of this sector are rather symbolic than economic (Braga & Sousa, 2021).

HHT is performed in manor houses—some of them dating back to the fifteenth and sixteenth centuries—which requires a big effort from the owners to preserve their original character. The HHT also conveys an ideology of the authentic and the singular. In other words, the success of the concept lies in the fact that it consubstantiates an alternative form of tourism, contrary to urban and sun, sea, and sand tourism massification, and enhances unique destinations driven by the pastoral-style ideas of returning to the countryside, but, essentially, by the exaltation of the singular and nonreplicable character of places and houses.

HHT hosts, in Portugal, are often members of the provincial elites, so they come from rural areas and originate from the place to which the manor house belongs. Sometimes, they are not only the owners of the house, but also of other buildings and agricultural lands that they operate in a heterogeneous way and which have been inherited. According to Cavaco (1999), the promoters of HHT are holders of high cultural and economic indices, and they dedicate themselves mainly to professional activities from the tertiary sector. They are businessmen, liberal professionals, managers, administrators, and senior public officers, although it is not uncommon that they are linked to agriculture and livestock. HHT units are often owned and operated by only one individual who, very often, assures their functioning. Nevertheless, family societies represent a form of property present in a significant number of houses.

The reason they became hosts was the recovery and profitability of properties obtained by inheritance. Therefore, HHT can offer a source of supplementary income to retired hosts or owners who are mainly engaged in agriculture or even young members of the family without permanent employment (Lorente et al., 2020).

One of the greatest motivations of tourists who choose HHT is to revisit the past, choose buildings that appeal to a sense of time travel and the experience of tradition, thus escaping the routine of modern life. These tourists also wish to have an in-depth knowledge of the country. The standard customer for this type of accommodation is the urban couple with a medium-to-high income, cultured, and with some interest or curiosity for traditional lifestyles (Pereiro, 2018). These individuals belong mostly to the middle and upper classes and choose alternative activities to mass tourism. The main reason for the choice of manor houses is their location and the quality–price ratio (Lemos & Farhangmehr, 2007). In fact, the location of the house is relevant for tourists who seek this type of accommodation to be in close contact with nature where they can practice outdoor activities. Tourists also favor the contact with the local population, both the owners and the employees.

HHT is an option for those who want to avoid the most popular and competitive tourist destinations. In this case, it is the adoption of a social differentiation strategy. Thus, those who select this type of tourism believe they will benefit from a more enriching, edifying, and authentic tourist experience than those who choose mass tourism. These HHT regulars can be considered anti-tourists since they deny their condition (Silva & Prista, 2016).

Demand, of course, varies according to the time of year. Three periods can be identified. The high season, from June to August, with a peak occupation in the latter month. Notwithstanding, this season, according to Cánoves et al. (2004), extends for about 12–14 weeks, and includes September. An extensive low season, from September to March, but with some busy periods, particularly, at the end of the year and during Carnival. Finally, a short intermediate season, which occurs from April to June (Silva, 2009).

4 Methods

The results of the present study emerged from the application of GT methods to the substantive area of HHT (Braga, 2016). GT comprises a set of methodological phases (Scott, 2009): (1) identification of the substantive area of the study; (2) collection of data relating to the substantive area; (3) open coding of the data as we collect it; (4) writing memos throughout the process; (5) selective coding and theoretical sampling; (6) sorting to find the theoretical code(s) that best organize the substantive codes; (7) reading and integrating literature into our theory through selective coding; and (8) Theoretical Writing.

Each of these phases of the HHT study will be addressed in detail below, in ascending order:

(1) HHT is the area of interest of the study (Braga & Silva, 2021). The current research focused on the perspective of hosts and/or owners of manors (substantive population), which have been converted into HHT enterprises.

(2) GT's principle that "all is data" (Glaser, 2001) means that data can be legitimately collected from any source. Possible primary sources include interviews, observation, and field notes. Among the secondary sources are any documents produced by institutions or by persons who participate in the study (textbooks, press releases, promotional brochures, newspapers, letters, websites, blogs, photographs, videos, etc.) that may be of investigative interest (Castellanos-Verdugo et al., 2010). In addition, this methodology enables the combination and integration of the various sources. The HHT GT used qualitative data from participant observation (5 sessions) and unstructured interviews (48 in person, 3 by telephone, and 2 by e-mail). Regarding the geographical location of the houses, the vast majority are in the old province of Minho (34). In contrast, a small number can be found in the former provinces Douro Litoral (4), in Trás-os-Montes and Alto Douro (3), Beira Litoral (4), Beira Alta (1), Ribatejo (1), Alto Alentejo (2), Baixo Alentejo (3), Algarve (1) and Azores (2). The interviews were based on two neutral questions of the grand tour type: (a) *can you please tell me about your experience as owner and host of the house X?* (b) *How do you feel as a host?* The purpose of these questions was to instigate the loquacity of the hosts or as Glaser (2001) refers to "instill the spill."

(3) The field notes of the participating observations and of unstructured interviews were expanded in the Word processor. Then, line-by-line coding was employed asking the following questions to the incidents (Glaser, 1998): (a) *what category does this incident indicate?* (b) *what property of a category does this incident indicate?* c) *what is the main concern of the participant?* Code assignment was conducted by using the Word processor's "comment" functionality. At the end of the open coding, followed the constant comparative analysis method (Glaser & Strauss, 2017[1967]). In this context, the identified incidents were "copied" and "pasted" into a new Word document under the appropriate substantive code title. Then, through the "highlight" feature of the Word, incidents were suppressed and only the codes were left visible. This mechanism makes the comparison of codes be easier. Next, we expanded the codes to see the incidents and thereby make the comparison of incident to incident within a code be easier. After that, we compared incidents through codes.

(4) Then, we wrote memos about the codes and about their relationships with other codes by creating a separate document (Scott, 2009). GT is an iterative process, so we analyzed each interview before we started the next one. At this stage, our analytical attention focused on the behavior of HHT house hosts in response to guest actions to distinguish guest and host properties. During the open coding phase, we aimed to interview as many hosts as possible. So, at this stage, we completed 25 interviews.

(5) After analyzing the field notes—those regarding the interviews and the participant observations we conducted—and the ideas we wrote down in the memos on the concepts and their interrelations, we deduced where the next interviews would take place (therefore applying theoretical sampling). Hence, the questions that were previously descriptive became more specific to collect data and saturate the missing categories. At this stage, we confronted the hosts with sensitive

information, removed the testimonies delivered by other hosts, and asked them: (a) *some people told me that...*; (b) *can you tell me about...?* (Braga, 2016). Thus, we improved our theoretical sampling, to theoretically saturate the categories. When we identified the core category and the main concern, the open coding ceased and the selective coding began, which only regard the core variable and the related categories (Scott, 2009). Therefore, the purpose of this first phase of the GT was to identify the main concern of the hosts and how they solved it. We realized that the main concern of the hosts was to preserve the history and the family connection to the HHT house and the way to resolve this concern was by "refunctionalizing" the manor house. This was the moment we started coding selectively for the core variable "refunctionalizing" (Braga & Sousa, 2021) as we believed it might interconnect all the main concerns of the participants. So, we stopped coding in a neutral way and only did it for the "refunctionalizing" code.

(6) Theoretical codes conceive how substantive codes can relate to each other as hypotheses to be integrated into theory (Glaser, 2005). Sorting literally involves distributing memos in conceptual piles. Sorting an important number of memos in an integrated theory is the culmination of months of conceptual elaboration. Sorting by theoretical codes enabled us to relate categories and their properties "into an integrated theory around a core category" (Glaser, 1992, p. 80). Our effort to classify resulted in a combination of theoretical codes: (a) a basic social process: "refunctionalize" composed of two stages ("improvising" and "professionalizing"); (b) a typology of hosts ("dedicated," "undedicated," "lineage," "by acquisition," "initiator," "continuous," "professional," "manipulative"); (c) three styles of refunctionalization ("classic," "hybrid," and "modern"); (d) an amplifying causal looping (Braga & Silva, 2022).

(7) In GT, the literature review is perceived as a source of additional data to be compared with existing data, so we integrated it into the constant comparative analysis process when the core category, its properties and the related categories emerged, and the basic conceptual development was already emerging.

(8) In the writing phase, we wrote the piles of memos obtained in (6) in the form of a coherent text (Braga & Silva, 2022).

5 Results

HHT is the solution found by the owners to preserve the manor house, thus avoiding selling, or degrading it. HHT does not seem to aim at profitability, since fixed charges are high, and the size of houses is small (Braga, 2016). Above all, it is intended to achieve self-sustainability which will make it possible to strengthen a financial structure that enables the recovery of the property.

Nevertheless, HHT can be economically unsustainable if the host does not devote enough time to the activity (Braga & Dionísio, 2021b).

In addition, HHT is a way to ensure a constant investment without the owner having to employ his own capital. If the house is self-sustaining, the income from

HHT can cover the fixed costs. However, economic sustainability needs to be increased to support the ongoing recovery that the house needs. Thus, HHT must be self-sustaining, therefore, equity should only be used if it is mandatory to make intensive recoveries that cannot be borne by HHT revenue.

In short, the owners' main concern seems not to be profit, but rather passing on the house in better condition to the next generation of the family (Braga, 2016).

In the first phase of implementation of HHT, which we called improvisation (Braga & Sousa, 2021), the capital with payment facilities, provided by the political structure, was considerable, so the pressure on the economic sustainability of the house was not so high.

In the next phase, the professionalization phase (Braga & Sousa, 2021), a negative financial structure will lead to less advantageous financing conditions and cause a slowdown in the pace of recovery of the manor house or a decrease in its intensity. This will, in turn, result in a lower economic sustainability of the house, as the recovery costs will be high, and the accommodation offer will be less satisfactory for the guest due to the accentuated degradation of the building (Braga & Dionísio, 2021b).

Thus, some houses can achieve self-sustainability and others are chronically unsustainable except in the period of high guest inflow. If the HHT is not self-sustainable, the owner will have to recover the house using equity. In fact, there are hosts who cannot stand the unpredictability of the market and end up with the HHT.

To raise sustainability, houses do their best to reduce fixed costs. Deferring non-priority expenses is one way they do this. Other means used are reducing the houses' visibility charges, or opting for free visibility, or even reducing travelling expenses from the house elsewhere.

In fact, HHT allows the political structure to renew the built heritage in low-density territories. In turn, for the owners, it allows them to solve problems related to the usefulness of house preservation, at a time when the traditional agricultural function of the property is no longer profitable (Lorente et al., 2020).

The continuity of the manor house presupposes that the descendants have sufficient equity for the maintenance of the property or that they endow it with a viable exploitation model from the point of view of its economic sustainability. Nevertheless, there are different degrees of satisfaction with the solution adopted for the preservation of the house, that is, HHT. There are, therefore, hosts fulfilled and hosts not fulfilled with the HHT (see interview no. 20).

> In summer, yes, in winter it is very weak. The showcase is Buçaco. From 2006/2007, the economy started to go down. Currently it is very difficult to sustain. This is a 19th century family house. There are a lot of maintenance expenses. The house comes from a Mr. Duparchy, who came to lead the construction of the railways. It was bought by my wife's family. Nowadays it is experiencing many difficulties of sustainability. Accommodation has decreased a lot. This type of tourism is one of the most important. It transmits values to foreigners. The activity is very much divided between luxury tourism and the rest. HHT has also been affected by this aspect. While there is life, there is hope. With the improvement of the economy, we hope to improve the activity. There are also country houses and village houses, where the presence of the owners is not compulsory. [Interview no. 20]

If, as we have seen, the owner's equity can be used when the economic sustainability of HHT is reduced and financing is insufficient to recover the house, the opposite can also happen, the owner's equity can be increased, even if modestly, by the economic sustainability of HHT.

Ultimately, the host's main concern being the restoration and maintenance of the house, he fights a permanent fight against time. The HHT must contribute, essentially, to this goal. If it contributes to the host's livelihood, then it will allow him to fully dedicate himself to the activity.

5.1 Raising the Economic Sustainability of HHT

The solution to raising the economic sustainability of the house may be to have a larger number of variable staff in periods of high inflow of guests. However, it is always necessary to instill the values of HHT in the staff. Another possibility is to reduce the number of variable staff while keeping the fixed staff and, in periods of high affluence, the host can dedicate more time to the HHT. In this case, the needs in terms of staff will be conditioned by the hosting capacity of the house.

The house's reputation feeds its economic sustainability. A reputable house in the eyes of guests is a house that can attract a continuous flow of bookings. On the other hand, the more economically sustainable the HHT is, the more intense the recoveries will be and the more apt the house will be to be passed on to descendants. Thus, hosting provides a financial motivation for recovery. Moreover, if the host knows recovery agents (i.e., construction workers) and hires them from the surrounding area, this can have positive effects on the intensity of the recovery undertaken and its costs for economic sustainability.

For HHT to be profitable, there must be a minimum occupancy to cover fixed costs. For this, a good part of the accommodation capacity must be open to occupation and the host must be fully dedicated to the activity. If the host wants to increase profitability, it will have to reduce staff by increasing the dedication it gives to the HHT. Thus, the host may choose to keep the staff he had from housekeeping in order not to incur fixed charges that burden the sustainability of HHT.

The HHT is not very sustainable by nature (see interview no. 50). One of the ways to raise profitability is to make the relationship with the guests more frequent, making them loyal to the house's accommodation proposal.

> It's the final straw! Keeping a house like this running is a cancer. Certifications are not significant, it's burying money. We keep the house running, it's expensive to keep the house running. We lose some quality of life. We have a different life. Since I am a teacher, I only have vacation in August, but I can't leave. We can still increase the number of guests, but everything is still very seasonal. If we can stop the winter bloodletting, it's already a lot. For that, you must have hotel quality, what's worth it's trying to capture tourism. [Interview no. 50]

Another way to monetize is to increase the energy and environmental sustainability of the HHT, whereby the host can use biological sources of energy and water.

Increasing sustainability also means not incurring fixed charges when there is no or low occupancy.

HHT is a way to preserve houses, to pass them on, so the profit margin should be immediately invested in recovery. To raise the economic sustainability, the host may also seek to increase the hosting capacity by recovering annexes.

One way to increase self-sustainability is to extend the hosting relationship with the clients. For that, it is critical to better articulate the house with its surroundings, as well as to formalize the HHT service. In fact, the attractiveness of the surroundings in which the house is inserted is also responsible for an increase in economic sustainability. However, it is up to the host to enhance this attractiveness, differentiating the accommodation proposal.

To raise economic sustainability, the host must make micromanagement decisions, concerning actions that seem of minor relevance, such as doing the laundry inside or outside the house.

In fact, it may be that other modality of lodging exploitation in which guests have total autonomy allow for greater economic sustainability of the house, modalities that fall within a less rigid legal framework than the one that falls on the HHT.

To fill the accommodation capacity that the house has available, the HHT can go for differentiated accommodation proposals (i.e., changing the product to be offered, such as, for example, half-board, 2-day stay with enhanced welcome drink, among others).

The annexes are a way to monetize the activity, escaping the functional constraints of the manor house. Thus, one way to raise economic sustainability is to create a hybrid style of HHT refunctionalization, in addition to the classic style typical of the improvisation phase (Braga, 2016).

As we saw above, capital from HHT can avoid intensive recoveries, which would require an investment project and equity or funds from the political structure, or even bank financing.

On the other hand, there seems to be an increasing tendency for houses to reach the guest through massive intermediaries (e.g., reservation systems like *booking.com*) who work with HHT-insensitive guests (Braga & Silva, 2021), as these will allow them to raise economic sustainability. However, this type of guest requires more dedication on the part of the host.

In the period of low affluence, there are circumstances in which isolated reservations can result in a loss of economic sustainability. Therefore, out of pure economic rationality, it may be better for the economic sustainability of the house if it is closed.

Thus, the host will have the goal of reducing the fixed charges to a minimum. Therefore, the host family, while it has no guests, should not use infrastructures that it would only use if it had guests because this will increase the fixed charges of the HHT.

The costs inherent to house restoration are, in HHT, higher than in other types of tourism, since in houses that have been recently built, the wear and tear on materials are much less significant. This is a major obstacle to lowering prices to compete with mass lodging. However, depending on the accommodation capacity of the house, the owner can raise the economic sustainability of the house by reserving accommodation capacity in the house for another model of exploring accommodation that does not burden the economic sustainability as much.

There are also add-ons to the HHT, such as wine tastings, merchandising sales, and paid tours of the house that help support the house economically, among others. By enhancing the HHT's hosting proposition, the host can recover the house at a higher rate.

The host should be dynamic and creative in the low affluence period to differentiate the lodging proposal, creating, if possible, a complementary economic modality to the HHT. The host will have to be more dedicated to the activity to increase its self-sustainability, and this dedication must be maximal in the period of high affluence, to allow the accommodation capacity to be filled.

Having more HHT-sensitive guests, because they spend more, as well as having guests with longer lodging relationships are other ways to raise economic sustainability.

It is by building up a good reputation that the house can increase the occupancy of the accommodation capacity and, therefore, its economic sustainability, since rival houses present very similar and, therefore, replaceable products.

Also, there may be occasional events in the house surroundings that make HHT more attractive at a certain time of the year and lead to greater economic sustainability during that period.

5.2 *Reducing the Economic Sustainability of the HHT*

As we have seen above, in the HHT it is more difficult to hire only variable staff, as is the case in many mass lodgings, which could raise its economic sustainability. There is a set of fixed staff always attached to the house who works for the HHT, but also for the host family. Thus, there seems to be less flexibility in hiring and firing staff at the HHT. Therefore, fluctuations in occupancy cannot be compensated for by laying off workers with the same turnover as in a mass accommodation. The cost of staffing is increased in the period of high affluence since more labor is needed to conveniently serve more guests.

When it comes to the recovery of the house, for it to be true to its identity, it is possible that the owner will have to hire recovery agents from outside the surroundings, which will be detrimental to the self-sustainability of the house. To do this, the host may have to apply his/her own capital.

On the other hand, political structures, in the professionalization phase, require a dedication from the host that cannot be diminished by hiring staff, as this will harm economic sustainability.

If the price depends on the service formalization achieved by the HHT and the articulation that the lodging proposal achieves with the surroundings, in the period of low affluence the lack of occupation of the lodging capacity will cause a huge lowering of prices which will make the HHT unsustainable from an economic point of view.

HHT's compliance with the legal framework decreases its economic sustainability. This is even more true because the legal framework is not adapted to the identity of the house, being undifferentiated irrespective of whether it applies to a mass lodging or HHT. Thus, HHTs must maintain prices, otherwise, they will suffer from competition from mass lodging and similar houses. On the other hand, the political structure can make service formalization requirements that are reductive to self-sustainability.

In short, HHT is more expensive than the other types of accommodation, precisely because there is a need to restore the house (see interview no. 39).

> My husband has the concept that if we rented the house it would be more cost effective. But it's my house, it's hard for me. This house has old things that I have feelings for. I don't know who's coming. As long as we can, we won't rent. It would be more profitable and worthwhile. There is a lot of work here. My husband sets the table, takes things out of the machine, we talk, I make dinner. If we rent the house it's easier, but it's not HHT anymore, it's conservation of old houses. [Interview no. 39]

Thus, HHT operates in amplifying causal loop (Braga & Dionísio, 2021b). The more economic sustainability is reduced, the less the host will be able to hire staff and may have to rely on existing permanent staff. Consequently, it is possible that the formalization of the HHT service will suffer, with damage to the reputation of the house.

Moreover, the reduction in the economic sustainability of HHT will be accompanied by a decline in the pace of recovery and a consequent decrease in its intensity since HHT does not generate equity. In turn, the resulting loss of reputation will reduce the economic sustainability of HHT, as it will increase the duration of low affluence periods.

Economic sustainability is also more difficult to achieve because the price is dependent on the weighting of fixed charges that are less controllable in the HHT than in its competitors (e.g., it is more difficult to lay off staff, more heating is needed in winter and maintenance of the pools in summer).

The peculiarity of being an establishment with little lodging capacity makes it difficult to achieve economies of scale in HHT (i.e., increased occupancy is not always likely to increase the burden of the lodging proposal by a smaller proportion).

In extreme cases, the reduction of the sustainability of the HHT can even lead to the breakdown of heritage transmission. In fact, some hosts close the hosting capacity of the house because the occupancy is not enough to make the HHT profitable, and the non-sustainability of the HHT can lead to making it impossible to pass on the house. Effectively, the HHT struggles with chronic low occupancy.

In summary, if there is a continued reduction in economic sustainability, HHT risks incurring a negative financial structure, which will lead to the pace of recovery becoming slower and continuity less assured.

6 Discussion and Concluding Remarks

The problem of the impacts of rural tourism in general and of HHT, in particular, have been addressed before by the tourism literature (Cavaco, 1999; Silva, 2005–2006). However, the present study presents relevant novelties regarding the knowledge of the behavior of the hosts and/or owners of HHT and the strategies used by them to ensure the self-sustainability of that tourism product and by metonymy of the manor house.

Some of the research findings are summarized in Fig. 1, in which "self-sustainability" is seen as a dependent variable of the properties "raising the economic sustainability" and "reducing the economic sustainability." Thus, this study allows us to recognize the impacts that the behaviors and actions of hosts, guests, and the political structure can have on the economic sustainability of the HHT.

In reality, a substantial part of these houses can no longer depend economically on agriculture, since this primary sector activity has lost relevance in the local economy. In this way, there has been a revaluation of the cultural and natural heritage in a logic of multifunctionality of rural areas.

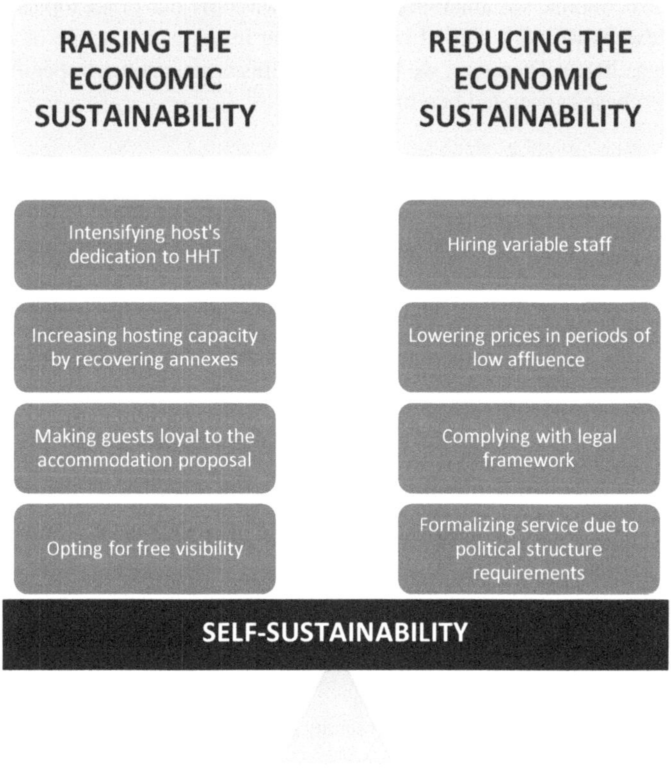

Fig. 1 HHT's self-sustainability scale. Source: Own elaboration

Owners who have houses with outbuildings should devise two differentiated hosting systems. Also, the HHT host must dedicate himself intensively to the activity, seeking to make an effective articulation of the house with its surroundings.

Manor houses should develop lodging proposals that appeal both to guests sensitive to the HHT and to guests insensitive to the same activity. Competition with rival houses should not be through price, but through formalization and differentiation of the lodging proposal. Nevertheless, hosts should not formalize HHT too much, otherwise, it will lose its differentiation from mass accommodation. In addition, the host should act as a host enhancer, being aware of the endogenous resources and events of the surroundings to facilitate the customer experience.

For its part, the public sector must ensure that the legal designation of HHT refers only to very high quality and highly homogeneous housing. In turn, the local authorities must ensure the attractiveness of the surroundings of the houses. Houses have a high bureaucratic burden that falls on HHT hosts, and the legal framework is constantly changing, which fuels insecurity amongst stakeholders. This instability is hindering the economic sustainability of HHT and should be changed. Finally, local administrations should stimulate tourist entertainment in the surroundings of the houses, not only in high season but also in low season.

The present study has limitations because the stated research problem, which concerns the economic sustainability of houses, was only one of the topics addressed in a more extensive work carried out within the investigative scope of a doctoral thesis (Braga, 2016). Therefore, we believe that this issue justifies a specific study to test the hypotheses generated herein.

References

Almeida, F., Almeida, J., & Mota, M. (2019). Perceptions and trends of booking online payments in tourism. *Journal of Tourism and Services, 10*(18), 1–15.

Bento, J. P. (2016). Tourism and economic growth in Portugal: An empirical investigation of causal links. *Tourism & Management Studies, 12*(1), 164–171.

Bote Gómez, V. (2001[1988]). *Turismo en espacio rural: Rehabilitación del patrimonio sociocultural y de la economía local.* 3ª ed. Editorial Popular, S.A.

Boyer, M. (2003). *História do turismo de massa.* EDUSC.

Braga, J. (2016). *Refuncionalizando a Casa Solarenga: uma Grounded Theory* [Ph.D. thesis]. Universidade de Santiago de Compostela, Santiago de Compostela. Retrieved from https://minerva.usc.es/xmlui/handle/10347/14805.

Braga, J., & Dionísio, M. (2021a). A typology of hosts in housing tourism. In A. Lubowiecki-Vikuk, B. Sousa, B. Dercan (Eds.), *Handbook of Sustainable Development and Leisure Services (World Sustainability Series).* Springer International Publishing.

Braga, J., & Dionísio, M. (2021b). An application of the "amplifying causal looping" model to manor house tourism. In S. Lokuge & D. Sedera (Eds.), *Rural entrepreneurship and innovation in the digital era* (pp. 102–125). IGI Global.

Braga, J., & Silva, O. (2021). Sense and sensibility in housing tourism: Uncovering patterns of behavior. In L. Oliveira, A. Amaro, & A. Melro (Eds.), *Handbook of research on cultural heritage and its impact on territory innovation and development* (pp. 276–296). IGI Global.

Braga, J. & Silva, O. (2022). Grounded theory. In D. Buhalis (Ed.), *Encyclopedia of tourism management and marketing* (in press). Edward Elgar.

Braga, J., Sousa, B. (2021). The basic social process of "re-functionalising" and its implications for housing tourism: a niche tourism perspective. In A. Abreu, D. Liberato, E. González & J. Ojeda (Eds.), *Advances in tourism, technology and systems: selected papers from ICOTTS20* (Vol. 2, pp. 125–141). Springer Nature Singapore.

Cánoves, G., Villarino, M., Priestley, G., & Blanco, A. (2004). Rural tourism in Spain: An analysis of recent evolution. *Geoforum, 35*(6), 755–769.

Castellanos-Verdugo, M., Caro-González, F., & Oviedo-García, M. (2010). An application of grounded theory to cultural tourism research: Resident attitudes to tourism activity in Santiponce. In G. Richards, W. Munsters (Eds.), *Cultural tourism research methods* (pp. 115–127). CAB International.

Cavaco, C. (1999). Turismo Rural e Turismo de Habitação em Portugal. In Cavaco, C. (Ed.), *Desenvolvimento Rural: Desafio e Utopia* (pp. 293–304). Centro de Estudos Geográficos da Universidade de Lisboa.

Chhabra, D., Healy, R., & Sills, E. (2003). Staged authenticity and heritage tourism. *Annals of Tourism Research, 30*(3), 702–719.

Comissão de Coordenação e Desenvolvimento Regional do Norte (CCDRN). (2011). *Programa Regional de Reformas Norte 2020. Public Consultation Document*. CCDRN. http://www.ccdr-n.pt/node/415.

Edwards, J., & Fernandes, C. (1999). Emigrants and Espigueiros – Tourism activities in peripheral area of Portugal. *International Journal of Tourism Research, 1*(5), 329–340.

Glaser (2001). *The grounded theory perspective: conceptualization contrasted with description*. Sociology Press.

Glaser, B. (1992). *Basics of grounded theory analysis: Emergence vs. forcing*. Sociology Press.

Glaser, B. (1998). *Doing grounded theory: Issues and discussions*. Sociology Press.

Glaser, B. (2005). *The grounded theory perspective III: Theoretical coding*. Sociology Press.

Glaser, B., & Strauss, A. (2017[1967]). *The discovery of grounded theory: Strategies for qualitative research*. Routledge.

Goethe, J. W. (2018). *Viagem a Itália*. Bertrand Editora.

Gössling, S., Scott, D., & Hall, M. (2021). Pandemics, tourism and global change: A rapid assessment of COVID-19. *Journal of Sustainable Tourism, 29*(1), 1–20. https://doi.org/10.1080/09669582.2020.1758708

Horta, A. (2020). *Tourism in Portugal expected to fall dramatically*. Retrieved 7 August, 2020, from https://www.theportugalnews.com/news/tourism-in-portugalexpected-to-fall-dramatically/53926

ILO. (2020). Importance of risk management for the sustainability of tourism. *Balıkesir University the Journal of Social Sciences Institute, 18*(33), 163–178.

Lemos, F., & Farhangmehr, M. (2007). Tourism marketing and management: The case of Pousadas vs. Solares de Portugal. *Revista Turismo e Desenvolvimento, 7*(8), 111–119.

Lopes, M. C., Liberato, D., Alén, E., & Liberato, P. (2020). Social tourism development and the population ageing: Case study in Portugal and Spain. In Á. Rocha et al. (Eds.), *Advances in tourism, technology and smart systems. Smart innovation, systems and technologies* (pp. 527–536). Springer.

Lorente, T., Braga. J., & Cardoso, A. (2020). The social problem of rural depopulation in Spain and Portugal. In F. Estrena-Durán, R. Soriano-Miras, & R. Duque-Calvache (Eds.), *Social problems in southern Europe: A comparative assessment (southern European societies series)* (pp. 143–156). Edward Elgar Publishing Limited.

Mahika, E. (2011). Current trends in tourist motivation. *Cactus Tourism Journal, 2*(2), 15–24.

Ministério da Economia e da Inovação (MEI). (2007). *Plano Estratégico Nacional do Turismo. Para o Desenvolvimento do Turismo em Portugal 2006–2015*. Ministério da Economia e da Inovação e Turismo de Portugal. https://travelbi.turismodeportugal.pt/pt-pt/Documents/Estratégia/PENT_2007.pdf.

Moreira, C. O. (2018). Portugal as a tourism destination. *Méditerranée*, 130. https://doi.org/10.4000/mediterranee.10402

Oliveira, E. (2014). The tourism potential of northern Portugal and its relevance for a regional branding strategy. *Advances in Hospitality and Tourism Research (AHTR), 2*(2), 54–78.

Park, Y. (2014). *Heritage tourism*. Routledge.

Pereiro, X. (2009). Turismo em Espaço Rural. In *Turismo Cultural: Uma visão antropológica* (pp. 253–286). PASOS, Revista de Turismo y Patrimonio Cultural.

Pereiro, X. (2018). Abordagem exploratória do turismo rural de Trás-os-Montes e Alto Douro (Portugal). *Análise Social, 53*(226), 58–87.

Pina, P. (1988). *Portugal, o Turismo no Século XX*. Lucidus Publicações.

Poria, Y., Butler, R., & Airey, D. (2003). The core of heritage tourism. *Annals of Tourism Research, 30*(1), 238–254.

PRESSTUR. (2019). Portugal é o 4° país onde o turismo tem mais 'peso' no emprego total. Retrieved 7 August, 2020, from http://www.presstur.com/mercados/portugal/portugal-e-o-4%2D%2Dpais-onde-o-turismo-tem-mais-peso-no-emprego-total/

Pulido Fernández, J., Calle Vaquero, M., & Velasco González, M. (2013). *Turismo Cultural*. Editorial Sintesis.

Responsible Tourism. (2020). *Travel and tourism after Covid-19*. Retrieved 10 August 2020, from https://responsibletourismpartnership.org/travel-and-tourism-after-covid-19

Ritchie, B., & Jiang, Y. (2019). A review of research on tourism risk, crisis and disaster management: Launching the annals of tourism research curated collection on tourism risk, crisis and disaster management. *Annals of Tourism Research, 79*, 1–15.

Roberts, L., & Hall, D. (2001). *Rural tourism and recreation: Principles to practice*. CAB International.

Scott, H. (2009). *How do you do grounded theory?* Grounded Theory Online. http://www.groundedtheoryonline.com/what-is-grounded-theory/

Silva, L. (2005–2006). Os impactos do turismo em espaço rural. *Antropologia Portuguesa, 22–23*, 295–317.

Silva, L. (2009). *Casas no Campo, Etnografia do Turismo Rural em Portugal*. Imprensa de Ciências Sociais.

Silva, L. (2010). Perspetiva antropológica do turismo de habitação em Portugal. *Pasos: Revista de Turismo e Património Cultural, 8*(1), 31–46.

Silva, L., & Prista, M. (2016). Social differentiation in the consumption of a pastoral idyll through tourist accommodation: Two Portuguese cases. *Journal of Rural Studies, 43*, 183–192.

Simková, E., & Holzner, J. (2014). Motivation of tourism participants. *Procedia – Social and Behavioral Sciences, 159*, 660–664.

Timothy, D. (2021). *Cultural heritage and tourism: An introduction* (2nd ed.). Channel View Publications.

Timothy, D., & Boyd, S. (2003). *Heritage tourism*. Pearson Education Limited.

Ural, M. (2015). *Tourism in Portugal expected to fall dramatically*. Retrieved 7 August, 2020, from https://www.theportugalnews.com/news/tourism-in-portugal-expected-to-fall-dramatically/53926

TravelBI. (2019). *Consumo Turístico representa 14,6% do PIB em 2018 | Conta Satélite do Turismo 2018*. Retrieved 9 August, 2020, from https://travelbi.turismodeportugal.pt/pt-pt/Paginas/conta-satelite-do-turismo-2018.aspx

Waitt, G. (2000). Consuming heritage: Perceived historical authenticity. *Annals of Tourism Research, 27*(4), 835–862.

WEF. (2018). *The travel & tourism competitiveness report 2018. Geneve, Switzerland: World Economic Forum Travel and Tourism after Covid-19*. Retrieved 10 August, 2020, from https://responsibletourismpartnership.org/travel-and-tourism-after-covid19

Wong, M. M. C., Cheung, R., & Wan, C. (2013). A study on traveler expectation, motivation and attitude. *Contemporary Management Research, 9*(2), 169–186.

Cooperation as a Basic Strategy for New Dynamics of a Cross-Border Destination: The Case of the Euroregion Galicia and Northern Portugal

Diana Pereira, Manuel Fonseca, Bruno Sousa, and Alexandra Correia

Abstract Cooperation is a strategy increasingly used in the tourism sector, as it can enhance cross-border tourism, including in regions characterised by historical singularities, social and institutional relations, which are presently facing challenges arising from the current pandemic situation. Thus, this study, based on the case of the Iberian Euroregion Galicia and Northern Portugal, aims to analyse stakeholders' perceptions of cooperation as a strategy to enhance the dynamism of a common tourist destination, consisting of two regions. To meet the objectives of the study, eight semi-structured interviews were conducted with Euroregion stakeholders from the public and private tourism sector, which allowed access to their vision on the need to create a common marketing strategy for the tourist destination Galicia and Northern Portugal. The results of this study have practical implications as they allow the identification of marketing strategies and actions for the enhancement of the Iberian tourist destination under analysis. At the end of the chapter, the limitations of the study will be presented, as well as lines of investigation for future research.

D. Pereira
IPVC – Polytechnic Institute of Viana do Castelo, Viana do Castelo, Portugal

M. Fonseca
IPVC – Polytechnic Institute of Viana do Castelo and UNIAG, Viana do Castelo, Portugal

B. Sousa (✉)
IPCA – Polytechnic Institute of Cávado and Ave, Barcelos, Portugal

CiTUR, Coimbra, Portugal
e-mail: bsousa@ipca.pt

A. Correia
IPVC – Polytechnic Institute of Viana do Castelo, Viana do Castelo, Portugal

CiTUR, Coimbra, Portugal

J. Leitão et al. (eds.), *Tourism Entrepreneurship in Portugal and Spain*, Tourism, Hospitality & Event Management, https://doi.org/10.1007/978-3-030-89232-6_6

1 Introduction

Cooperation has been the basis for the development of several European projects encompassing cross-border territories and involving different stakeholders. As a strategy, cooperation not only makes it possible to combat inequalities in the most disadvantaged areas, mainly interior territories but also brings competitive advantages of using the location of two cities in two different countries (Kurowska-Pysz et al., 2018).

The flexibility of the cross-border cooperation strategy in direct engagement between regional and local agencies, but also to many other entities, institutes and organisations, and even businesses from neighbouring cities (Kurowska-Pysz et al., 2018), allows the development of new products based on creativity (Richards and Wilson, 2007), and co-creation experiences, which add value to both the visitor and the resident while contributing to the uniqueness and authenticity of the destination (Binkhorst and Dekker, 2009). In this context, the places where the experience takes place should be related to the destination's history, ethnography, traditions, among others. During the creative process of developing tourism experiences, both the resident's environment and (future) visitors play a key role (Ratten et al., 2019).

The COVID-19 pandemic was declared in March 2020 by the World Health Organisation as a global pandemic, leading to restrictions, border closures and compulsory confinement orders in several countries, and has devastatingly affected the travel and tourism sector, putting it on pause. This context has proved to be particularly acute given that tourism was one of the fastest-growing sectors in 2019, accounting for one in four new jobs created worldwide over the past 5 years.

The change in tourists' motivations, namely in the search for travel/holidays was one of the main changes resulting from the pandemic, looking for places that prove to be reliable, safe, and low risk. Despite the uncertainty brought about by this whole situation, the truth is that this panorama brings opportunities to rethink strategies and to enhance partnerships with local communities, a determining factor of tourist destinations, as may be the case of the dynamisation of border territories through the cooperation strategy for the enhancement of creative tourism experiences.

Given the current trends of dynamization of the territory and this being a time that requires disruptive attitudes, in the field of tourism, the disruption, although timid (Zhang et al., 2014), is already happening, both on the supply side and on the demand side. Therefore, creativity combined with cooperation can be viewed as an opportunity. Creativity can be consumed in countless ways, by different types of audiences, between visitors and residents, and should be seen as an opportunity for local communities (Melo et al., 2019).

With the changes in the criteria for choosing a destination, mainly due to the restrictions imposed, the following stand out: proximity to the area of residence, the search for outdoor spaces and nature that do not force agglomerations, greater planning and research before the trip, the valorisation of review platforms and real-time research on the reality of the destination, the evaluation of compliance with the imposed distance and agglomeration rules by local companies and providers. It is noteworthy that WTTC points to young people (millennials) and professionals as being the first segments to travel again, the former because they constitute a segment that essentially seeks nature and the latter due to the need to establish partnerships with face-to-face meetings.

The same criteria are pointed out by the report "European Tourism: Trends & Prospects", released by the European Travel Commission (Commission, 2021), which underlines that consumer sentiment will be subject to the evolution of the pandemic and the lifting of restrictions.

Through the "European Tourism 2020—Trends & Prospects" of the European Travel Commission, it is estimated that in 2023 business travel between European countries will begin and in 2024 there will be a recovery in business travel outside the European area, which may present pre-covid numbers.

In terms of economy, the tourism sector will have to rely on the digitalisation of processes and prepare for an economic recession by rethinking business models. Communication is the key, with the application of marketing strategies for its enhancement at the promotion level.

Based on the arguments addressed, the objectives of this study are: (1) the identification of cooperation as a strategy for the dynamization of territories; (2) the analysis of the potential of the Euroregion Galicia and Northern Portugal in the pre- and post-COVID-19; (3) to identify the perceptions of public and private stakeholders on the creation of a common marketing strategy for the Euroregion destination; (4) to identify potential marketing actions for the dynamization of the territory.

1.1 Cross-Border Cooperation as a Strategy for the Dynamization of Territories

Buhalis (2000) mentions that destinations are an amalgamation of tourism products, which offer an integrated experience to consumers. The author recognises that destination is a subjective concept, by the interpretation of the consumer, the itinerary of the trip, the cultural background, the reason for visiting, the educational level and previous experience. In a more administrative view, Buhalis (2000) mentions that often destinations are artificially divided by geographical and political barriers. For Kotler (2011), tourist destinations are places with real or imaginary border: the physical borders that delimit an island, the political borders and those created by the market, that is, the travel agent delimits the destination when promoting a certain place. For him, the tourist destination has a life cycle like any other product. He identifies two types of destinations: macro-destinations (example of a country) and micro-destinations (regions, cities, villages, or tourist attractions).

However, and even with the geographical and political barriers, tourists tend to perceive the Euroregion as a common destination. For this reason, the territories need the elaboration of an applicable and simple model for monitoring tourism demand and supply within the Euroregion and, at the same time, one that is acceptable to the two administrations involved in the management of the two regions. Only with coordinated actions can visitors perceive the Euroregion as a single destination. Rojo (2010) considers cross-border cooperation to be one of the main typologies of the current era.

Cross-border cooperation has become not only a way to promote the development of territories but also a way to use the competitive advantages of cities located on the borders of two countries (Kurowska-Pysz et al., 2018).

Kozak and Buhalis (2019) present politics and political support as points for a sustainable cooperative relationship, sometimes being the beneficial and/or problematic element in cross-border cooperative relations. It is also identified that trust and this political support together with coordinated marketing actions ensure comprehensive and continuous experiences for the market. Another interesting point made was the need to develop the business ecosystem in both destinations to ensure that bridges can be developed to facilitate cross-border tourism marketing and management.

Kurowska-Pysz et al. (2018) identified five aspects for the success of strategic alliances in cross-border cooperation projects: (i) well defining the objectives of the alliance; (ii) ensuring the participation of various stakeholder groups in the alliance; (iii) involving both partners with extensive experience in cooperation; (iv) ensuring the coherence of the main objective and (v) ensuring that the alliance benefits both parties. These aspects can contribute to the achievement of more effective results in projects of this scope.

Kurowska-Pysz et al. (2018) pointed out that cross-border cooperation includes direct relations between local governments (this being the most frequent form), but also gradually extends to many other entities, institutes, and organisations, and even companies from nearby cities. Collaboration between all stakeholders should ensure that experiences are co-created (Kozak e Buhalis, 2019). Euroregions play a big role in cross-border cooperation, creating networks involving a wide range of local and regional actors on both sides of the borders.

It is important to stress that the strategies developed, and the cross-border governance structures should be limited to promoting the active involvement and mobilisation of local and regional actors in the cooperation process and in the implementation of various cross-border projects of local/regional importance.

In cross-border marketing it is necessary to be bolder than in local marketing, branding and signage need to demonstrate the benefits of the combined experience and show how value will be co-created, even knowing that each destination will want to stand out. However, commitments and innovations should bring a unique proposition that does not compete with existing destinations, but rather a dynamic product and experience proposition based on the synergies of the combined offer (Kozak e Buhalis, 2019).

To enhance the cross-border cooperation strategy it is also necessary to consider cross-border travel trends, as indicated by the World Tourism Organisation:

- Travel facilitation allows more tourists to cross borders.
- Few cross-border formalities.
- Flexible and backup transportation networks.
- Shorter trips.
- Search for more varied experiences.
- Availability of information through technology.
- Collaboration between international partners and development of cooperation ecosystems.
- Last-minute bookings and special offers available.
- Increase in long-distance travel.
- Increase in middle-class travel to 'once in a lifetime' destinations. Need for travellers to feed social media with as much content as possible.

The development of creative tourism experiences from a cooperation strategy can be used in various ways, namely through (i) existing tourism products and experiences; (ii) revitalising existing products; (iii) enhancing cultural and creative assets; (iv) providing economic outcomes for creative development; (v) using creative techniques to enhance tourism experiences, and finally, (vi) creating a creative environment (Richard, 2012).

The element of creativity can be used in creating experiences from other elements such as active participation, co-creation, contact with the locals and skills development (Binkhorst, 2007; Richards & Raymond, 2000).

The Euroregion Galicia and the North of Portugal, two destinations with geographical and political barriers, will then be framed.

1.2 Brief Analysis of the Euroregion Galicia and North of Portugal's Potential

The concept of "Euroregion" refers to a cooperation structure between two or more nearby territories located in two or more border countries on the European continent. A Euroregion is generally not part of a legislative or governmental institution, nor does it have direct political power, but regions try to integrate and cooperate through different ways (Otocan, 2010).

The Euroregion Galicia and Northern Portugal is located in the northwest of the Iberian Peninsula. It is made up of 12 territorial units of NUTS III (Nomenclature of Territorial Units for Statistics), that is, four provinces of the Autonomous Community of Galicia in Spain—A Coruña, Lugo, Ourense and Pontevedra—and eight intermunicipal communities of the Northern Region of Portugal—Alto Minho, Cávado, Ave, Porto Metropolitan Area, Alto Tâmega, Tâmega e Sousa, Douro and Terras de Trás-os-Montes. With 1232 km of border, of which 100 km (about 8%, in the north of Portugal), along the international stretch of the river Minho, between Caminha and Melgaço, more people live than in the remaining 92%. Cross-border cooperation is particularly relevant given the existence of a single land border, one of the oldest and most stable in Europe, which was established with the creation of the Galicia-Northern Portugal Working Community in 1991.

The following year, the Atlantic Axis of the Iberian Northwest was born, an association of cross-border urban municipalities. The first EGTC (European Grouping of Territorial Cooperation) on the Iberian Peninsula was created in 2008, more precisely in Galicia and in the North of Portugal, through an agreement between the Xunta de Galicia and the Northern Regional Coordination and Development Commission (CCDR-N).

The Euroregion Galicia and Northern Portugal has been promoting several cooperation strategies that promote its regional development. The proximity and the history of existing relations between the two regions make this cooperation solid and efficient.

Through the various European Union cross-border cooperation programmes, support was provided for the development of territorial cohesion, fostering proximity relations and the sharing of solutions to problems and obstacles encountered by citizens and local administrations in their daily lives. Initiatives were then developed at the micro-level, the Eurocities, which within their possibilities manage to present strategies for development and territorial cooperation based on the sharing of facilities and services, collaboration, and complementarity. However, despite the existence of a "Europe without borders", there are invisible barriers that do not strengthen these territories, namely in issues of mobility and use of goods and services, as well as the differences in the daily lives of citizens, which lead these territories to be known as having less economic and social development (Campos &Pardo, 2008).

Currently, in the Euroregion Galicia and Northern Portugal, there are four projects that are called Eurocities: Chaves-Verín (the pioneer, with approval of its POCTEP project in 2007), Valença-Tui (twinning protocol in 2012, POCTEP project approved in 2019), Monção-Salvaterra (signing of twinning protocol in 2015) and Vila Nova de Cerveira-Tomiño (signing of project constitution and approval of POCTEP in 2018).

The Euroregion Galicia and Northern Portugal, in terms of tourism, is not jointly promoted (Trillo-Santamaría, 2014). There are entities that promote it separately, such as the Galicia Tourism Agency and the Regional Authority of Porto and North Portugal (Sousa & Simões, 2018). There are cooperation projects between the two entities, namely the enhancement of a cooperation agenda in the tourism sector and a POCTEP project—"Facendo o Camiño"—led by the Porto and North Entity and, as partners, Turismo de Galicia and EGTC Galicia and North Portugal.

Galicia and the North of Portugal are united by an identical geography that can be seen in parks, spaces, areas and natural landscapes that occupy many kilometres of both regions. The contiguous landscape and material and immaterial culture between the North of Portugal and Galicia are elements of cohesion between two regions where, progressively, various types of conflict have been deactivated in order to move towards a horizon of progress and joint development (Sousa & Simões, 2018).

Locally, besides the projects, twinning and Eurocities, there are other projects focused on a common characteristic, as is the case of EGTC Rio Minho, which has as partners the municipalities integrated in the Deputación de Pontevedra (Galicia) and the CIM Alto Minho (Northern Portugal). The Euroregion Galicia and Northern Portugal is endowed with a set of infrastructures and accessibilities that allow a good connection between the two regions, a characteristic that is not so present in other Euroregions.

In terms of land transport, there is a network of motorways connected around a main axis running along the coast from Ferrol (Galicia) to Vila Nova de Gaia (Northern Portugal), passing through the Galician cities of A Coruña, Santiago de Compostela, Vigo, and the northern cities of Valença, Viana do Castelo and Porto, and linking Braga and Guimarães. From there it continues with a network of roads and motorways to the interior, connecting to the rest of Spain and Portugal. The railway network follows a similar structure, with a main line along the coast. However, the only existing connection is between Porto and Vigo, which is not in the best condition and is currently being modernised (Ramos et al., 2008).

The Euroregion also has several commercial ports and airports. Galicia has six ports: San Cibrao, Ferrol, A Coruña, Vilagarcía de Arousa, Marín and Vigo. In the North of Portugal there are also, but in a smaller number: Viana do Castelo, Leixões and Douro. As for air transport, the Euroregion has four international airports: Lavacolla in Santiago de Compostela (SCQ), Peinador in Vigo (VGO), Alvaredo in A Coruña (LCG) and Francisco Sá Carneiro in Porto (OPO).

Although the North of Portugal has only one airport, this one surpasses the four Galician ones in passenger numbers. According to official statistics, the Francisco Sá Carneiro airport (OPO) had 11,942,333 passengers in 2018 and 13,112,453 in 2019, while the Galician airports (Vigo, Lavacolla and Alvaredo) obtained in total 5,082,076 passengers in 2018 and 5,269,133 passengers in 2019.

Galicia and the North of Portugal have great tourism potential, both together and individually.

In terms of UNESCO World Heritage, the Euroregion has 11 classified heritage elements:

- Five in the region of Galicia: Dry stone construction, historical centre of Santiago de Compostela, Caminos de Santiago, Lugo Wall, and Hercules Tower.
- Six in the North region of Portugal: Alto Douro Wine Region, Guimarães historic centre, Porto city centre, Côa Valley Rock Art, Bom Jesus do Monte Sanctuary (Braga) and Podence Carnival.

The two regions also have a remarkably similar and complementary natural, cultural, and historical heritage. This was built on a common history that allows them to present themselves as a single destination.

In terms of natural heritage, namely national and nature parks, this Euroregion has:

- Two national parks: Gerês National Park and Atlantic Islands National Park.
- 10 nature parks: Alvão, North Coast, Montesinho and International Douro (North Portugal); Carrubedo Dunar Complex and Carregal and Vixán Lagoons, Lower

Limia and Xurés Mountains, Eume Fragas, Aloia Mountain, Invernadeiro, Enciña da Lastra Mountain (Galicia).

There are similar aspects in the behaviour of the tourists visiting the Euroregion (Ladeiras et al., 2015). For tourists, the main attractions for visiting the territory are history, culture, heritage, and nature, together with landscape and gastronomy.

According to Pardellas and Padín (2017), there are several studies that show the potential for the Euroregion to promote itself together and become competitive in Europe and the world.

The same authors mention the Minho area as the Euroregion's tourist potential, not only because of its great proximity but also because of the set of common tourist attractions: the Minho River, the Gerês-Xurês National Park. These constitute sufficient resources for the creation of a cross-border tourist destination.

The two regions have different legislation regarding the categories and regulation of tourist accommodation establishments, which leads to a great disparity between them, as shown by the statistics made available by Turismo de Galicia and the National Statistics Institute (INE).

Galicia has a higher number of tourist accommodation establishments than Northern Portugal. The concept of local accommodation does not exist in the Galician region; however, it is in the same class as tourist flats, villas for tourist use, pensions, and tourist hostels, because they all have similar regulations. In terms of tourist resorts, in 2018, Galicia had 9091 units in operation. In Northern Portugal, there were 1785 units in operation (Sousa & Simões, 2018).

Regarding Galicia's tourism demand in 2018 recorded a total of 4,773,496 tourists (Galicia, 2019) and 5,107,273 tourists the following year. In the same period, the North of Portugal reported a total of 5,285,297 guests and, in 2019, it reached 5,873,000 guests. In addition, both have an average stay of approximately 2 days.

In 2020, the year in which COVID-19 was declared a world pandemic, the Euroregion had potential growth, but everything changed. The North of Portugal, which ended 2019 with positive results, ends 2020 with −63% of overnight stays, −36% from residents, and −74.3% from non-residents. It should be noted that in this period 50% of the hotels were closed or had no guest movement.

In the case of Galicia, the situation is the same. As soon as the order of containment was declared, most establishments closed and reopened in the high season, with—60.9% of overnight stays, 89% of which were by residents and 11% by non-residents.

As reflected in the 2019 data, tourism growth was expected to continue in both Northern Portugal and Galicia. The Galician region was preparing for the arrival of millions of pilgrims to celebrate the holy year, Xacobeo 2021, which would create an economic and tourist boost in this Euroregion, via the Camino de Santiago (Casais & Sousa, 2020). However, with the declaration of the pandemic, everything was changed, restricted and delayed.

2 Methodology

Within the cooperation strategy, the Euroregion Galicia and Northern Portugal is an example, due to the protocols and European cooperation projects developed.

Considering the objective of this study, eight exploratory semi-structured interviews were conducted with stakeholders from Galicia and Northern Portugal, with the intention of knowing and understanding their vision of the Euroregion, both at the institutional and business level. The instrument used of qualitative analysis for this work consists of interview surveys, which according to Brinkmann (2007), allows knowledge to be produced through the experience, wishes and opinions of the interviewees.

In particular, the interview aimed to analyse the opinion of the main stakeholders in Galicia and Northern Portugal and understand in more detail the following points: (i) availability of cooperation for the creation of a destination brand; (ii) identification of key tourism products; (iii) identification of the target audience; (iv) evaluation of current communication strategies; (v) contribution to the structuring of a common communication strategy. Given the need to listen to the stakeholders of Tourism and the Euroregion (due to their direct contact with the phenomenon), with the aim of knowing their characteristics and opinions, a qualitative approach was chosen. Knowledge of these visions makes it possible to analyse not only the strategies that have already been developed but also to understand how they work to propose alternative strategies.

The interview was divided into four parts to obtain the stakeholders' perceptions about the Euroregion, through several dimensions: Availability, Identification, Tourism Products and Analysis. In the first part of the interview, the aim is to find out about the reality of stakeholder cooperation and whether there is already a dynamic of cooperation between agents in the two regions. Then, it is intended that through their experience, they indicate key tourism products that in their opinion positions the destination, as well as the identification of the target audience, with an analysis of the current and potential audiences for Galicia and Northern Portugal. The third part refers to the evaluation of current marketing strategies developed by or for stakeholders. In the last part, it is intended that the stakeholders indicate potential competitors and what differentiates the Euroregion from them.

The interviews, conducted between July 2019 and March 2020, with three public and five private entities, were transcribed and analysed according to the referenced dimensions. Public entities have a history of cooperation. The European Grouping of Territorial Cooperation Galicia and Northern Portugal is an agent of dynamization and interconnection of institutions, companies, and associations of the Euroregion. The Municipalities of Valencia and Tui form the Eurocity Tui-Valencia, a project of local dynamization of cross-border scope. The companies involved present both in their offer and demand, cooperation initiatives and are based in the Euroregion. The definition of the sample was based on the proximity between the entities (Table 1), their knowledge of European projects and cooperation, activities and cross-border

Table 1 List of participants

1	Agrupamento Europeu de Cooperação Territorial – Galiza e Norte de Portugal	Galicia—Spain
2	Concello de Tui	Galicia—Spain
3	Município de Valença	North of Portugal—Portugal
4	Hostel Ideas Peregrinas	Galicia—Spain
5	Hotel Fábrica de Chocolate	North of Portugal—Portugal
6	Elos da Montanha	North of Portugal—Portugal
7	Verdant Experiences	Galicia—Spain
8	Galaxipotential	North of Portugal—Portugal

Source: Own elaboration

cooperation projects developed and cross-border business relations over the last 10 years.

3 Discussion of Results

First insights into the availability of cooperation for the development of a common marketing strategy show that there is cooperation at the institutional level, on a local basis, between municipalities and between the various bodies, which at the public level have an influence on the tourism sector, but also at business and associative level. When referring to associations, the relationship between citizens is considered: "Non-institutional cooperation or social cooperation is very intense, as a joke, I usually say that between Galicia and Northern Portugal there is cooperation beyond the institutional, I mean there is collaboration between chambers of commerce, universities and municipalities, but citizens cross the border as if there was no border" [Interviewee 1]. Regarding the availability and type of cooperation, Galicia and Northern Portugal is taken as an example of cooperation between the various agents of society, seeing these as an opportunity for economic growth. "It makes perfect sense to have more business cooperation, it is an added value and I think that a partnership is just that" [6].

In assessing the current communication strategies, namely the current communication strategies of regional tourism entities, the interviewees highlight the disparities in the promotion and communication of destinations.

Both businesses and municipalities say that tourism entities focus on destinations that are already marketable, "massacring" them [7], leaving aside potential alternative destinations to boost tourism.

For the interviewees, the tourism products that most identify this Euroregion are "Religious Tourism with the Camino de Santiago, Nature Tourism, valuable

Heritage. There is such a differentiated offer with a huge potential for the various types of public. Bringing all the points together we would present quality and diversified products for each public" [3], which mainly attract a target public that seeks local experiences with interaction with culture and traditions, walking trails, hiking [6].

Regarding the potential of a common marketing strategy, the interviewees refer that there are conditions to exist and that this should be led by the regional tourism entities and coordinated by a destination marketing organization that already develops the issues of territorial cooperation, allowing not only a sectoral scope but also a local one.

To understand the potential of the Euroregion, a SWOT (Strengths, Weaknesses, Opportunities and Threats) analysis was designed. Based on the research carried out on the tourism strategies of each region and the interviews carried out, it is intended to demonstrate the strengths, weaknesses, opportunities, and threats of the object under study. As exemplified in Table 2, briefly, the destination Galicia and Northern Portugal presents itself as a unique destination due to its proximity, its common identity, the history of social, institutional, and business relations. Good accessibility means that there is no real frontier between the two regions and that travel is free, demonstrating their complementarity. However, there are still bureaucratic difficulties in the implementation of actions (such as authorisations for the use of equipment or simply the use of the banks of the River Minho for nautical activities, since the legislation is different in both countries), and in the communication between municipalities in the more technical field (such as the example of tourist shops that do not have access to promotional material). In this way, the destinations' communication strategies are not coordinated and sometimes confuse tourists. The COVID-19 pandemic that shook the world has become a threat, causing a spirit of distrust and fear, which in the case of this Euroregion has not been experienced for a long time, but this new "new beginning" should also be seen as an opportunity, because it will allow the Euroregion to reinvent itself and to align strategies in order to boost the internal market, which in other strategies was placed in third place. To understand the opportunity of this Euroregion as a tourist destination, through the creativity of the actions to be implemented, one should understand what exists and "seize" the opportunities and strengths and promote authentic and creative tourism products, to meet the current trends and consumers in the post-covid.

The following are defined as strategic objectives for the valorisation of the Euroregion of Galicia and the North of Portugal as a creative tourist destination based on its complementarity: (i) to valorise the cross-border territory and its communities; (ii) to boost endogenous resources; (iii) to create networks between companies in the Euroregion; and finally, (iv) to project the "Galicia and the North of Portugal" brand-destination.

To develop these objectives, it is necessary to enhance institutional relations with the creation of a destination marketing organisation (DMO), based on an existing entity that encompasses the various stakeholders and has a track record in cross-border cooperation and in developing joint projects. By defining this entity, it is more accessible to redefine the objectives and the strategies to apply, through the

Table 2 SWOT analysis

Strengths	Weaknesses
• Its own cultural identity. • Diversity of cultural, sports and natural attractions. • Proximity to the main tourist destinations. • Good accessibility of connections between the two regions. • Long history of institutional relations. • High number of projects financed within the scope of cross-border cooperation. • Ease of communication (similar languages). • Material and immaterial heritage with links between the two regions. • Privileged geographical location.	• Even though there are four airports, there is only one international airport (Francisco Sá Carneiro airport—Porto). • A deficient bus and railway network in terms of timetables and quantity. • Institutional bureaucracies. • Too many public entities for the same territory. • Outdated cultural equipment. • Lack of joint tourism promotion. • Lack of major joint events. • Weak evidence of R&D in tourism. • Significant offer of non-classified accommodation. • Outdated and unattractive communication strategies (social media, relational marketing, among others). • Poor coordination between the tourist information offices.
Opportunities	**Threats**
• Cooperation between cross-border municipalities at local and institutional levels. • Recognition of the importance of tourism for the economy. • Considered as safe and economic destinations. • Diversity of endogenous resources for the creation of unique experiences. • Growth of smart cities and destinations. • Technological knowledge of the population.	• Strong competitiveness with other Portuguese and Spanish Euroregions. • Economic and political situation in Europe. • Closure of borders. • COVID-19 pandemic. • Economic crisis. • Vulnerability of tourism to external factors. • Relevance of price in decision-making.

Source: Authors

knowledge of the territory, its potentialities and the projects that exist, to create a relationship between all of them and to potentialize benchmarking actions to boost coordination between the various micro-destinations in the Euroregion. With the knowledge of the territory, both in terms of resources and business, it is necessary to create actions of mutual knowledge, that is, that the business agents get to know each other and start working together, potentializing synergies to obtain greater profitability (monetary and temporal) from their business. As there is knowledge among business entities and agents, the development of marketing actions should be based on what makes them stronger: cooperation.

The cooperation strategy becomes an impulse for the dynamization of the territory at the tourist level, through the meeting of the singularities and strengths of each one to combat common problems, as well as to take advantage of the gains for the territory.

The creation of tourism products and tourism experiences based on cooperation will allow for differentiation from other destinations, but also for sustainability, because there will be a greater concern for cost reduction, environmental, cultural and heritage concerns, through better use of resources and equipment.

When thinking about creative products and experiences for this destination it is inevitable to mention the use of technologies, thinking about the development of a smart Euroregion, through the combination of several financed projects that already exist and/or are in the final stages of completion. This action would bring together all the smart destinations and add value through the creation of a platform where all the stakeholders can provide information so that the tourist can create their own experience in the Euroregion.

4 Conclusions, Limitations, and Future Lines of Research

Galicia and the North of Portugal have always defined themselves as one, with similar characteristics that stood out for their strategic position, sometimes far from the central power (Madrid and Lisbon).

To combat this distance, they became partners and allied, thus promoting regions, something that Europe promotes among its constituent countries through the creation of the European Groupings of Territorial Cooperation.

During this research new elements were always appearing to highlight and support the idea and the objectives that are presented, such as the Iberian Summit on October 10, 2020, and the various strategies that have already been presented with a view to the future (2030), joint investment plans for the Euroregion (2021–2017), among others that have been presented and discussed between the two countries, highlighting the border territory.

Through the interviews carried out it was possible to verify that stakeholders see as an added value the potential of a cooperation strategy for a common destination for Galicia and Northern Portugal, through the coordination of a DMO (Destination Marketing Organization) with experience in cross-border cooperation.

The results of the study allow us to state that, for marketing strategies, the experience of being in two different countries in the same destination should be highlighted, enjoying creative experiences based on the singularities of this Euroregion. With the creation of a uniform communication strategy to communicate the destination as equal "without borders", promoting the complementarities of destinations, mainly micro-destinations that are not so massified, and thus giving relevance to the proximity between nearby localities. It is important that these strategies identify the territories that are or need a boost for tourism. They should be identified based on their common singularities and strengthened through cooperation with creative experiences.

Two factors that can further contribute to the growth or boost of these strategies is the valorisation of cooperation projects that already exist and their enhancement through digital tools, using social networks, but with the potential of smart tourism, using solutions that energise the territory with co-creation experiences with the use of smart technologies.

The stakeholders' perceptions highlighted the willingness to cooperate to create a tourist destination of reference and excellence through cooperation practices.

The economic, cultural, and touristic potentialities for the creation of the destination and the vast existing resources that potentialize the creation of a joint strategy for tourism development and the launch of a cooperative destination brand, even in the face of the COVID-19 situation, are notorious.

In several studies, the issues of tourism potential of cross-border cooperation and marketing strategies that enhance this cooperation are addressed and referenced with examples to follow. However, this study addresses a theme in the boosting of creative tourism experiences through the cooperation strategy and as a boost to energize the territory.

The limitations relate to the fact that the perspective of public and private entities was analysed, leading to results based on the perceptions of local stakeholders, both institutional and business. Thus, it will be important in the future to consider the perceptions of residents, tourists, and visitors to the destination, in order to obtain a more comprehensive view of the destination and whether the strategy is necessary or likely to succeed. Finally, it is important to stress the need to know the European directives and legislation which can further enhance the implementation of more cooperation strategies.

In the possible limitations to the implementation of the communication actions presented, it is highlighted the fact that there are several public entities that assume the role of promoting the territory. However, they do not do it in totality, but by regions, or in this case, municipalities that integrate that entity, as is the case of EGTC-Rio Minho and the Atlantic Axis. The entity that has the prospect of moving towards the creation of the DMO should encompass the entire territory and all agents, without the need to create another entity for this purpose. Through the research carried out, the entity that best fits in is the EGTC Galicia and Northern Portugal, due to existing protocols with various entities and organizations in this Euroregion, both at the public and associative level.

References

Binkhorst, E. (2007). Creativity in tourism experiences: The case of Sitges. In G. Richards & J. Wilson (Eds.), *Tourism creativity and development* (pp. 125–144).

Binkhorst, E., & Den Dekker, T. (2009). Agenda for co-creation tourism experience research. *Journal of Hospitality Marketing & Management, 18*(2–3), 311–327.

Brinkmann, S. (2007). The good qualitative researcher. *Qualitative Research in Psychology, 4* (1–2), 127–144.

Buhalis, D. (2000). Marketing the competitive destination of the future. *Tourism Management, 21* (1), 97–116.

Campos, A. L., & Pardo, M. C. (2008). Eurocidade Chaves-Verín, um instrumento de desenvolvimento económico e social. *Revista da Eurorrexión Galiza-Norte de Portugal, 13*, 81–99.

Casais, B., & Sousa, B. (2020). Heterogeneity of motivations and conflicts in pilgrim-to-pilgrim interaction: A research on the way of Saint James. *Tourism Management Perspectives, 36*, 1–9. https://doi.org/10.1016/j.tmp.2020.100748

Commission, E. T. (2021). *European tourism: Trends & prospects*. European Travel Commission.

Galicia, A. d. (2019). *Balanço Anual do Turismo de Galicia*. Turismo de Galicia.

Kotler, P., Bowen, J. T., Makens, J. C., Garcia, J., & Flores, J. (2011). *Marketing turístico*. Prentice Hall.

Kozak, M., & Buhalis, D. (2019). Cross–border tourism destination marketing: Prerequisites and critical success factors. *Journal of Destination Marketing & Management, 14*, 100392.

Kurowska-Pysz, J., Castanho, R. A., & Loures, L. (2018). Sustainable planning of cross-border cooperation: A strategy for alliances in border cities. *Sustainability, 10*(5), 1416.

Ladeiras, A., Mota, A., & Pardo, M. C. (2015). A comparative study of thermal legislation in the Galicia–North Portugal Euroregion. In *Health and wellness tourism* (pp. 1–20). Springer.

Melo, C., Correia, A. I., Cardoso, M., & Marques, V. (2019). Creative tourism experiences in Guimarães : A twofold analysis o f visitors ' and suppliers' perspectives. *Revista Portuguesa de Estudos Regionais, 51*, 73–91.

Otocan, O. (2010). *Euroregion as a mechanism for strengthening transfrontier and interregional co-operation: opportunities and challenges*. Venice Commission.

Pardellas, X., & Padín, C. (2017). La Eurorregión Galicia-Norte de Portugal y los modelos de cooperación trans fronteriza, Polígonos. *Revista de Geografía, 29*, 11–35.

Ramos, R. A., Carballo Cruz, F., & Ribeiro, J. C. (2008). *Efeitos económicos da melhoria da ligação ferroviária Porto-Vigo na euroregião Norte de Portugal-Galiza*.

Ratten, V., Braga, V., Álvarez-García, J., & Del Río, M. D. L. C. (Eds.). (2019). *Tourism innovation: Technology, sustainability and creativity*. Routledge.

Richards, G. (2012). Tourism, creativity and creative industries. In *Conference Creativity and Creative Industries in Challenging Times*, NHTV Breda.

Richards, G., & Raymond, C. (2000). Creative tourism. *ATLAS News, 23*(8), 16–20.

Richards, G., & Wilson, J. (Eds.). (2007). *Tourism, creativity and development* (Vol. 10). Routledge.

Rojo Salgado, A. (2010). Cross-border cooperation and its consequences: Towards the territorial restructuring in Europe. *Investigaciones Regionales-Journal Of Regional Research, 18*, 141–152.

Sousa, B., & Simões, C. (2018). An approach on place attachment, involvement and behavioural intentions in iberian marketing contexts: The case of Galicia-north Portugal Euroregion: An abstract, In P. Rossi & N. Krey (Eds.), *Finding new ways to engage and satisfy global customers, proceedings of the 2018 academy of marketing science (AMS) world marketing congress (WMC)* (p. 617). Springer.

Trillo-Santamaría, J. M. (2014). Cross-border regions: The gap between the elite's projects and people's awareness. Reflections from the Galicia-North Portugal Euroregion. *Journal of Borderlands Studies, 29*(2), 257–273.

Zhang, H., Fu, X., Cai, L. A., & Lu, L. (2014). Destination image and tourist loyalty: A meta-analysis. *Tourism Management, 40*, 213–223.

The Presence of Rural Accommodation in Extremadura on Facebook: An Approach to Its Intensity of Use Through a Grouping Analysis

Juan Ignacio Rengifo Gallego, José Manuel Sánchez Martín, Marcelino Sánchez Rivero, Luz María Martín Delgado, and María Cristina Rodríguez Rangel

Abstract Facebook has acquired a notable presence in many business activities as a social medium and networking service. Tourism has not been oblivious to this trend and consequently numerous accommodation establishments have joined this network. However, when analysing a particular type of accommodation, i.e. rural accommodation, certain deficiencies can be observed at least in the context analysed. For this reason, this article aims to ascertain the activity of rural accommodation in Extremadura based on the hypothesis that the potential of this network to promote the visibility of its establishments is not always exploited. At the same time we investigate the possible existence of a territorial pattern in certain characteristics of the social network that could lead to an improvement in the visibility of the accommodation establishments. To corroborate this assumption, a Grouping Analysis is applied both with and without spatial restriction in order to determine the presence of spatial clusters. The results clearly show that a higher performance is achieved when this technique is applied based on spatial relationships than without them. At the same time, it can be observed that most of the rural accommodation establishments do not stand out in the use of the social network analysed and very few of them play a prominent role, both positively and negatively, with respect to the average situation of the accommodation establishments. In the first case, they should act as models when their behaviour can be replicated by other establishments, and in the second case they should serve as a basis for the implementation of corrective measures to involve the establishments in the management and control of this widely distributed social network. Among the conclusions drawn from this study is the need to actively incorporate Facebook as a basic tool for the positioning of accommodation.

J. I. R. Gallego (✉) · J. M. S. Martín · M. S. Rivero · L. M. M. Delgado · M. C. R. Rangel
University of Extremadura, Badajoz, Spain
e-mail: irengifo@unex.es

J. Leitão et al. (eds.), *Tourism Entrepreneurship in Portugal and Spain*, Tourism, Hospitality & Event Management, https://doi.org/10.1007/978-3-030-89232-6_7

1 Introduction

The traditional role of tourism destination management, consisting of disseminating information sources for tourists, has been changing in recent years owing to the emergence of online communication tools such as Facebook (Hays et al., 2013; Li et al., 2017). These platforms have therefore become essential tools for positioning brands of tourism (Li et al., 2017), for attracting potential tourists (Schroeder & Pennington-Gray, 2015) and for adapting the offer to tourists' needs and being more competitive (Mariani et al., 2016), among other matters.

Facebook is currently the social network with the largest number of users in the world. According to the WeAreSocial and Hootsuite reports (2021), Facebook had 2.74 billion users in January 2021, which exceeded the figures for YouTube by more than 400 million users and for WhatsApp by more than 700 million.

Social media have now become perhaps the main source of information about destinations for tourists (Zeng & Gerritsen, 2014) and their role is likely to continue to grow in the future (Phillips et al., 2010). However, social networks are not merely a source of information but also a powerful tool for tourism business management. They help tourists make the decision to visit the destination (Phillips et al., 2010; Marchiori & Onder, 2015; Önder & Marchiori, 2017; Marder et al., 2019), promote tourism establishments much more effectively than websites or paid media advertisements (Lee et al., 2012; Moise & Cruceru, 2014), develop a destination brand image (Kim et al., 2016; Marine-Roig & Clavé, 2015), and help clarify misperceptions of the destination (Joyner et al., 2018; Kotsi et al., 2018).

Despite the abundance of studies analysing the importance of social networks in the management and marketing of hotels and tourist destinations, there is a clear lack of applied research on social networks (social media) referring to rural tourism and agritourism (Bowman et al., 2020; Cho et al., 2014; Zeng & Gerritsen, 2014). For this reason, this paper aims to fill this gap by determining the degree of use of Facebook profiles by rural accommodation establishments in one of the Spanish regions with the greatest development of this segment.

The aim of this chapter is to demonstrate that in the case of rural accommodation establishments in the region of Extremadura, Spain, the potential of Facebook to promote their visibility has not yet been sufficiently exploited. Likewise this research aims to determine whether those establishments that are more aware of the importance of being present and active on this social network are grouped in certain areas of Extremadura, i.e. whether there are spatial patterns justifying the existence of groups of establishments according to their greater or lesser presence and activity on Facebook.

To this end, the second section of this paper gives a brief overview of the uses of the social network Facebook as a marketing and promotion tool for any type of economic activity, paying special attention to the interaction established between the company and its customers (real or potential) through the messages published by both parties. Subsequently the third section establishes the relationship between social networks and tourism, highlighting the positive and negative impacts of the use of social networks by any tourist company in general and rural and agritourism accommodation establishments in particular. The case study to be analysed, the data to be processed and the methodology used in this work are presented below. The most relevant results of the application of this spatial statistical technique are mentioned and discussed in the fifth section of the paper, which ends with a presentation of the main conclusions and their implications for the management of the presence of rural accommodation establishments in the region on social networks and more specifically on Facebook.

2 Facebook as a Tool for Marketing and Promotion

Social media is currently a unique marketing communication channel (Eagleman, 2013) which enables the mass dissemination of information about products, goods and services (Dahnil et al., 2014) and the building of brand loyalty among customers (Castronovo & Huang, 2012). Moreover, from the point of view of consumers social networks allow the latter to have more control over what they are willing to buy or reject (Tynan et al., 2010).

One of the greatest potentialities of social networks is the possibility of interaction with profile followers. The content of the publications (posts) on social networks determines the greater or lesser degree of interaction of the followers (Kim, 2018; Kim & Yang, 2017; Saxton & Waters, 2014). In addition, the length of the posts (in number of words) and the number of images, videos and links included in the posts also determine the interaction of profile followers (Pino et al., 2018; Sabate et al., 2014). Other factors which also condition the interaction between the profile and the followers are the use of hashtags (Sevin, 2013; Usakli et al., 2017) and the tagging of friends (Oeldorf-Hirsch & Sundar, 2015).

As established by Muntinga et al. (2011), forms of social media behaviour can be divided into three categories: consuming, contributing and creating. Consuming is the lowest level of participation in social networks, as it does not involve contributing or creating content. The intermediate level is contributing, which involves interaction between users and content or between users and users. Finally, the highest level is creating, which involves producing and publishing content.

In the specific case of the Facebook social network, the lowest level (consuming) represents liking, the intermediate level (contributing) represents commenting and the highest level (creating) represents sharing (Kim & Yang, 2017). This gradation of forms of behaviour on Facebook is organised around the basic unit of communication on this social network, which is the publication of messages (posts). However, not all Facebook posts generate the same amount of attention among Facebook users (Brafton, 2014). Therefore, certain aspects of the publication must be taken into account so that it generates as much interest and attention as possible. According to Kim and Yang (2017), the aspects to consider are the strategy of the message, the form of the message, the posting type and the interactivity of the message.

Every message on Facebook must correspond to a certain strategy, which takes into account aspects such as the type of product, the target audience, etc., since the response of users of this social network will be highly conditioned by the strategy chosen. Several studies have demonstrated the relationship between message strategy and user response (Ashley & Tuten, 2015; Chauhan & Pillai, 2013; Saxton & Waters, 2014; Swani et al., 2013).

On the other hand, the form of the message (text, photo, audio, video, etc.) is an element which may increase its usefulness (Kent et al., 2003), with the most common being the combination of different message forms to reach a greater number of users (Men & Tsai, 2012; Waters et al., 2009). Images (photos and videos) thus generate more emotions than text (Brantner et al., 2011). Furthermore, Kim and Yang (2017) show that photos contribute towards an increase in the number of likes but a decrease in the number of comments. These authors also show that visual elements (photos and videos) are positive predictors of the number of shares.

Type posting can be classified into two main groups: created posting and shared posting (Kim & Yang, 2017). According to these authors, in the same way that a shared post can save time and money for the organisation owning the Facebook profile, its usefulness may be less than that of a created post. In fact these authors show that a created post generates a higher number of likes, comments and shares than a shared post.

Finally, given that Facebook attributes to each profile the same level of structural interactivity (Lovejoy & Saxton, 2012), the perceived interactivity of the posts of an organisation will depend on the extent to which it wishes to use that interactive structure (Saffer et al., 2013) and the degree to which it initiates two-way communications (Lee & Park, 2013). Kim and Yang (2017) show that posts soliciting responses encourage users to comment on the post, although they do not increase the number of likes or shares.

3 Social Media and Tourism

The current impacts of social media on tourism are beyond doubt and this is so to such an extent that they are changing tourism information search habits (Roth-Cohen & Lahav, 2019) and modifying existing promotion and communication structures (Zeng & Gerritsen, 2014). Although these impacts are usually positive, some negative impacts have also been identified in the literature, although these are smaller than the positive impacts.

The positive impacts of social media include their usefulness to tourists in evaluating tourist destinations in real time, providing data to managers for decision-making (McCreary et al., 2019) and generating expectations (Narangajavana et al., 2017). In the specific case of the social network Facebook, it is being used to recommend a tourism brand or destination (Ben-Shaul & Reichel, 2018), to post complaints about tourism products or services (Pantano & Di Pietro, 2013), to provide real-time information which is not available in the media (Ketter,

2016), to enhance destination reputation (De Moya & Jain, 2013), to promote hotel brands or make bookings (Mariani et al., 2016), to model tourism demand (Önder et al., 2020) and so on.

However, despite the above some studies have revealed the existence of some negative impacts of social networks on tourism, although they are scarce. Zhang et al. (2016) therefore show that tourists' excessive exposure to social media can affect their decision-making process due to information overload. De Lencastre and Corte-Real (2013) show that the social exchange of information resources between residents and tourists on social networks can generate a reputation in the management of the destination's brand image.

The publication of photographs on the profile of tourism companies on social networks has become an essential element of their online marketing strategies. This is so much so that several studies (Choi et al., 2007; Dwivedi, 2009; Govers & Go, 2005; Stepchenkova & Morrison, 2008) have shown that online images of tourist destinations are a powerful tool for their management and marketing. The emotional response which certain photographs inspire in tourists can help or hinder their willingness to visit certain areas or book certain accommodation establishments. For example, in a study of agritourism businesses Joyner et al. (2018) show that tourists prefer photographs showing the farmer and animals together, or photographs showing farmers interacting with tourists, or photographs showing children with animals. Conversely, photographs showing fences or animals with ear tags generate negative emotions among tourists.

Social networks have also become a selection tool for tourists in view of the large number and variety of destinations from which they can choose (Roth-Cohen & Lahav, 2019), which is due among other reasons to the confidence they have in the comments made by other tourists (Casalo et al., 2010).

To turn to the specific case of rural tourism, despite the fact that well-planned commercial communication is a fundamental issue for companies operating in this segment (Simková, 2007), the truth is that the vast majority of establishments lack the means of carrying out major marketing campaigns (Lane, 2009). However, these small establishments always have the possibility of promoting themselves through social networks and also of providing customers with information about the establishment and the environment in which it is located, and of commenting on their experience during the trip (Joo et al., 2020). Moreover, the importance that customer satisfaction and behaviour is acquiring on social networks is causing tourism establishments in general and rural accommodation establishments in particular to give increasing prominence to new issues such as big data, online comments, eWom and digital marketing (Nusair, 2020; Nusair et al., 2019).

In addition, social media also represent an opportunity for rural residents to participate in the rural tourism development process (Lundgren & Johansson, 2017; Ryu et al., 2020; Senyao & Ha, 2020) and to define the image of the rural destination through their online behaviour on these social networks (Uchinaka et al., 2019).

However, despite the current importance of Facebook as an online marketing tool, this social network is little used by rural accommodation establishments. A study by Pesonen (2011) analyses the Facebook tools used by rural accommodation establishments in Finland. The author concludes that most of the establishments analysed do not update their Facebook page on a weekly basis, with an average number of posts per week of only 0.62. In contrast to the more intense Facebook activity of other larger tourism businesses, Pesonen (2011) found that small rural tourism businesses do not devote much effort to their Facebook presence, which leads him to question whether they should even have Facebook pages if they do not plan to update their content on a regular basis.

4 Materials and Method

4.1 The Case Study

The study area chosen to carry out the research is the autonomous region of Extremadura. This region is located on the periphery of peninsular Spain and also shares a border with Portugal. It is an eminently rural territory as can be seen from the articles of Law 45/2007 of 13th December on the sustainable development of the rural milieu. This Law includes in its Article 3 different definitions of the concept of rural areas. In this sense, section a of the same law classes as rural the geographical area formed by the sum of municipalities or smaller local entities with a population of less than 30,000 inhabitants and a population density of less than 100 inhabitants per km^2. Paragraph c specifies that a small rural municipality is one with a resident population of less than 5000 inhabitants and which is an integral part of the rural environment (Law 45/2007 of 13th December on the sustainable development of the rural milieu). Based on the articles of this Law, most of the 388 municipalities in Extremadura are rural, as only Badajoz, Plasencia and Almendralejo meet both requirements.

A large part of the surface area of Extrenadura is dotted with numerous tourist attractions that are of interest to the demand for rural tourism in its broadest conception (Sánchez-Martín et al., 2020a) (Fig. 1). Natural landscapes of great richness and with an exceptional state of conservation stand out, which allows them to be included among protected natural areas (Sánchez-Martín et al., 2018).

There are also numerous bathing areas in the form of natural pools and a smaller number of river beaches located in the vicinity of the areas of upland relief (Sánchez-Martín et al., 2020a) or on large reservoirs, which are ideal for practising water sports. On the other hand, historical-artistic sites or assets of cultural interest constitute an important historical legacy (Sánchez-Martín et al., 2020c) and are accompanied by attractive gastronomy based on high quality products produced in the region itself, at times under the protection of a Protected Designation of Origin or Geographical Indication (Ortega-Rosell et al., 2012).

The great wealth of heritage present in Extremadura, whether natural or cultural, has been transformed into a resource and integrated into tourism products in certain areas, but not in all. It can be observed that there are areas in which the tourism

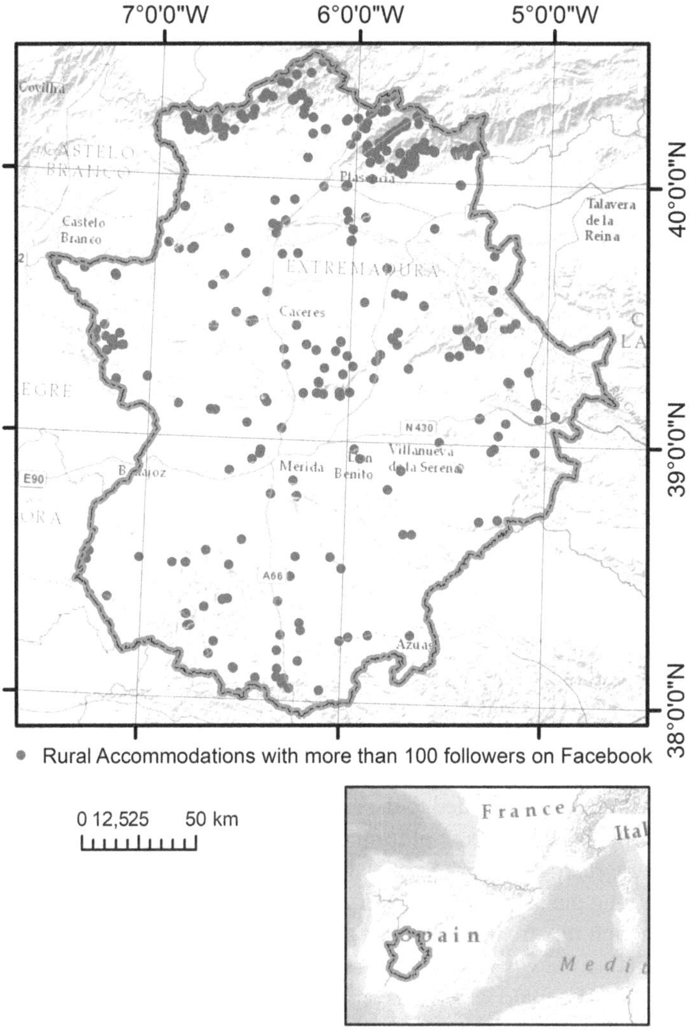

Fig. 1 Location. Source: Own material

potential of their main attractions is not exploited, as is the case, for example, with large reservoirs of water or emblematic cultural landscapes, such as the *dehesas*. This is despite the fact that they should serve as a basis for promoting an increase in rural tourism in the area analysed and contributing to the socio-economic development of the territory.

The reasons which may explain this differentiated behaviour are varied, one of the most common being the absence of the main attractions sought by the demand (Sánchez-Martín et al., 2020b). However, when analysing the supply of rural accommodation establishments in the territory, numerous studies agree on their good online reputation (Martín-Delgado et al., 2020; Sánchez-Martín et al., 2019a).

Despite the significant number of attractions and the good reputation of the rural accommodation offered on different social networks, Extremadura has a demographic dynamic which shows a secular trend in decline, especially in rural areas which make up most of the territory. In addition, we find processes of the masculinisation of the population, ageing and a lack of generational replacement. This peculiar demographic situation is both a cause and a consequence of its clear economic backwardness, as it is one of the regions making the lowest contribution to the country's GDP, which was estimated according to 2019 data at 20,677 million euros (Datos Macro, https://datosmacro.expansion.com). In addition, communication routes which up until a few decades ago were in very poor condition conditioned limited accessibility to the main economic centres of the Iberian Peninsula, which contributed to its isolation. This is still the case even today for rail or air transport (Sánchez-Martín et al., 2019b).

However, these limitations have led to a population of only 1,063,987 inhabitants in 2020 according to official figures obtained from municipal registers on 1st January (https://www.ine.es/jaxiT3/Datos.htm?t=2853); the environment is well preserved and its amalgam of landscapes can be considered a tourist attraction of great interest to those making up the demand, who are eager for knowledge and new experiences. From this we can infer the great potential for the development of activities such as rural tourism, especially if they are accompanied by effective tourism policies based on knowledge of the tourism system, which is not always the case.

As a consequence of the socio-economic backwardness, numerous initiatives have been launched to diversify production, thus giving substance to Article 20 of Law 45/2007 which is committed to the economic diversification of the rural world, with tourism specifically in its rural form occupying a privileged position, as can be seen from paragraph e of the aforementioned article. Since the 1990s therefore, projects involving the implementation of actions to support rural tourism have been initiated, although this trend has accelerated since the beginning of this century. It was during this period when there was a strong commitment to the development of this type of tourism, a trend which was prevalent even during the economic crisis and has continued until now. This circumstance has led to mismatches between the accommodation capacity in rural areas and the demand for accommodation in these establishments (Sánchez-Martín & Rengifo-Gallego, 2019). On the other hand, the commitment to rural tourism has concentrated on the creation of a significant rural accommodation pool as mentioned in the Rural Tourist Accommodation Occupancy Survey (*Encuesta de Ocupación en Alojamientos de Turismo Rural*, EOTR). This increase in the supply of accommodation was not initially accompanied by the implementation of other types of actions linked to the complementary offer in the form of activities, which would have meant a notable improvement in the enjoyment of the tourist experience. Fortunately, today there is a proliferation of tourist activity companies which enhance the tourist experience.

4.2 Data

The information supporting the research has been obtained from different sources. These include the Spanish National Institute of Statistics (*Instituto Nacional de Estadística*, INE) which provides all information collected at a regional level on parameters such as travellers, overnight stays, beds available and the number of estimated establishments, in addition to others such as the average stay or the level of occupancy. However, this information has been complemented with much more detailed data from the Register of Tourism Enterprises, which is accountable to the Regional Ministry of Culture, Tourism and Sports of the Regional Government of Extremadura, updated to 31 December 2019; these make up the database on the supply of tourist accommodation. These data, broken down by establishment, were georeferenced using the Google Maps application and corroborated with Street View to ensure that the locations were reliable. In addition, this database has been complemented in the form of consultations made to each of the establishments, indicating where appropriate whether they had a Facebook social network.

An initial approach to this database has shown that of the total number of rural tourism accommodation establishments in Extremadura (846), only 509 had the social network analysed. This figure represents 60.2% of the establishments, which is a significant proportion giving statistical solidity to the analyses carried out, as can be seen on the data sheet of the survey. It was however decided to use only establishments with at least 100 followers so as to make the values extracted from this sample even more reliable (Table 1).

In all the rural tourism establishments which had Facebook, the information available was complemented by 21 items which sought to find out the characteristics of the accommodation. These included the number of visitors, their reaction, the insertion of graphic material such as photos or videos, and also other information such as the interaction between owners and customers as shown in Table 2.

In addition, the alphanumeric information was supplemented with cartographic data from the National Geographic Institute (*Instituto Geográfico Nacional*, IGN) of 2001 under a Creative Commons CC-BY 4.0 International licence, which protects its free and unrestricted use for legitimate purposes with the sole obligation of acknowledging and mentioning its origin and ownership. Among the different alternatives, it

Table 1 Data sheet

Universe:	846 rural accommodation establishments on 31st December 2019
Sampling Size:	447 (those with more than 100 followers on Facebook)
Degree of confidence:	95%
Sampling error for the most unfavourable (pq=50%) and most favourable (pq=90%) scenario	3.19; 1.91
Date of completion	15th March to 15th June 2020

Source: Own material

Table 2 Information collected on each rural establishment using Facebook

Parameters
1. Does the hotel establishment have a Facebook profile?
2. Date of creation of the Facebook profile:
3. Number of people who have indicated that they like the page:
4. Number of people who follow the page:
5. Number of visits the page has received:
6. Type of content. Number of cover photos:
7. Type of content. Number of photos in the biography:
8. Type of content. Number of videos:
9. Type of content. Photos uploaded by mobile phone:
10. Use of hashtags:
11. Time (in days) since the most recent publication (in the "publications" tab) [Indicate the time fraction if <24 h has elapsed]
12. Time (in days) since the second most recent publication (in the "publications" tab) [Indicate the time fraction if <24 h has elapsed] 13.
13. Time (in days) since the third most recent publication (in the "publications" tab) [Indicate the time fraction if <24 h has elapsed] [Indicate the time fraction if <24 h has elapsed]
14. Time (in days) since the fourth most recent publication (in the "publications" tab) [Indicate the time fraction if <24 h has elapsed]
15. Time (in days) since the fifth most recent publication (in the "publications" tab) [Indicate the time fraction if <24 h has elapsed]
16. Interaction with followers: What is the score given by the followers to the accommodation establishment?
17. Interaction with followers: How many views have been used to calculate the above score?
18. Use of surveys:
19.1. Conducting sweepstakes: [Discounts]
19.2. Conducting prize draws: [Promotions] 19.3.
19. Conducting prize draws: [Giveaways]
20.1. Interacting with followers/users (do you respond to comments in the "community" section?): [Respond to positive comments]
20.2. Interaction with followers/users (do you respond to comments in the "community" section?): [Reacts to positive comments (likes)]
20.3. Interaction with followers/users (do you respond to comments in the "community" section?): [Responds to negative comments]
20.4. Interaction with followers/users (do you respond to comments in the "community" section?): [Reacts to negative comments (I don't like it, it makes me angry)]
21. Inclusion of the "Book Now" button for direct booking from the profile:

Source: Own material

was decided to use the National Topographical Base at a scale of 1:100,000 (BTN100), which provided sufficient spatial resolution for the purposes of this research (20 m). The information provided by this mapping is varied, although it has been complemented by the Spatial Data Infrastructure of Extremadura (*Infraestructura de Datos Espaciales de Extremadura*, IDEEX) of 2021, which is also under the same CC-BY 4.0 International licence (Table 3).

Table 3 GIS design project

Data type	Source	Cartographic information	Alphanumerical information
Cartographic	IGN	Administrative units	Area
	IDEEX	Population centres Natural protected areas Bathing areas	Population Area Location
Alphanumeric	Extremadura Turismo	Georeferencing information on Google Maps	Type of accommodation Address Municipality Beds available
	Authors by means of Facebook enquiries		Information collected in Table 3

Source: Own material

All alphanumeric information has been implemented in a Geographic Information System based on the aforementioned cartography, using the ArcGIS 10.5 software.

4.3 Methodology

The methodology of analysis proposed in this research consists of six stages (Fig. 2).

During the first stage the specialised literature on the subject was reviewed. At the same time the alphanumeric information which served as the basis for the research and the cartographic information was compiled. In the second stage the two databases were purified. They were then implemented in a GIS after the corresponding designing of the relevant project. From this point only those rural establishments with Facebook were selected and specifically those with more than 100 followers. The third stage focused on the application of spatial statistics, taking advantage of the specific module integrated in the GIS software which was used. In this case, and given the wide variety of geostatistical techniques, Mapping Clusters and more specifically a Grouping Analysis were selected. This has been used to carry out numerous tests, the main results of which are presented in the research. During the fourth stage the results were studied in order to decide on the most suitable model and to focus on an exhaustive analysis of the most representative group.

In thc fifth stage the results and the technique used were discussed, while in the sixth stage the main conclusions were drawn.

4.4 Analysis Techniques

The spatial statistics present on the ArcGIS software (10.5 version) have been used. Among all the Mapping Clusters tools included we specifically chose to use a

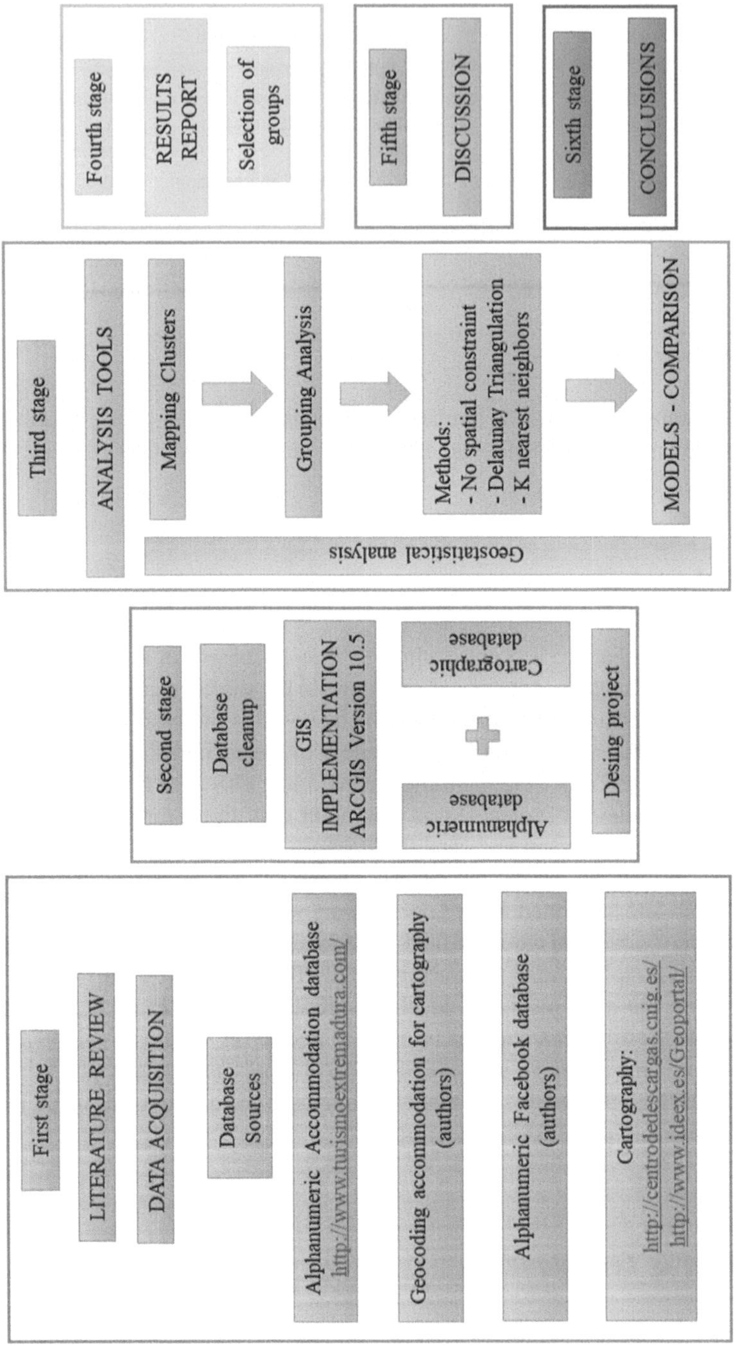

Fig. 2 Methodological scheme. Source: Own material

Grouping Analysis, the functionality of which aims to create groupings of a set of data represented cartographically considering numerical variables. For this purpose, the tool itself allows the use of several options, not only referring to the use of different variables but also to the selection of a spatial criterion which determines precisely the configuration of the neighbourhood criterion. In this way it is possible to differentiate between the use of a spatial criterion, based on the neighbourhood, or its omission. In this sense, and in order to make the results presented more rigorous, different tests have been carried out to combine both variables and types of spatial restriction, including two different models. The first did not use the territorial component, i.e., it established the groups by taking only the numerical attributes as a reference. On the other hand, the second was chosen to include the neighbourhood criterion according to two different models, Delaunay triangulation and K nearest neighbours. This double application aims to determine which type of technique obtains the best results for establishing clusters of properties according to their common criteria and at the same time those which are different from the rest.

The number of accommodation establishments used to carry out the analysis was 447, precisely those that had a Facebook social network and more than 100 followers. They were taken as specific entities, given that they were georeferenced in order to be able to carry out a more rigorous analysis. After extensive testing, it was decided to select the variables which were suitable for differentiating the groups. These variables concern three key aspects: the success of the page of the establishment, the content published and the interaction with followers. Logically other variables were discarded because their contribution to the creation of differentiated groups was low, as was deduced from the R2 coefficient calculated. Furthermore, according to the literature the inclusion of a greater number of variables does not imply greater rigour, and prior selection of these variables is advisable if suitable results are to be obtained (Mitchell, 2005; Venkatramanan et al., 2015). Grouping analysis explores similar values, and if a spatial criterion is also used it can consider spatial proximity to configure different groupings. Each of the groups obtained presents the greatest similarity between its members and at the same time maximises its differences with the remainder of the groups (Cardona et al., 2007).

The technique has been applied following three basic references, taking into account that the geometry used is of the point type and not a polygon, which would allow other options, although this possibility was ruled out given that they did not share any segments. In view of the above, three ways of obtaining the groups are used. The first dispenses with the spatial relationship to use exclusively the numerical values of the parameters included; the second uses a Delaunay triangulation as a spatial restriction criterion. It is therefore obvious that the configuration of the groups is constructed taking into account three lodgings; the third uses its eight nearest neighbours as a spatial criterion. Its operation, which is described in numerous bibliographies (ESRI, 1999), is based on the calculation of a coefficient of determination for each of the variables analysed and the value they maintain after the construction of the group. However, both parameters vary according to the type of spatial restriction used. The result obtained also depends on the number of clusters to be achieved, although the best way of calculating the number of ideal clusters is to use the F-Statistic index (Calinski & Harabasz, 1974).

The optimal number of clusters is determined by using the following equation:

$$R^2 = \frac{(TSS - ESS)}{TSS}$$

TSS corresponds to the total sum of squares obtained by squaring and then summing the deviations from the overall mean value of a variable. In contrast, ESS is the sum of squares explained and is calculated in a similar way, although in this case the deviations are grouped by each cluster. In other words, each value is subtracted from the average of the group to which it belongs and squared and summed, thus showing its degree of similarity. It can be deduced from this that the technique aims to obtain the greatest possible similarities between the components of the group and at the same time the most marked differences with regard to the rest of the groups.

On the other hand, the parameters which make up the above function are obtained from the following equation:

$$\text{TTS} = \sum_{i=1}^{n_c} \sum_{j=1}^{n_i} \sum_{v=1}^{n_v} \left(V_{ik}^{k} - \overrightarrow{V}^{k} \right)^2,$$

$\text{ESS} = \sum_{i=1}^{n_c} \sum_{j=1}^{n_i} \sum_{v=1}^{n_v} \left(V_{ij}^{k} - \overrightarrow{V}_i^{k} \right)^2$,in which n is the number of characteristics; ni the number of characteristics in group i; n_c the number of classes or groups; n_v the number of variables used, V_{ij}^{k} the value which the variable takes on kth in the jth characteristic in the ith group, $\overrightarrow{V}^{k}$ the average of the kth variable; and $\overrightarrow{V}_i^{k}$ the mean value of the kth variable in the i group.

5 Results

The clustering analysis has been performed by taking into account the parameters affecting both the type of spatial constraint and the starting method (Table 4), since both the number of accommodation establishments used and the starting method remain unchanged. This makes it possible to assess whether using a configuration considering spatial relationships has substantial advantages over one which disregards them and focuses exclusively on numerical values. It also makes it possible to decide whether the clusters obtained are strengthened by including a higher volume of lodgings in the spatial configuration, as three and eight have been used, depending on the spatial constraint method used, either Delaunay triangulation or K nearest neighbours, respectively.

Table 4 Parameters used in cluster analysis

Parameter name	Variable
Features	• Rural establishments with more than 100 followers on Facebook.
Fields of analysis	• People who follow the page (P4).
	• Visits to the page (Q5)
	• Cover photos (P6).
	• Photos in biography (P7).
	• Photos uploaded by mobile phone (P9).
	• Accommodation rating given by followers (P16).
Spatial constraints	• No spatial constraint. • Delaunay triangulation. • K nearest neighbours ($K = 3$ & $K = 8$).
Distance method	• Euclidean.
Initialisation method	• Find seed locations.

Source: Own material

Table 5 Group-wise summary

Variable	Mean	Std. Dev	Min.	Max.	No spatial constraint	Delaunay triangulation	K nearest neighbours (K=8)
P4	879.13	1346.09	101	14,271	0.865	0.123	0.609
P5	101.81	366.67	0	4629	0.898	0.477	0.865
P6	7.05	18.04	0	228	0.898	0.342	0.632
P7	102.97	216.46	0	2128	0.839	0.044	0.717
P9	84.95	209.78	0	1968	0.831	0.058	0.285
P16	3.83	1.98	0	5	0.939	0.087	0.072
Number of groups					15		

Source: Own material

When calculations are made according to these three methods (Table 5), certain conclusions can be drawn concerning the number of groups obtained by each of these methods.

The summary of the main parameters on which the technique is based offers sufficient variability as can be seen from the descriptive statistical analysis. In this sense it should be remembered that the values of each of the fields are standardised so that the variables reaching higher figures do not condition the formation of the groups. It is also significant that any of the methods recommend the creation of 15 groups using F-Statistic.

However, the first differences arise when R2 is analysed as the coefficient of determination is in general considerably higher when no spatial restriction is used. This is logical since we are only looking for numerical relationships and not for their conditioning on any other criterion. On the other hand, when using a Delaunay triangulation or a number of neighbours, 8 in particular, certain variables experience a notable reduction in the coefficient referred to; this decrease is more pronounced when the number of dwellings used as a spatial criterion is reduced.

In addition, the model that does not take into account the restriction considers that the score given to the accommodation establishment by visitors to the Facebook pages is the most effective parameter for classifying the groups. Despite this, the remainder of the variables also attain very significant values, which means that they are highly effective when it comes to establishing the groups. On the other hand, when the spatial restriction is applied following the Delaunay triangulation, apart from lowering the coefficient of determination in all the variables, the parameter which best discriminates the groups is the number of visits to the pages. Furthermore, if we take the number of neighbours as the neighbourhood criterion, the coefficients of determination are higher than in the previous case, although the same variable remains as the one best discriminating the groups. This circumstance makes it possible to observe an interesting differentiation: if only a numerical criterion is used, the most important factor for establishing differences between groups is the score on the Facebook page of the establishment. Meanwhile, if a spatial parameter is included, whatever it may be, the number of visits is the most important element.

Despite the differences between the three alternatives compared, it is worth noting that in any of the cases poorly defined clusters are obtained, which affects a variable number of accommodation establishments. At the same time, if we resort to spatial autocorrelation analysis to detect whether the groups are formed according to different spatial criteria, we can observe that when no spatial constraint is required the sets appear grouped as revealed by the spatial autocorrelation, which does not occur when any of the selected types of constraint are used.

This comparison between different methods of approaching cluster analysis establishes that there are two valid options, one disregarding the spatial relationship and the other taking into account a neighbourhood criterion based on the consideration of eight nearest neighbours. Meanwhile, the option of using Delaunay triangulation as a way of defining the neighbourhood criterion is discarded, as is the option of considering three nearest neighbours, as the results obtained are very similar.

Nevertheless, if we resort to obtaining the spatial autocorrelation of the results obtained in the form of clusters, Moran's Index reflects that it is clustered when there is no spatial constraint and random when this relationship is configured by means of the neighbourhood criterion as is also demonstrated by the log p-value. However, there are notable differences between the clusters obtained by Delaunay triangulation and the remainder of the spatial restrictions. In this sense, the highest z-score (standard deviation) is obtained using only numerical criteria, while it occupies an intermediate position when 8 nearest neighbours are considered. Similarly, the p-value (the probability of randomness) also considers these two forms as the most prominent when establishing groupings; there is some advantage in omitting the spatial restriction by considering that there is a prior grouping of the results, which may perhaps be explained by the distribution of the accommodation establishments.

Given this alternative, it is proposed to develop two options to illustrate the goodness of one versus the other, discarding the constraint imposed by Delaunay triangulation or with a number of neighbours of less than 8.

5.1 Cluster Analysis Without Spatial Restriction

As has been pointed out, the fact that no spatial restriction is used implies that the way of establishing the groups is exclusively numerical. The variables listed in the methodological section (P4, P5, P6, P7, P9 and P16) are used to establish the groups and the Pseudo F-Statistic analysis is used to verify the ideal number of groups. This statistic shows that the division into 15 groups is the most effective (Table 6).

However, the summary of the contribution of the groups (Table 7) shows that of the 15 groups obtained only 7 manage to bring together a significant number of establishments. These are groups 1, 3, 8, 9, 10, 13 and 14, although there are large differences between them, with group 9 being the most numerous with 229 establishments followed by group 10 with 88 and group 13 with 55. These three groups together account for 83.22% of the sample, although all the groups with more than five cases have been considered for a more detailed analysis in order to detect some kind of hidden territorial pattern.

Schematically these groups are identified by a significant contribution in some of the variables considered.

- Group 1 (Table 8) is made up of 16 establishments, the main defining characteristics of which are that they outperform the average for the whole sample in all the variables considered. Particularly noteworthy is the excellent rating of 4.82 out of 5 given by the followers of the page. It is therefore a fairly dynamic group as far as Facebook actions are concerned and in addition is recognised by users.

- Group 3 (Table 9) is made up of 23 establishments sharing the dynamism and recognition which characterised the previous group, although there is a substantial difference in the number of photos uploaded by mobile phone (P6). In this parameter its values attain only 50% of the average of all rural accommodation establishments, this is the aspect which most needs improving.
- Group 8 (Table 10) is made up of only 10 accommodation establishments, the main distinguishing feature of which is that in all parameters it is above the average of all the accommodation establishments, except for the cover photos (P6). Among the positive aspects, it also stands out for having a very good rating (4.77) by the followers of the page, which is particularly relevant if one takes into account that it has a large number of them (P4).

Table 6 Pseudo F-statistic summary

	Pseudo F-statistic (max)			**Pseudo F-statistic (max)**		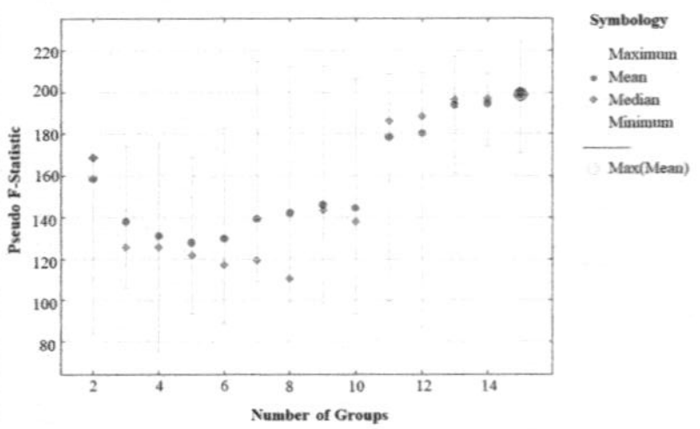
Number of groups	2	169.91	Number of groups	9	212.65	
	3	174.26		10	206.53	
	4	176.16		11	208.62	
	5	169.25		12	209.45	
	6	182.75		13	217.17	
	7	214.58		14	209.25	
	8	212.25		15	224.39	

Source: Own material

Table 7 Group-wise summary

	Variable								
Share Groups	P4	P5	P6	P7	P9	P16	Count	Std. distance	SSD
1	0.21	0.19	0.25	0.10	0.17	0.12	16	856.56	30.04
2	0.31	0.09	0.31	0.38	0.25	0.98	3	1874.57	29.77
3	0.13	0.15	0.14	0.24	0.21	0.10	23	606.65	24.98
4	0.43	0.28	0.01	0.14	0.01	0.04	2	3134.21	17.68
5	0	0	0	0	0	0	1	0	0
6	0	0	0	0	0	0	1	0	0
7	0	0	0	0	0	0	1	0	0
8	0.20	0.21	0.04	0.20	0.31	0.10	10	1082.40	21.57
9	0.09	0.06	0.09	0.11	0.08	0.46	229	240.54	40.30
10	0.24	0.04	0.08	0.20	0.26	0.00	88	531.48	50.59
11	0	0	0	0	0	0	1	0	0
12	0.06	0.01	0.18	0.03	0.06	0.98	2	422.05	5.94
13	0.08	0.27	0.11	0.18	0.17	0.2	55	632.53	54.25
14	0.36	0.19	0.12	0.32	0.18	1.00	13	1356.67	49.01
15	0.03	0.03	0.03	0.03	0.03	0.03	2	197.53	0.09

Source: Own material

Table 8 Group 1

Variable	Mean	Std. Dev	Min	Max	Share
P4	1190.13	809.05	292	3303	0.21
P5	193.94	228.06	16	886	0.19
P6	37.25	13.09	23	80	0.25
P7	150.88	112.97	11	374	0.10
P9	108.94	119.13	0	335	0.17
P16	4.82	0.18	4	5	0.12

Source: Own material

Table 9 Group 3

Variable	Mean	Std. Dev	Min	Max	Share
P4	995.52	561.45	239	2053	0.13
P5	134.22	168.04	0	705	0.15
P6	7.70	7.25	0	32	0.14
P7	430.09	130.62	241	743	0.24
P9	42.30	86.26	0	418	0.21
P16	4.87	0.17	5	5	0.10

Source: Own material

- Group 9 (Table 11) is the largest group, comprising 229 establishments. The main characteristic defining it is that in most of the established criteria it is below the average for all accommodation establishments. Nevertheless, it achieves the best rating by the followers of the page with an average score of 4.85.
- Group 10 (Table 12) is made up of 88 establishments in which there is no rating of the page (P16) and the remainder of the variables analysed are well below the

Table 10 Group 8

Variable	Mean	Std. Dev	Min	Max	Share
P4	1609.00	1025.05	178	3020	0.20
P5	221.80	279.57	0	949	0.21
P6	4.10	3.62	1	11	0.04
P7	149.40	122.30	1	417	0.20
P9	675.10	166.53	430	1034	0.31
P16	4.77	0.16	5	5	0.10

Source: Own material

Table 11 Group 9

Variable	Mean	Std. Dev	Min	Max	Share
P4	408.33	227.88	101	1339	0.09
P5	35.04	49.61	0	292	0.06
P6	2.92	3.00	0	21	0.09
P7	36.72	47.24	0	235	0.11
P9	26.27	35.06	0	157	0.08
P16	4.85	0.32	3	5	0.46

Source: Own material

Table 12 Group 10

Variable	Mean	Std. Dev	Min	Max	Share	Variable
P4	487.56	513.89	101	3538	0.24	P4
P5	9.73	26.43	0	172	0.04	P5
P6	3.03	3.23	0	18	0.08	P6
P7	48.08	84.63	0	426	0.20	P7
P9	48.20	102.55	0	507	0.26	P9

Source: Own material

general average. This is one of the groups which should be most strongly supported, given that its values are well below average. In addition, it presents notable differences between the minimum and maximum values, which results in significant variability as is reflected in the standard deviation.

- Group 13 (Table 13) is made up of 55 rural accommodation establishments in which the followers give a very high score (4.8), with the remainder of the variables being well above the averages obtained by all the establishments, except for the number of followers of the page (P4) which is much lower.
- Group 14 (Table 14) is made up of 13 establishments, the rating of which is close to the average for all the establishments although it is low (3.75), which contrasts with the large number of followers. Moreover, it has very different values, which corroborates the high standard deviation in all the variables analysed.

The geographical distribution of the groups described shows that there are overlaps between some groups and others, which means that no general conclusion can

Table 13 Group 13

Variable	Mean	Std. Dev	Min	Max	Share
P4	6.33	5.33	1	20	0.08
P5	166.69	210.13	3	1244	0.27
P6	73.91	62.00	0	238	0.11
P7	137.07	112.50	0	359	0.18
P9	1238.38	582.59	183	2643	0.17
P16	4.7982	0.2324	4	5	0.2

Source: Own material

Table 14 Group 14

Variable	Mean	Std. Dev	Min	Max	Share
P4	4698.77	1312.91	3031	8148	0.36
P5	237.15	272.91	0	857	0.19
P6	11.38	9.23	0	28	0.12
P7	317.69	168.19	13	685	0.32
P9	102.92	118.18	0	357	0.18
P16	3.75	2.06	0	5	1.00

Source: Own material

be drawn as to the possible intervention of a territorial pattern which acts in an exclusive manner. A somewhat contradictory situation can be detected at a territorial level, since as can be seen in the north of the province of Cáceres there are very different groups interspersed throughout the territory. This fact can be observed in groups 9, 10 and 13, which bring together the largest volume of accommodation establishments and which are basically made up of rural lodgings with good ratings on Facebook profiles (groups 9 and 13). Their situations are however very different, as the remainder of the parameters of the first group are below average for accommodation and those of the second are above average. Group 10 deserves a separate mention as it has no ratings (Fig. 3).

5.2 *Spatially Constrained Cluster Analysis Using K Nearest Neighbours* (K = *8)*

The use of a restriction criterion based on the consideration of the eight nearest neighbours following the Euclidean distance implies that the territory acquires significant importance in the configuration of the groups. As indicated in the methodology, the variables used are naturally the same (P4, P5, P6, P7, P9 and P16). On the other hand, by using the Pseudo F-Statistic analysis it can be verified that the most effective number of groups is 15 (Table 15).

However, the summary of the contribution of the groups (Table 16) shows that of the 15 groups constructed only two which manage to gather a significant group of cases can be distinguished. These are groups 12 and 15 which comprise 109 and

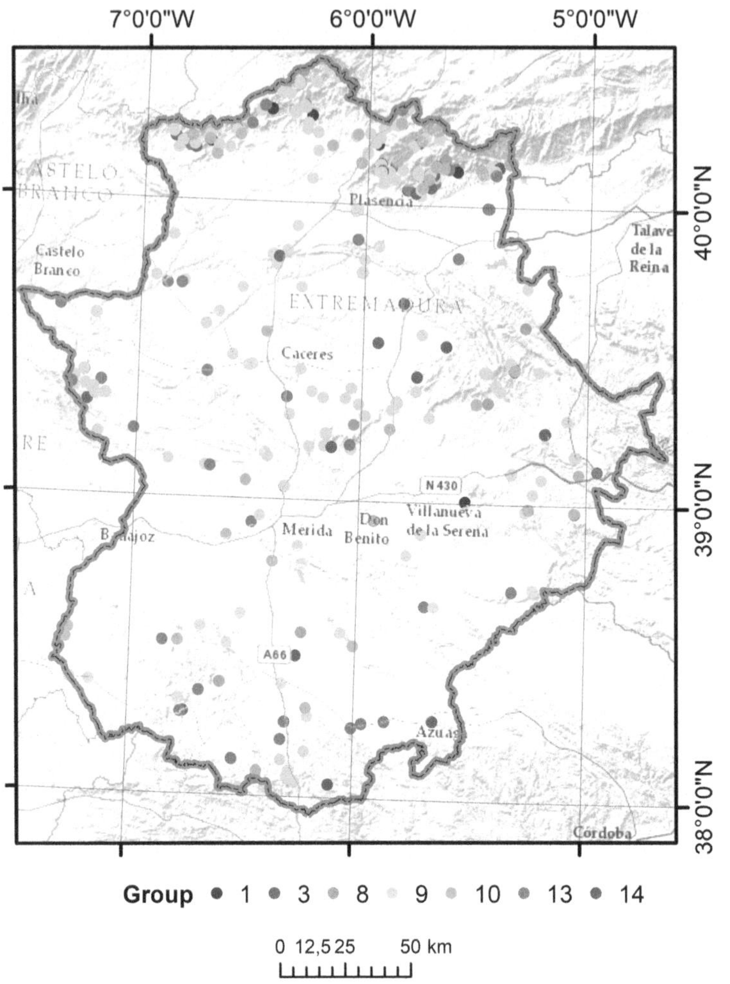

Fig. 3 No spatial constraint distribution. Source: Own material

318 members, respectively. However, it was decided to also include group 10 despite its containing five establishments owing to its type of grouping; the remaining groups were discarded owing to their low representativeness as in most cases they consisted of only one establishment.

A more detailed analysis of each of the selected groups shows quite interesting patterns, which is further supported by their territorial proximity.

- Group 10 (Table 17) is composed of only five establishments. These are characterised by being well below the average of the total number of rural accommodation establishments with more than 100 followers on Facebook.

Table 15 Pseudo F-Statistic Summary

	Pseudo F-Statistic			Pseudo F-Statistic	
Number of groups	2	4.092	Number of groups	9	26.019
	3	18.323		10	24.510
	4	24.173		11	22.920
	5	28.473		12	21.565
	6	29.750		13	20.190
	7	29.339		14	26.621
	8	27.785		15	34.800

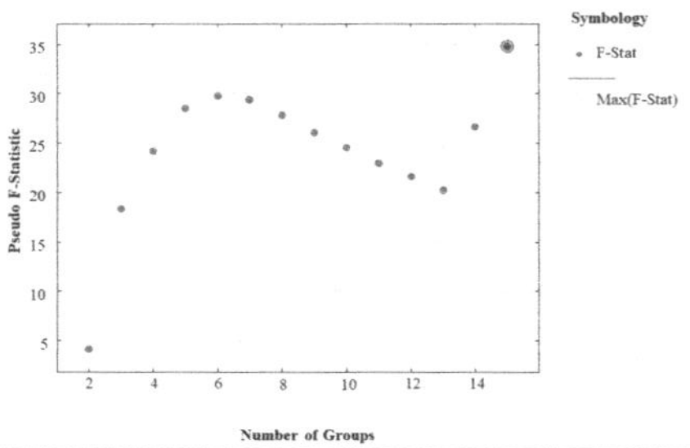

Source: Own material

Table 16 Group-Wise Summary

Share Groups	Variable P4	P5	P6	P7	P9	P16	Count	Std. distance	SSD
1	0	0	0	0	0	0	1	0	0
2	0.184	0.069	0.057	0.164	0.102	0.080	3	1111.67	4.758
3	0	0	0	0	0	0	1	0	0
4	0	0	0	0	0	0	1	0	0
5	0	0	0	0	0	0	1	0	0
6	0	0	0	0	0	0	1	0	0
7	0	0	0	0	0	0	1	0	0
8	0	0	0	0	0	0	1	0	0
9	0	0	0	0	0	0	1	0	0
10	0.034	0.001	0.013	0.056	0.022	0.000	5	184.759	0.352
11	0	0	0	0	0	0	1	0	0
12	0.296	0.120	0.180	0.334	0.379	1.000	109	694.216	185.889
13	0	0	0	0	0	0	1	0	0
14	0.027	0.024	0.000	0.005	0.001	0.020	2	197.628	0.088
15	0.474	0.269	0.693	0.349	1.000	1.000	318	952.628	1069.382

Source: Own material

Table 17 Group 10

Variable	Mean	Std. Dev	Min	Max	Share
P4	318.80	178.51	133	619	0.03
P5	1.20	2.40	0	6	0.00
P6	2.20	1.17	1	4	0.01
P7	32.60	44.80	1	121	0.06
P9	11.60	15.98	0	43	0.02
P16	0.00	0.00	0	0	0.00

Source: Own material

Moreover, the contribution made by most of the variables to the formation of the group is very small.

- Group 12 (Table 18) is made up of 109 establishments which share lower values than the general average in most of the variables, although they stand out for having an above-average rating. To be precise, the followers of their Facebook profiles give them an average rating of 4.34. However, all the variables show very disparate values with significantly high standard deviations. In this regard, it should also be noted that the variables which contribute most to the formation of the group are the rating (P16) and to a lesser extent the number of photos (P9 and P7) and the number of followers of the page (P4).
- On the other hand, group 15 (Table 19) is made up of 318 rural accommodation establishments in which once again the averages are evidently lower than those of the establishments as a whole. In addition, it is also noteworthy that the rating of

Table 18 Group 12

Variable	Mean	Std. Dev	Min	Max	Share
P4	670.50	668.59	101	4296	0.30
P5	40.34	69.70	0	557	0.12
P6	5.34	7.99	0	41	0.18
P7	73.74	106.95	0	710	0.33
P9	75.91	136.25	0	745	0.38
P16	4.34	1.48	0	5	1.00

Source: Own material

Table 19 Group 15

Variable	Mean	Std. dev	Min	Max	Share
P4	755.28	911.80	101	6816	0.47
P5	76.85	154.09	0	1244	0.27
P6	5.72	12.08	0	158	0.69
P7	77.63	120.20	0	743	0.35
P9	74.73	194.37	0	1968	1.00
P16	3.67	2.09	0	5	1.00

Source: Own material

these establishments is below that of the remainder of the groups at 3.67. It is worth noting in this respect that the rating of the establishments is lower than that of the remainder of the groups. In fact the contribution to the configuration of this group of the score given by the followers and the photos uploaded by mobile phone (P16 and P9) is striking in this group, although they are also very relevant in the remainder of the variables in their contribution to the group.

The geographical distribution of the groups described reflects well-defined territorial patterns. At first sight, it is striking that as general criteria there is a predominance of differentiated patterns in the different groups. In this sense, the north-western area of Extremadura stands out; within it, a small cluster with distinct characteristics can be observed. It is also evident that despite having above-average ratings of Facebook profiles these need to improve as to the remainder of the variables analysed. On the other hand, cluster 15 requires further action as it has lower than average ratings for the average number of establishments (Fig. 4).

6 Discussion

The aim of this study is to determine the degree of presence and the use that rural tourism is making of these new tools, focusing attention on what is undoubtedly the social network par excellence, Facebook, in a destination which is firmly committed to strengthening rural tourism as the main means of tourism development, i.e. the region of Extremadura.

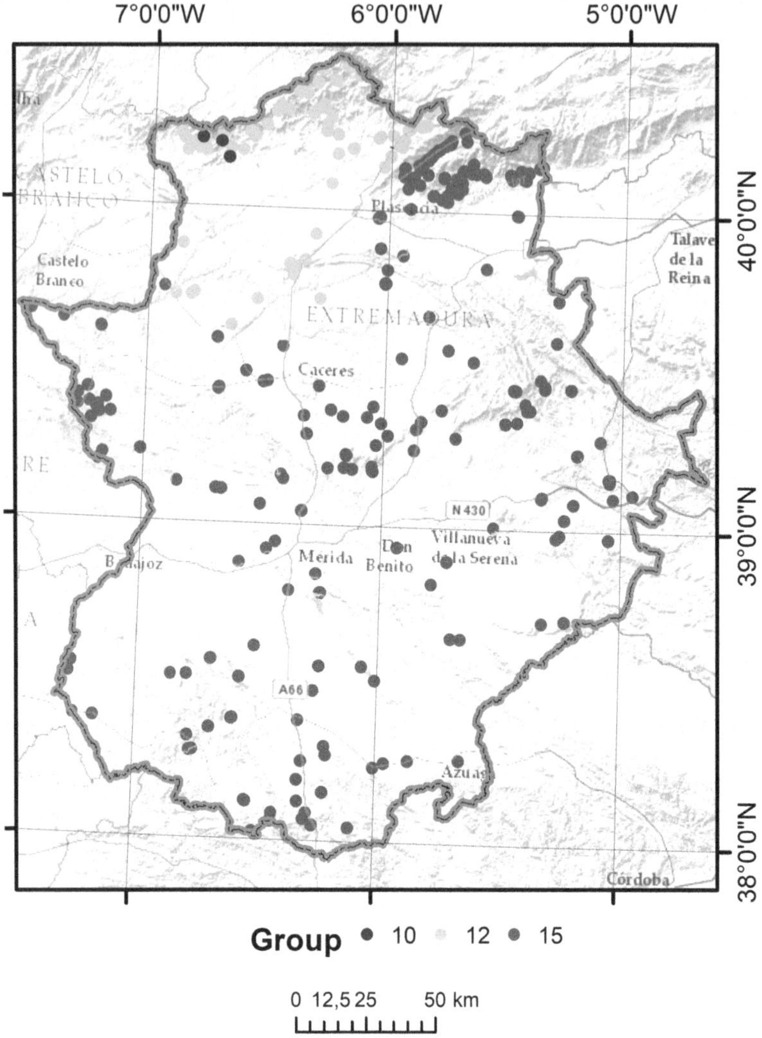

Fig. 4 K nearest neighbours (8) distribution. Source: Own material

The expansion of the use of social media has reshaped the way in which tourism activity is managed, which is why the presence and use of these tools is essential in the current context of the tourism sector. The low cost associated with the use of these social tools represents an opportunity for those accommodation establishments which do not have the means of carrying out advertising campaigns, as is the case of rural tourism and agritourism (Lane, 1994). However, the scarcity of studies on the subject, especially in comparison with the hotel sector, raises doubts as to whether this sector is really taking advantage of the development potential offered by social media.

On numerous occasions and in the most varied disciplines, statistical analysis is used to attempt to describe or explain a complex reality (Frías Guzmán, 2017). Tourism is no stranger to this trend in which the statistical analysis of data has been used for decades (Coenders & Ferrer-Rosell, 2020; Majewska & Truskolaski, 2019), although the literature specialising in geostatistical analysis has also developed considerably (Adamiak et al., 2019; Lisiak-Zielińska & Ziernicka-Wojtaszek, 2021; Rodríguez-Rangel et al., 2020; Rodríguez Rangel & Sánchez Rivero, 2020; van der Zee et al., 2020). Given these two ways of analysing tourism data and the difficulties deriving from the use of spatial criteria, this article has focused on solving the problem in both ways by using a grouping analysis. The numerical and spatial criteria have been used to carry out the analysis in such a way that the results allow the comparison of the two conceptualisations of the tool.

Perhaps the most striking aspect of the results is that regardless of the type of spatial relationship selected (none or the neighbourhood) the number of groups obtained is similar. This leads us to the conclusion that it is a very complex reality by placing its ideal number at 15, according to the Pseudo F-Statistic criterion. However, the groups obtained by means of both conceptualisations of the spatial criterion have nothing to do with each other, neither from a numerical nor from a territorial point of view. In view of this circumstance, the advantages or disadvantages deriving from each of them must be analysed.

When no spatial restriction of any kind is used it becomes clear that the groups show greater internal similarities and at the same time greater divergences with the remainder, which is certainly the aim of the technique. In this sense it can be argued that from a strictly numerical point of view they are more effective. However, when the results are reflected on the territory, no pattern can be observed which would make it possible to consider them as a homogeneous group. On the contrary, the predominant situation is one of plausible heterogeneity with the interspersion of accommodation establishments belonging to various groups.

On the other hand, the results obtained with a spatial restriction, taking into account the eight nearest neighbours following a Euclidean distance, show at least two antagonistic groups which are very well differentiated at a territorial level. In this sense it is shown to be a much more efficient way of conceptualising the groups, as it is easy to obtain groups with similar characteristics. This favours the implementation of tourism policies promoting training and the updating of the Facebook profiles of rural accommodation establishments.

Despite everything, the tool used also has certain limitations when it comes to specifying the groups. Perhaps the most outstanding among them is the use of few variables to make the clustering analysis work more efficiently. However, the use of Euclidean distance instead of real distance or more accurately travel time to specify the number of neighbours can also be mentioned. This leads to the continuation of this line of geostatistical analysis, albeit combined with the use of even more efficient tools to establish models or clusters, such as Geographically Weighted Regression (GWR) and Multi-scale Geographically Weighted Regression (MGWR).

With regard to the results obtained in this study, the only precedent to the authors' knowledge in the literature which allows us to compare the results obtained with previous data in order to be able to compare the evolution of the phenomenon under study is the work carried out by Rodríguez and Sánchez (2015). In this study they analyse the degree of presence of tourist accommodation establishments in Extremadura in 2013, with the results obtained showing that 67.6% of rural accommodation establishments had a Facebook profile in that year. This study reflects a certain decline or stagnation in the use of this network given that of the total of 846 accommodation establishments in the region only 60.2% are currently present on it. Moreover, this percentage is reduced to 52.84% when we consider those accommodation establishments which reach the significant number of 100 followers. It can therefore be concluded that the age of these tools, and consequently their increasingly widespread use among users, is not accompanied by an increase in the interest and implementation of rural tourism accommodation establishments in the region.

Finally, the fact that the largest group of accommodation establishments (group 15) is made up of more than half of the establishments analysed and that it is characterised by lower than average values of the remaining groups in all the variables of presence and activity on Facebook is a clear indication that at both public and private level this social network is undervalued in the region as a promotional and marketing tool. In this sense, the results obtained are in line with those obtained by Pesonen (2011).

7 Conclusions

From a generic point of view the usefulness of social networks is highly transverse; one of its purposes is to generate information. One of these networks is Facebook, which as has been shown in various reports is currently that with the largest number of users worldwide. In this sense Facebook is a communication channel of prime importance which has proved its usefulness as a valid tool for the management of tourism companies. However, despite the existence of a wide range of studies addressing the importance of social networks in the management and marketing of hotels and tourist destinations, there is a lack of studies to highlight the role played by Facebook in the rural tourism and agritourism sector. For this reason, in this study we have tried to provide knowledge in this regard, starting from the objective of finding out how rural accommodation establishments in Extremadura, of which there are many, use this social network. From the results obtained, after the relevant processing of the information using the spatial statistics techniques explained above the following conclusions can be drawn:

- Spatial statistics is a very useful tool for the detection of territorial groupings as it allows us to determine the areas that stand out owing to their greater or lesser use of Facebook, establishing notable differences between the use of this tool and traditional statistics.

- After using both techniques, the existence of 15 groups of rural accommodation establishments with common characteristics has been determined. However, in those groups obtained without spatial restriction the most important variable is the rating given to the Facebook page, while in the case of those deriving from the application of geostatistics this variable is the number of visits received by the profile of this social network.
- Among the groups developed by traditional statistics, there are seven groups which can be classified at the same time as follows:
  - Those with a large presence on Facebook and high scores in the rating of their profile (Groups 1, 3, 8 and 13).
  - Those with a low presence on Facebook, Groups 9 and 10, although the former has the highest score in terms of profile ratings.
  - Finally, there is Group 14 in which the accommodation establishments making up the group have a common characteristic: the lowest score with respect to the profile rating, despite having a large number of followers.
- The results deriving from the application of spatial restriction, taking eight close neighbours as a reference, determine the existence of 15 groups as was the case after using the statistics without spatial restriction. However, in this case only the coefficients obtained in groups 10, 12 and 15 stand out. The first of these (Group 10) stands out for presenting values below the average of the remainder of the accommodation establishments in the study sample. As for Group 12, it is made up of a wide range of accommodation establishments totalling 109, in which most of the variables present values below the general average except in the case of the assessment of the Facebook profile. Finally, Group 15 has a common characteristic: it shows the lowest values in all the variables analysed with regard to the remainder of the accommodation establishments.

On the other hand, the results obtained in this study have generated some very useful conclusions in terms of the management of the presence of rural accommodation establishments in Extremadura on social networks in general and on Facebook in particular.

Firstly, the fact that the number of accommodation groups identified was so high (15, both when using spatial restrictions and when not using them) shows that the behaviour of the profiles of rural accommodation establishments in the region on Facebook is highly heterogeneous. It is therefore not possible to define a clear pattern of behaviour of these establishments since both the regular publication of messages, photographs, videos, etc., on the social network and the degree of interaction with customers (real or potential) are so varied that it is not possible to define a single Facebook presence strategy for rural tourism in Extremadura for the whole region. It is clear that the degree of professionalisation of the managers of these establishments is the cause of the very high heterogeneity detected.

Another significant result is the fact that almost 40% of rural accommodation establishments in the region do not have a Facebook profile. In the highly connected social networking society of today, not being present on Facebook is practically equivalent to being out of the tourism market. In response to the complaint of some owners that Extremadura is not sufficiently promoted as a rural tourism destination, it should be argued that this promotion is not only the responsibility of the public tourism administration but also that of private business initiative. Consequently, many of the rural tourism entrepreneurs should be more aware of the potential of Facebook as a tourism promotion tool.

The identification of different groups of rural accommodation establishments with different levels of presence and activity on Facebook in specific geographical areas of the region is a factor which will make it considerably easier for the tourism administration of the region to launch different campaigns with different content (raising awareness of the need to be present on Facebook, strategies to define more active profiles, assessment of customer opinions and comments, etc.), rather than a single regional campaign with monographic content.

Finally, interaction with customers (real or potential) of rural accommodation establishments with a Facebook profile requires certain language skills, especially English and Portuguese. The Facebook social network is an online platform which reaches every corner of the world. For this reason, it does not make much sense for the Regional Government of Extremadura to carry out campaigns to promote rural tourism at an international level (Lisbon, London, Berlin, New York, Beijing, etc.) without these campaigns being accompanied by a policy of training tourism entrepreneurs in the region to communicate with foreign customers in English and Portuguese. The use of the latter language is essential if we want to take advantage of the Portuguese market which, given its geographical proximity, should be where the main efforts to promote rural tourism in Extremadura are directed at an international level.

Acknowledgements This work was supported by the Regional Government of Extremadura and co-financed by the European Regional Development Fund through the IB18015 research project and aid GR18052 (DESOSTE) and GR18124 (AEA).

References

Adamiak, C., Szyda, B., Dubownik, A., & García-Álvarez, D. (2019). Airbnb Offer in Spain—Spatial Analysis of the Pattern and Determinants of Its Distribution. *ISPRS International Journal of Geo-Information, 8*, 155. https://doi.org/10.3390/ijgi8030155

Ashley, C., & Tuten, T. (2015). Creative strategies in social media marketing: an exploratory study of branded social content and consumer engagement. *Psychology & Marketing, 32*(1), 15–27. https://doi.org/10.1002/mar.20761

Ben-Shaul, M., & Reichel, A. (2018). Motives, modes of participation and loyalty intentions of Facebook tourism brand page consumers. *Journal of Travel Research, 57*(4), 453–471. https://doi.org/10.1177/0047287517704087

Bowman, B., Settle, Q., Riggs, A., Tomas, S., & King, A. (2020). Characteristics of Oklahoma Agrotourism Facebook Posts. *Journal of Applied Communications, 104*(1) https://doi.org/10.4148/1051-0834.2294

Brafton. (2014). *Zuckerberg answers questions about organic Facebook reach in Q&A*. Retrieved February 15, 2021, from https://www.brafton.com/news/zuckerberg-answers-questions-organic-facebook-reach-qa/

Brantner, C., Lobinger, K., & Wetzstein, I. (2011). Effects of visual framing on emotional responses and evaluations of news stories about the Gaza conflict 2009. *Journalism & Mass Communication Quarterly, 88*(3), 523–540. https://doi.org/10.1177/107769901108800304

Calinski, T., & Harabasz, J. A. (1974). Dendrite method for cluster analysis. *Communications in Statistics, 3*(1), 1–27.

Cardona, J. C., Suriñach, J., & Ramos, R. (2007). Supervised Regionalization Methods: A Survey. *International Regional Science Review, 30*(8), 195–220.

Casalo, L. V., Carlos, F., & Miguel, G. (2010). Determinants of the intention to participate in firm-hosted online travel communities and effects on consumer behavioral intentions. *Tourism Management, 31*, 898–911. https://doi.org/10.1016/j.tourman.2010.04.007

Castronovo, C., & Huang, H. (2012). Social media in an alternative marketing communication model. *Journal of Marketing Development and Competitiveness, 6*(1), 117–134.

Chauhan, K., & Pillai, A. (2013). Role of content strategy in social media brand communities: a case of higher education institutes in India. *Journal of Product & Brand Management, 22*(1), 40–51. https://doi.org/10.1108/10610421311298687

Cho, M., Schweickart, T., & Haase, A. (2014). Public engagement with nonprofit organizations on Facebook. *Public Relations Review, 40*, 565–567. https://doi.org/10.1016/j.pubrev.2014.01.008

Choi, S., Lehto, X. Y., & Morrison, A. M. (2007). Destination image representation on the web: content analysis of Macau travel related websites. *Tourism Management, 28*(1), 118–129. https://doi.org/10.1016/j.tourman.2006.03.002

Coenders, G., & Ferrer-Rosell, B. (2020). Compositional Data Analysis in Tourism: Review and Future Directions. *Tourism Analysis, 25*(1), 153–168. https://doi.org/10.3727/108354220X15758301241594

Dahnil, M. I., Marzuki, K. M., Langgat, J., & Fabeil, N. F. (2014). Factors influencing SMEs adoption of social media marketing. *Procedia-Social and Behavioral Sciences, 148*, 119–126. https://doi.org/10.1016/j.sbspro.2014.07.025

De Lencastre, P., & Corte-Real, A. (2013). Brand response analysis: a Peircean semiotic approach. *Social Semiotics, 23*(4), 489–506. https://doi.org/10.1080/10350330.2013.799005

De Moya, M., & Jain, R. (2013). When tourists are your 'friends': exploring the brand personality of Mexico and Brazil on Facebook. *Public Relations Review, 39*(1), 23–29. https://doi.org/10.1016/j.pubrev.2012.09.004

Dwivedi, M. (2009). Online destination image of India: a consumer based perspective. *International Journal of Contemporary Hospitality Management, 21*(2), 226–232. https://doi.org/10.1108/09596110910935714

Eagleman, A. N. (2013). Acceptance, motivations, and usage of social media as a marketing communications tool amongst employees of sport national governing bodies. *Sport Management Review, 16*(4), 488–497. https://doi.org/10.1016/j.smr.2013.03.004

ESRI. (1999). *The ESRI guide to GIS analysis. Volume 1: Geographic patterns & relationships* (p. 188). Redlands: ESRI Press.

Frías Guzmán, D. H. (2017). *Estudio experimental, geoestadístico y numérico sobre los procesos de fracturamiento superficial* [Tesis Doctoral]. Universidad Autónoma de Aguascalientes. Centro de Ciencias del Diseño y de la Construcción. Departamento de Geotecnia e Hidráulica. http://hdl.handle.net/11317/1429.

Govers, R., & Go, F. M. (2005). Projected destination image online: website content analysis of pictures and text. *Information Technology & Tourism, 7*(2), 73,89. https://doi.org/10.3727/1098305054517327

Hays, S., Page, S. J., & Buhalis, D. (2013). Social media as a destination marketing tool: its use by national tourism organisations. *Current Issues in Tourism, 16*(3), 211–239. https://doi.org/10.1080/13683500.2012.662215

Joo, Y., Seok, H., & Nam, Y. (2020). The moderating effect of social media use on sustainable rural tourism: a theory of Planned Behavior Model. *Sustainaibility, 12*(10), 4095. https://doi.org/10.3390/su12104095

Joyner, L., Kline, C., Oliver, J., & Kariko, D. (2018). Exploring emotional response to images used in agrotourism destination marketing. *Journal of Destination Marketing & Management, 9*, 44–55. https://doi.org/10.1016/j.jdmm.2017.10.004

Kent, M. L., Taylor, M., & White, W. J. (2003). The relationship between Web site design and organizational responsiveness to stakeholders. *Public Relations Review, 29*(1), 63–77. https://doi.org/10.1016/S0363-8111(02)00194-7

Ketter, E. (2016). Destination image restoration on Facebook: the case study of Nepal's Gurkha earthquake. *Journal of Hospitality and Tourism Management, 28*, 66–72. https://doi.org/10.1016/j.jhtm.2016.02.003

Kim, C., & Yang, S. U. (2017). Like, comment, and share on Facebook: how each behavior differs from the other. *Public Relations Review, 43*, 441–449. https://doi.org/10.1016/j.pubrev.2017.02.006

Kim, J. W. (2018). The liked and shared: effects of social media virality metrics on perceptions of message influence and behavioral intentions. *Computers in Human Behavior, 84*, 153–161. https://doi.org/10.1016/j.chb.2018.01.030

Kim, W. G., Li, J., & Brymer, R. A. (2016). The impact of social media reviews on restaurant performance: the moderating role of excellence certificate. *International Journal of Hospitality Management, 55*, 41–51. https://doi.org/10.1016/j.ijhm.2016.03.001

Kotsi, F., Balakrishnan, M. S., Michael, I., & Ramsoy, T. Z. (2018). Place branding: aligning multiple stakeholder perception of visual and auditory communication elements. *Journal of Destination Marketing & Management, 7*, 112–130. https://doi.org/10.1016/j.jdmm.2016.08.006

Lane, B. (2009). Rural tourism: An overview. In T. Jamal & M. Robinson (Eds.), *The SAGE handbook of tourism studies* (pp. 354–371). Sage Publications.

Lane, B. (1994). Sustainable rural tourism strategies: a tool for development and conservation. *Journal of Sustainable Tourism, 2*(1–2), 102–111. https://doi.org/10.1080/09669589409510687

Lee, H., & Park, H. (2013). Testing the impact of message interactivity on relationship management and organizational reputation. *Journal of Public Relations Research, 25*(2), 188–206. https://doi.org/10.1080/1062726X.2013.739103

Lee, W., Xiong, L., & Hu, C. (2012). The effect of Facebook users' arousal and valence on intention to go the festival: applying an extension of the technology acceptance model. *International Journal of Hospitality Management, 31*, 819–827. https://doi.org/10.1016/j.ijhm.2011.09.018

Ley 45/2007, de 13 de diciembre, para el desarrollo sostenible del medio rural. Publicado en BOE núm. 299 de 14 de diciembre de 2007.

Li, S., Robinson, P., & Oriade, A. (2017). Destination marketing: the use of technology since the millennium. *Journal of Destination Marketing and Management, 6*(2), 95–102. https://doi.org/10.1016/j.jdmm.2017.04.008

Lisiak-Zielińska, M., & Ziernicka-Wojtaszek, A. (2021). Spatial Variation in Tourism and Investment Potential in the Context of Sustainable Development—A Case Study of Staszowski County. *Sustainability, 13*, 3. https://doi.org/10.3390/su13010003

Lovejoy, K., & Saxton, G. D. (2012). Information, community, and action: how nonprofit organizations use social media. *Journal of Computer-Mediated Communication, 17*(3), 337–353. https://doi.org/10.1111/j.1083-6101.2012.01576.x

Lundgren, A. S., & Johansson, A. (2017). Digital rurality: producing the countryside in online struggles for rural survival. *Journal of Rural Studies, 51*, 73–82. https://doi.org/10.1016/j.jrurstud.2017.02.001

Majewska, J., & Truskolaski, S. (2019). Cluster-mapping procedure for tourism regions based on geostatistics and fuzzy clustering: example of Polish districts. *Current Issues in Tourism, 22*(19), 2365–2385. https://doi.org/10.1080/13683500.2018.1467883

Marchiori, E., & Onder, I. (2015). Reframing the image of a destination: a pre-post study on social media exposure. In *Information and Communication Technologies in Tourism. Proceedings of the International Conference in Lugano, Switzerland, February 4–6* (p. 2015).

Marder, B., Archer-Brown, C., Colliander, J., & Lambert, A. (2019). Vacation posts on Facebook: a model for incidental vicarious travel consumption. *Journal of Travel Research, 58*(6), 1014–1033. https://doi.org/10.1177/0047287518786465

Mariani, M. M., Di Felice, M., & Mura, M. (2016). Facebook as a destination marketing tool: evidence from Italian regional destination management organizations. *Tourism Management, 54*, 321–343. https://doi.org/10.1016/j.tourman.2015.12.008

Marine-Roig, E., & Clavé, S. A. (2015). A detailed method for destination image analysis using user-generated content. *Information Technology & Tourism, 15*(4), 341–364. https://doi.org/10.1007/s40558-015-0040-1

Martín-Delgado, L. M., Sánchez-Martín, J. M., & Rengifo-Gallego, J. I. (2020). An analysis of online reputation indicators by means of geostatistical techniques - the case of rural accommodation in Extremadura, Spain. *ISPRS International Journal of Geo-Information, 9*(4), 208. https://doi.org/10.3390/ijgi9040208

McCreary, A., Seekamp, E., Davenport, M., & Smith, J. W. (2019). Exploring qualitative applications of social media data for place-based assessments in destination planning. *Current Issues in Tourism, 23*(1), 1–17. https://doi.org/10.1080/13683500.2019.1571023

Men, L. R., & Tsai, W.-H. S. (2012). How companies cultivate relationships with publics on social network sites: evidence from China and the United States. *Public Relations Review, 38*(5), 723–730. https://doi.org/10.1016/j.pubrev.2011.10.006

Mitchell, A. (2005). *Guide to GIS analysis. Volume 2: Spatial measurements & statistics. Spatial an measurements* (p. 249). Redlands: ESRI Press.

Moise, D., & Cruceru, A. F. (2014). An empirical study of promoting different kinds of events through various social media networks websites. *Procedia – Social and Behavioral Sciences, 109*, 98–102. https://doi.org/10.1016/j.sbspro.2013.12.426

Muntinga, D. G., Moorman, M., & Smit, E. G. (2011). Introducing COBRAs: exploring motivations for brand-related social media use. *International Journal of Advertising, 30*(1), 13–46. https://doi.org/10.2501/IJA-30-1-013-046

Narangajavana, Y., Fiol, L. J. C., Tena, M., Artola, R. M. R., & García, J. S. (2017). The influence of social media in creating expectations. An empirical study for a tourist destination. *Annals of Tourism Research, 65*, 60–70. https://doi.org/10.1016/j.annals.2017.05.002

Nusair, K. (2020). Developing a comprehensive life cycle framework for social media research in hospitality and tourism: a bibliometric method 2002–2018. *International Journal of Contemporary Hospitality Management, 32*(3), 1041–1066. https://doi.org/10.1108/IJCHM-09-2019-0777

Nusair, K., Butt, I., & Nikhashemi, S. R. (2019). A bibliometric analysis of social media in hospitality and tourism research. *International Journal of Contemporary Hospitality Management, 31*(7), 2691–2719. https://doi.org/10.1108/IJCHM-06-2018-0489

Oeldorf-Hirsch, A., & Sundar, S. S. (2015). Posting, commenting and tagging: effects of sharing new stories on Facebook. *Computers in Human Behavior, 44*, 240–249. https://doi.org/10.1016/j.chb.2014.11.024

Ortega-Rossell, F. J., Sánchez-Martín, J. M., & Hernández-Mogollón, J. M. (2012). *La gastronomía de alta gama, de lujo, o de calidad como eje potenciador del turismo cultural y de sensaciones de Extremadura* (Gastrolujo Project). Fundación Caja de Extremadura.

Önder, I., & Marchiori, E. (2017). A comparison of pre-visit beliefs and projected visual images of destination. *Tourism Management Perspectives, 21*, 42–53. https://doi.org/10.1016/j.tmp.2016.11.003

Önder, I., Gunter, U., & Gindl, S. (2020). Utilizing Facebook statistics in tourism demand modeling and destination marketing. *Journal of Travel Research, 59*(2), 195–208. https://doi.org/10.1177/0047287519835969

Pantano, E., & Di Pietro, L. (2013). From e-Tourism to f-Tourism: Emerging issues from negative tourists' online reviews. *Journal of Hospitality and Tourism Technology, 4*(3), 211–227. https://doi.org/10.1108/JHTT-02-2013-0005

Pesonen, J. (2011). Tourism marketing in Facebook: comparing rural tourism SME's and larger tourism companies in Finland. In R. Law, M. Fucks & F. Ricci (Eds.), *Information and communication technologies in tourism*. Springer. Retrieved March 2, 2021, from, https://link.springer.com/chapter/10.1007%2F978-3-7091-0503-0_43.

Phillips, M., Thilmany-McFadden, D. & Sullins, M. (2010). Possible roles for social networking in agrotourism development. *Colorado State University Extension Market Report.* Retrieved February 12, 2021, from http://dare.colostate.edu/pubs.

Pino, G., Peluo, A. M., del Vecchio, P., Ndou, V., Passiante, G., & Guido, C. (2018). A methodological framework to assess social media strategies of event and destination management organizations. *Journal of Hospitality Marketing & Management, 28*(2), 189–216. https://doi.org/10.1080/19368623.2018.1516590

Rodríguez Rangel, C., & Sánchez Rivero, M. (2015). Presencia de alojamientos extremeños en redes sociales. *Gran Tour, Revista de Investigaciones Turísticas, 12*, 4–22.

Rodríguez Rangel, M. C., & Sánchez Rivero, M. (2020). Spatial imbalance between tourist supply and demand: The Identification of Spatial Clusters in Extremadura, Spain. *Sustainability, 12*, 1651. https://doi.org/10.3390/su12041651

Rodríguez-Rangel, M. C., Sánchez-Rivero, M., & Ramajo-Hernández, J. (2020). Spatial Intensity in Tourism Accommodation: Modelling Differences in Trends for Several Types through Poisson Models. *ISPRS International Journal of Geo-Information, 9*, 473. https://doi.org/10.3390/ijgi9080473

Roth-Cohen, O., & Lahav, T. (2019). Going undercover: online domestic tourism marketing communication in closed and open Facebook groups. *Journal of Vacation Marketing*. https://doi.org/10.1177/1356766718796054

Ryu, K., Roy, P. A., Kim, H., & Ryu, H. B. (2020). The resident participation in endogenous rural tourism projects: A case study of Kumbalangi in Kerala, India. *Journal of Travel & Tourism Marketing, 37*(1), 1–14. https://doi.org/10.1080/10548408.2019.1687389

Sabate, F., Berbegal-Mirabent, J., Cañabate, A., & Lebherz, P. R. (2014). Factors influencing popularity of branded content in Facebook fan pages. *European Management Journal, 32*, 1001–1011. https://doi.org/10.1016/j.emj.2014.05.001

Saffer, A. J., Sommerfeldt, E. J., & Taylor, M. (2013). The effects of organizational Twitter interactivity on organization-public relationships. *Public Relations Review, 39*(3), 213–215. https://doi.org/10.1016/j.pubrev.2013.02.005

Sánchez-Martín, J. M., Blas-Morato, R., & Rengifo-Gallego, J. I. (2019b). The *dehesas* of Extremadura, Spain: a potential for socio-economic development based on agritourism activities. *Forests, 10*, 620. https://doi.org/10.3390/f10080620

Sánchez-Martín, J. M., Gurría, J. L., & García-Berzosa, M. J. (2020c). The cultural heritage and the shaping of tourist itineraries in rural areas - the case of Historical Ensembles of Extremadura, Spain. *ISPRS International Journal of Geo-Information, 9*(4), 200. https://doi.org/10.3390/ijgi9040200

Sánchez-Martín, J. M., Gurría, J. L., & Rengifo-Gallego, J. I. (2020b). The distribution of rural accommodation in Extremadura, Spain - Between the randomness and the suitability achieved by means of regression models (OLS vs. GWR). *Sustainability, 12*(11), 4737. https://doi.org/10.3390/su12114737

Sánchez-Martín, J. M., Martín-Delgado, L. M., & Rengifo-Gallego, J. I. (2019a). La reputación online de los alojamientos rurales en Extremadura desde una óptica geoestadística. *Boletín de la Asociación de Geógrafos Españoles, 82*(2758), 1–36. https://doi.org/10.21138/bage.2758

Sánchez-Martín, J. M., & Rengifo-Gallego, J. I. (2019). Evolución del sector turístico en la Extremadura del Siglo XXI: auge, crisis y recuperación. *Lurralde: Investigación y espacio, 42*, 19–50. http://www.ingeba.org/lurralde

Sánchez-Martín, J. M., Rengifo-Gallego, J. I., & Martín-Delgado, L. M. (2018). Tourist mobility at the destination toward protected areas: The case-study of extremadura. *Sustainability, 10*, 4853. https://doi.org/10.3390/su10124853

Sánchez-Martín, J. M., Sánchez-Rivero, M., & Rengifo-Gallego, J. I. (2020a). Water as a tourist resource in Extremadura: assessment of its attraction capacity and approximation to the tourist profile. *Sustainability, 12*(4), 1659. https://doi.org/10.3390/su12041659

Saxton, G. D., & Waters, R. D. (2014). What do stakeholders like on Facebook?. Examining public reactions to nonprofit organizations' informational, promotional, and community-building messages. *Journal of Public Relations Research, 26*(3), 280–299. https://doi.org/10.1080/1062726X.2014.908721

Schroeder, A., & Pennington-Gray, L. (2015). The role of social media in international tourist's decision making. *Journal of Travel Research, 54*(5), 584–595. https://doi.org/10.1177/0047287514528284

Senyao, S., & Ha, S. (2020). How social media influences resident participation in rural tourism development: a case study of Tunda in Tibet. *Journal of Tourism and Cultural Change*. https://doi.org/10.1080/14766825.2020.1849244

Sevin, E. (2013). Places going viral: Twitter usage patterns in destination marketing and place branding. *Journal of Place Management and Deveopment, 6*(3), 227–239. https://doi.org/10.1108/JPMD-10-2012-0037

Simková, E. (2007). Strategic approaches to rural tourism and sustainable development of rural areas. *Agricultural Economics-Czech, 53*(6), 263–270. https://doi.org/10.17221/979-AGRICECON

Stepchenkova, S., & Morrison, A. M. (2008). Russia's destination image among American pleasure travellers: revisiting Etchner and Ritchie. *Tourism Management, 29*(3), 548–560. https://doi.org/10.1016/j.tourman.2007.06.003

Swani, K., Milne, G., & Brown, B. P. (2013). Spreading the word through likes on Facebook: evaluating the message strategy effectiveness of Fortune 500 companies. *Journal of Research in Interactive Marketing, 7*(4), 269–294. https://doi.org/10.1108/JRIM-05-2013-0026

Tynan, C., McKechnie, S., & Chhuon, C. (2010). Co-creating value for luxury brands. *Journal of Business Research, 63*(11), 1156–1163. https://doi.org/10.1016/j.jbusres.2009.10.012

Uchinaka, S., Yoganathan, V., & Osburg, V. S. (2019). Classifying residents' role as online place-ambassadors. *Tourism Management, 71*, 137–150. https://doi.org/10.1016/j.tourman.2018.10.008

Usakli, A., Koç, B., & Sönmez, S. (2017). How 'social' are destinations?. Examining European DMO social media usage. *Journal of Destination Marketing and Management, 6*(2), 136–149. https://doi.org/10.1016/j.jdmm.2017.02.001

van der Zee, E., Bertocchi, D., & Vanneste, D. (2020). Distribution of tourists within urban heritage destinations: a hot spot/cold spot analysis of TripAdvisor data as support for destination management. *Current Issues in Tourism, 23*(2), 175–196. https://doi.org/10.1080/13683500.2018.1491955

Venkatramanan, S., Chung, S. Y., Rajesh, R., Lee, S. Y., Ramkumar, T., & Prasanna, M. V. (2015). Comprehensive studies of hydrogeochemical processes and quality status of groundwater with tools of cluster, grouping analysis, and fuzzy set method using GIS platform: a case study of Dalcheon in Ulsan City, Korea. *Environmental Science and Pollution Research, 22*, 11209–11223. https://doi.org/10.1007/s11356-015-4290-4

Waters, R. D., Burnett, E., Lamm, A., & Lucas, J. (2009). Engaging stakeholders through social networking: how nonprofit organizations are using Facebook. *Public Relations Review, 35*(2), 102–106. https://doi.org/10.1016/j.pubrev.2009.01.006

WeAreSocial & Hootsuite. (2021). *Digital 2021. Global Overview Report.* Retrieved February 22, 2021, from https://wearesocial.com/digital-2021

Zeng, B., & Gerritsen, R. (2014). What do we know about social media in tourism?. A review. *Tourism Management Perspectives, 10*, 27–36. https://doi.org/10.1016/j.tmp.2014.01.001

Zhang, L., Wu, L., & Mattila, A. S. (2016). Online reviews: the role of information load and peripheral factors. *Journal of Travel Research, 55*(3), 299–310. https://doi.org/10.1177/0047287514559032

Place-Linked Products and Creative Tourism in Iberian Regions

Celso Lopes, João Leitão, and Juan Rengifo Gallego

Abstract Local agri-food products are one of the most important differentiating assets of the rural territories of Portugal and Spain. They are often considered part of the representative cultural capital of their home territories, country, society and history. They are also a tourist resource, with the potential to promote creative tourism activities that offer visitors the opportunity to develop their creative potential, through active participation in courses and learning experiences specific to the destination where they take place.

This chapter aims to understand to what extent local agri-food products serve as a brand of their places of origin and local drivers of creative tourism. To this end, four regional agri-food products were taken as case studies, together with their places of origin in Portugal and Spain: *Cereja do Fundão* (Cherries of Fundão) and *Queijo Serra da Estrela* (Serra da Estrela cheese) from the Portuguese Central Region, and *Cereza del Jerte* (Cherries from Jerte) and *Torta del Casar* (Torta del Casar cheese) from Spanish Extremadura.

The research focuses on the communication channels used most by local, municipal, supra-municipal or regional public and private entities, whose scope of activity includes the promotion and valorisation of tourism, namely their official websites and Facebook pages. The qualitative and quantitative exploratory method was used to achieve the objective defined. After completing an exhaustive database of entities,

C. Lopes (✉)
Department of Art and Science of the Territory, University of Extremadura, Badajoz, Spain

NECE—Research Center in Business Sciences, University of Beira Interior, Covilhã, Portugal

J. Leitão
NECE—Research Center in Business Sciences, University of Beira Interior, Covilhã, Portugal

ICS, Instituto de Ciências Sociais, University of Lisbon, Lisbon, Portugal

CYFE—Center for Young and Family Enterprise, Università Degli Studi di Bergamo, Bergamo, Italy
e-mail: jleitao@ubi.pt

J. R. Gallego
Department of Art and Science of the Territory, University of Extremadura, Badajoz, Spain
e-mail: irengifo@unex.es

J. Leitão et al. (eds.), *Tourism Entrepreneurship in Portugal and Spain*, Tourism, Hospitality & Event Management, https://doi.org/10.1007/978-3-030-89232-6_8

as a data collection procedure, differentiated content analysis was carried out, through metrics created and applied to websites and Facebook pages.

This investigation revealed that the four agri-food products studied are catalysts for tourism activities and events in their places of origin. These products are also drivers of a range of creative tourism activities.

1 Introduction

Agriculture is one of the oldest sectors of activity in the global economy, aiming to meet a series of basic needs linked to the subsistence and development of different peoples. In turn, tourism is one of the economic activities with the greatest and fastest expansion, based on international mobility. Given the global problems related to climate change, food shortages, increased food prices, financial crises and, more recently, public health crises, the link between agriculture and tourism should be prioritised as a way to search actively for new solutions for countries' sustainable development.

However, major challenges arise in strengthening this link. First of all, greater understanding of the processes of restructuring the rural world and the new, principally social and environmental, functions of agriculture, and secondly, understanding of the multiple dimensions of tourism and its development (Torres & Momsen, 2011).

In the agricultural and agri-food sector, regional and local agri-food products are some of the most important differentiating assets of rural parts of Portugal and Spain (Cava-Jiménez et al., 2019; Folgado-Fernández et al., 2017; Rodrigo et al., 2015; Tibério, 1998). For this reason, they are frequently considered as part of the representative cultural capital, not only of their places of origin but even of the country, its society and history (Richards, 2014). Some examples are *Vinho do Porto* (port) and *Pêra Rocha* (pears) in Portugal or *Jamón Ibérico* (ham) and *Queso Manchego* (cheese) in Spain.

So it is not surprising that these products have the potential to bring social and economic benefits for their places, regions or countries of origin, besides their producers and commercial agents. Consequently, these products are also a tourism resource, and currently a wide range of tourist activities is stimulated mainly by elements related to agriculture and the rural world (Armesto-López & Gómez-Martín, 2006). Recognising this potential, in many places a growing number of authorities and policy-makers identify the development of creative agri-food product clusters as a way to stimulate the creative economy and the tourist industry (Lee et al., 2015; OECD, 2014; Richards & Raymond, 2000).

The expansion of this type of tourism promotes the emergence of creative tourism activities, i.e., a type of tourism that gives visitors the opportunity to develop their creative potential through active participation in courses and/or learning experiences associated with the place they are held in (Richards & Raymond, 2000).

The study presented in this chapter is part of wider research aiming to understand the impact of regional agri-food products of differentiated quality (with PDO, GPI and TSG certifications) on the sustainable development of their places of origin. Following this line, the main objective of this study is to understand to what extent regional agri-food products are considered tourism assets in their places of origin and encourage creative tourism initiatives.

The research focuses on four regional agri-food products from two neighbouring border regions in Portugal and Spain: *Cereja do Fundão* (cherries) and *Queijo Serra da Estrela* (cheese) from the Central Region of Portugal, and *Cereza del Jerte* (cherries) and *Torta del Casar* (cheese) from the province of Cáceres in Extremadura, Spain. The analysis is of the territories of production, more precisely, the tourist initiatives related to these products announced on websites and Facebook by local bodies with responsibilities for promoting and publicising tourist activities, and mainly creative tourism initiatives.

Following the introduction, there is a review of the literature on some of the main concepts related to use of the internet as a means to promote rural areas with quality agri-food products, the role of creativity and creative tourism in rural areas and the potential of using agri-food products as the reason for holding creative tourism initiatives. This section ends with formulation of the research hypotheses and a proposed conceptual model, which will be used to organise, explain and discuss the results of the research. Then the empirical study is presented, followed by discussion of the main results obtained. Finally, the conclusions are presented, together with the limitations and suggestions for future research.

2 Literature Review

2.1 ICT and Promotion of Territories with Agri-Food Products

In recent years, the commercialisation of tourist products, destinations and services has undergone important changes in trends, mainly due to the development and application of communication processes (Jiménez Morales & de San Eugenio Vela, 2009). This strengthens the idea that information and communication technology (ICT), and more specifically, the internet, has played a fundamental role in stimulating the leisure industry (Piñeiro-Naval & Serra, 2018). These means of communication are currently essential tools to promote and publicise tourist destinations (Túñez López et al., 2016), and are prioritised by entities with responsibility in this area (Cafiero et al., 2020).

A recent report from Eurostat about the use of ICT in tourism demonstrates that 40% of Europeans use the internet to look for information related to travel (Eurostat, 2016). In addition, the internet is the main source used by tourists who intend to explore less conventional destinations (Cafiero et al., 2020).

On the one hand, the development of ICT, and particularly the internet, gave the "new" tourist the power to become an expert in seeking exceptional value for their money and time (Piñeiro-Naval & Serra, 2018). New tourists are less interested in following the crowd and much keener to follow their own preferences and timetable (Buhalis & Law, 2008). On the other hand, the internet allows organisations with responsibilities for promoting tourism, obviously including local authorities, to develop a broad informative structure. This structure can cover all the topics, places and actors in the territory; from its history and heritage to basic information about accommodation and eating out (de Moragas Spà, 2015). That is how, in the local context, the entities that publish information, consciously or unconsciously, create a certain image of a place. They act as an antenna broadcasting concepts, attributes, values, impressions, sensations and visual impacts that form a place's image (Jiménez Morales & de San Eugenio Vela, 2009).

Therefore, websites have become a very relevant element of organisations' communication, particularly for local authorities, considerably modifying the way of spreading information and relating to their publics (Cabeceiro et al., 2019). Through this channel, local authorities and regional tourism organisations try to attract potential visitors by building an attractive leisure and cultural provision (Piñeiro-Naval & Serra, 2018). This communication tool helps a destination's managers to attract tourists to their area, based on information, establishing a dialogue between the parties involved or systematisation of the tourist service provision, and including access to hyperlinks to hotels and tourism operators (Alencar de Farias et al., 2011).

Another communication channel much used by local entities to promote tourism in their areas is Facebook (Cabeceiro et al., 2019). Unlike municipal websites, where access requires more effort by potential tourists, Facebook pages are more visible and accessible (Lev-On & Steinfeld, 2015).

Local authorities' use of the Facebook environment and complexity has shown continuous, incremental improvement (Deakin, 2010; Mossberger et al., 2013; Oliveira & Welch, 2013). Consequently, it has become an essential means of communication between local authorities and residents in Europe and the USA (Bonsón et al., 2012; Norris & Reddick, 2013).

Due to the great potential related to the speed of conveying messages, the capacity to spread information and mainly the ease of access, both the websites and Facebook pages of local entities with responsibilities in promoting tourism are tools of destination marketing, and as such, must be well managed, in order to create sustainable competitive advantages for destinations in the competitive scenario of global tourism (Pike & Page, 2014).

The major boom in world tourism in recent years is due to a wide set of factors. Piñeiro-Naval and Serra (2018) have no hesitation in pointing out the use of ICT as one of the main factors, as this has allowed tourists to plan their journeys and select their leisure experiences autonomously, based on their own interests and motivations.

2.2 *Creativity and Creative Tourism in Rural Areas*

Professionals in creative areas are located in places where they can benefit from technology, talent and tolerance (Florida, 2003). So it is not surprising that creativity continues to be a concept associated with urban life, since its main source of economic growth, creative people, live and work in cities (Jarábková & Hamada, 2012).

Creativity is therefore an extremely important element of knowledge-based economies (OECD, 2014). This, together with competences, has become central to creating value in the economy as a whole (Richards, 2011), and so creative industries in particular have become a significant economic, cultural and social force. They promote growth and employment, contribute to innovation, entrepreneurship and the development of competences, support urban and rural renewal, stimulate exports, help to maintain cultural identity and increase cultural diversity (OECD, 2014). Therefore, as highlighted by Richards (2011), creativity has been widely applied to tourist facilities such as "design hotels", iconic museums, art galleries and even vineyards.

Becoming aware of this opportunity, many rural areas have redefined themselves as places for consumption, where the commercial possibilities of nature, heritage and traditions have joined agricultural production as key elements of such areas. This leads to the question of the impacts, if any, of this "*creative turn*" on reproduction of the rural area (Cloke, 2007). Creativity can be a response to the social and economic viability of rural areas (Roberts & Townsend, 2016). Creative individuals and creative communities contribute to building cultural capital in the rural world and to their communities' resilience (Gibson & Gordon, 2018; Roberts & Townsend, 2016). Cultural and creative practices express identity and social cohesion, contributing to the quality of life of people living in rural areas (Cabeça et al., 2020).

A strong awareness of the resources available in the community, together with the sense of community (Roberts & Townsend, 2016), plays a fundamental role in developing future paths of sustainable development, besides allowing local communities to develop the capacity to adapt to changes observed in their environment (Cabeça et al., 2020).

The growing articulation between tourism and creativity has been stimulated by the search for alternative models of tourism development and the expansion of the creative economy (Long & Morphet, 2016; OECD, 2014). Consequently, associating creativity with tourism has become a common diversification strategy, principally in the field of cultural tourism, leading to the development of new events and festivals, restoration of old buildings or animation of static attractions being everyday activities nowadays (Richards, 2020). Today, different places, areas, regions and even countries seek rapid ways to differentiate themselves in the global economy, leading them to direct their attention to the creative industries (Turok, 2009).

Cabeça et al. (2020) mention various studies by specialists in the area of tourism, which indicate the emergence of new trends and a change in how people plan and find their destinations. According to these studies, today, a great many tourists seek usefulness rather than novelty, leading them to plan their trips according to the desire for personal growth (Tan et al., 2013). This type of tourist tends to seek authenticity, interaction, cultural immersion, unique experiences, traditional crafts and knowledge (Ohridska-Olson & Ivanov, 2010), wishing to partake in a service that provides them with memorable experiences (Ferreira et al., 2018), involving them deeply in the areas they visit and participating actively in activities that involve the local community (Cabeça et al., 2020).

Tourists are more aware of, and sensitive to the context, and so give more value to the culture of the place and the feeling of being in special, unique destinations (Moscardo, 1996). This means giving greater value to emotional aspects (Jensen, 1999), as well as the possibility of having unique experiences characteristic of a certain place which cannot be reproduced elsewhere (Gu & Ryan, 2008). This type of tourist seeks memorable events (Pine & Gilmore, 1999), engaging experiences (Richards, 2016) and co-created experiences that involve local communities and arouse their creativity (Richards, 2010).

Responding to the psychological needs expressed by this type of tourist, such as inspiration and creativity (Binkhorst, 2008), requires a change of perspective (Cohen & Cohen, 2012), as more than observing, they want to become involved, learn, intervene and experience (Cabeça et al., 2020). The territorial assets of a given destination and the different activities it can provide are therefore increasingly important at the time of deciding on a destination (Martin & Woodside, 2008).

As a result, creative tourism can be a response to this new paradigm, since it depends on the tourist's willingness to engage in learning activities connected to the art, heritage and culture of a destination, based on interaction with the local community (Cabeça et al., 2020). In truth, this is a new form of tourism, which assumes a link between visitors and residents in the same experience (UNESCO, 2006), giving each visitor the opportunity to develop their creative potential and learn, through active participation in the experiences characterising the destination (Richards & Raymond, 2000). Tourists and residents are involved in creative experiences in which both play active roles. Tourists learn, participate, interact and experience the daily life of the destination (Ilincic, 2013), while the resident becomes a key element of the tourist provision (Cabeça et al., 2020).

For Richards (2010), the main characteristics of creative tourism can be expressed as follows:

1. The development of creative potential: tourists need to be provided with the necessary tools to develop their creativity and so that when the time comes to return home they take more than just "souvenir".
2. The creation of active participation: tourists should be involved in the creative process. This promotes the generation of authentic exchange committed to the local population and culture.
3. The production of creative experiences: creativity can arise anywhere. What is important is the union between the creative process and the destination.

4. The development of co-creation: experiences can only be considered unique when tourists, besides interacting and participating, create, design, select and reflect on the experience.

As indicated by Cabeça et al. (2020), this type of tourism is inseparable from the places where it takes place, allying creativity to the places and providing unique cultural experiences. It requires the use of local resources and immersion in the culture of the destination.

Besides generating new cultural heritage (Ohridska-Olson & Ivanov, 2010), according to Richards (2001), creative tourism can potentialise a different type of value from that created by cultural tourism, aiming to differentiate destinations. This involves innovating destinations so as to create products more quickly, designing a process where renewable creative resources are also sustainable, and giving mobility to artistic shows or events, as the last-named can take place anywhere.

For all this to occur, it is enough for there to be a "place" which, in its broadest sense, includes the countryside and all its geographical, cultural (Hildreth & Bailey, 2014) and social specificities, as well as inter-connected institutions. The place serves as a source of inspiration to project distinct, creative tourist provision, with local resonance and meaning (Bakas et al., 2020), and in this sense, a rural territory is undoubtedly a unique place.

Creative tourism is therefore an experience of co-creation, which promotes the effective exchange of experiences, knowledge and competences. It provides the link between people and places, as well as between participants and experiences, valorises competences, knowledge, practices and creative avenues, simultaneously promoting investment in local potential, knowledge, competences and traditions.

2.3 Linking Local Agri-Food Products and Creative Tourist Activities

Regional agri-food products of recognised, differentiated quality and reputation are much more than mere agricultural products. Due to these differentiating characteristics, resulting from their strong connection to the land, they guarantee competitive advantage for their places of origin in a great variety of sectors and activities (Pacciani et al., 2001).

Conscious of this fact, through its quality policy, the European Union (EU), protects geographical indications (GI),[1] in order to promote the unique characteristics associated with geographical origin and traditional production methods (Belletti

[1] GIs are defined by Article 22 of the Agreement on Aspects of Intellectual Property Rights Related to Commerce (TRIPS) as "...indications that identify produce as originating in the territory of a member, or a region or locality of that territory, where a certain quality, reputation or other characteristic of the product is essentially attributable to its geographic origin" (WTO, 1994). This definition suggests that GIs guarantee not only a product's geographic origin but also some

et al., 2017; Cei et al., 2018; Tregear et al., 2007). In Europe, that quality usually refers to the presence of various factors in the place of origin (environmental and human), which are generally summed up in the French word *terroir* (Cei et al., 2018).

This type of product based on its origin is the result of a diversified set of technical, social and economic interactions that include the mobilisation of specific local resources, local know-how and cultural traditions and form the quality of the product within commercialisation chains, between producers and consumers (Belletti & Marescotti, 2011).

Beyond the strict and direct economic component, local and regional agri-food products can stimulate the emergence of other activities in their places of origin (Cei et al., 2018). Tourism is probably the best known example, through various initiatives such as wine trails and gastronomy (Gatti & Incerti, 1997; Millán-Vazquez de la Torre et al., 2017) or the supply of products and/or local dishes to agro-tourism structures (Sidali, 2011), being clearly linked to the presence of local food specialities. In addition, tourists, through activities related to the purchase and consumption in local specialised shops, grocers and restaurants, also represent an economic bonus for their places of origin (Bessière, 1998).

Therefore, local and regional agri-food products are also a tourism resource. Besides the increasing number of tourists who consciously seek new food experiences (Richards, 2012a), a growing number of tourists are willing to devote part of their money and free time to getting to know the world of wine, olive oil and other agri-food products. Their history, the landscape they belong to, their production methods and the people involved, besides visiting reserved spaces, oil presses, wine cellars or orchards (Millán-Vázquez de la Torre et al., 2013). Associating regional agri-food products with creative tourism initiatives can produce yet another interesting characteristic of creative tourism. Principally because, as stated by Richards (2014), these products combine different forms of culture: traditional culture, historical culture, popular culture, contemporary culture and mass culture. So they are another important advantage for peripheral areas, as preserving only the tradition is not enough to create vibrant communities.

These products can be the anchor for a diverse range of tourist experiences. Consequently, stimulating their development can be an attractive strategy, since this type of tourism is perceived as high-income tourism and can increase tourist expenditure (Bertella, 2011; Du Rand et al., 2003; Hall, 2012). Nevertheless, tourist activity, and particularly creative tourism initiatives related to local and regional agri-food products, and agricultural production in general, is still at a very early stage of research, with the literature on this topic being very limited.

A search of the literature on this subject quickly reveals there is not even a clear designation for this type of tourism. It usually emerges associated with food in

type of quality, which together with the product's reputation are essential conditions for GI protection (Lopes et al., 2018).

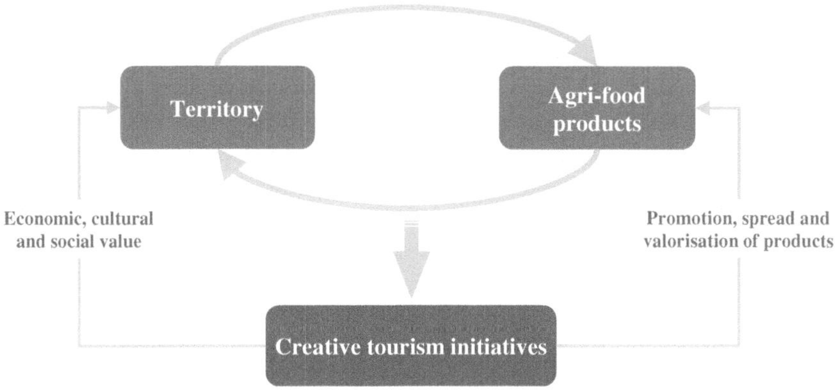

Fig. 1 Creative tourism initiatives based on the territory and agri-food products. Source: Own elaboration

general (Rachão et al., 2019) or gastronomy (Hjalager, 2002; Richards, 2012b, 2014).

From the above, it is logical to assume there is a synergistic relation between regional agri-food products and their places of origin, regarding the tourism sector. Firstly, as already presented, these types of products owe their principal qualitative characteristics to the territory and all the social, economic and technical interactions occurring there (Spielmann & Charters, 2013). Secondly, through the reputation acquired, the economic income generated, the landscape created and the traditions formed in their environment, these products add competitive value to their places of origin. This cultural blend results in the potential emergence of creative tourism initiatives (cf. Fig. 1).

As explained by Cabeça et al. (2020), developing creative tourism initiatives according to the supply of local resources, through the involvement of the destination's communities, can be an effective response to the challenges of sustainable development faced by areas of low population density, adding economic and social value to communities, local institutions and territories and regions as a whole. This type of initiative can add the effects of dissemination and economic valorisation to the whole range of regional agri-food products.

Creative tourism is an emerging form of tourism, on both the supply and demand side, allowing tourists to have deeper contact with, and knowledge of the local culture, through direct participation in cultural and creative activities that create experiences and emotions through involvement in the destination's life (Almeida, 2019).

The rural world is usually centred on particular forms of nature-society relations, in which people cooperate or struggle against the non-humanity of nature (Szerszynski, 2003). Creative spaces can represent the simple multi-functionality of a village hall, rather than extended artistic quarters of metropolitan areas and creative activities can consist of apparently simple learning to appreciate local products, rather than acquiring more ostentatious skills (Richards & Wilson,

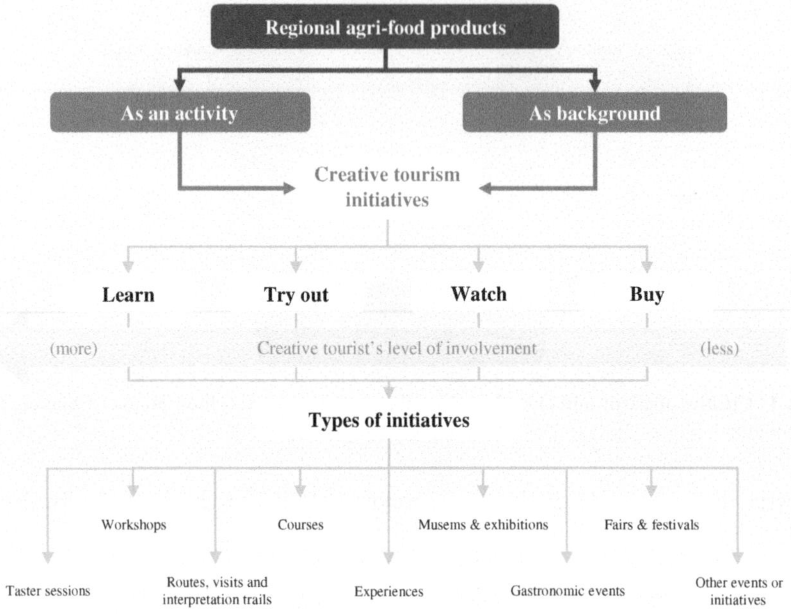

Fig. 2 Creative tourism initiatives based on the territory and agri-food products. Source: Own elaboration, based on Richards (2009, 2010)

2006). Alternatively, creativity in rural tourism can suggest new ways of understanding creativity, for example, in terms of hybrid nature-society activities based on eco-experience or adventure. Clearly, however, various practices giving tourists the opportunity to develop their creative potential through forms of active participation in rural contexts can, in turn, lead to new reproductions of the rural space (Cloke, 2007).

Therefore, creative tourism covers a vast range of experiences. These can imply more active involvement in more formal settings (such as courses or workshops) and more passive forms of creative consumption, such as cultural itineraries or even design shops (Richards, 2010).

Creative tourism initiatives developed in rural areas supported by their endogenous resources follow exactly the models of creative tourism proposed by Richards (2010, 2020). Many different types of experiences and products can be offered to the creative tourist. Once again, as mentioned by Richards (2009), these experiences vary from more active forms of involvement, such as learning a specific activity, to engaging in visits to galleries and shops in search of creative products. The same happens for activities and initiatives involving regional agri-food products that can be provided to the creative tourist. However, due to the shortage of means and critical mass with the capacity and availability to develop them, it is more common to hold events including a variety of creative tourism initiatives or activities (Fig. 2).

2.4 Central Research Question, Proposition and Conceptual Model

The progress in its technological capacity and generalised adoption has made ICT, and particularly the internet and social networks, an effective, low-cost instrument, both to spread information and to gather ideas from increasingly informed publics (Bonsón et al., 2015). This study focuses on the information contained on the websites and Facebook pages of local entities responsible for promoting and raising awareness of tourism, in relation to the main objective defined. These two forms of communication were chosen because nowadays they are powerful tools to market destinations (Pike & Page, 2014), being used to create sustainable competitive advantages for destinations in the competitive scenario of global tourism (Piñeiro-Naval & Serra, 2018).

As this study is part of a broader line of research aiming to understand the effects of regional agri-food products of differentiated quality (for example, with PDO, PGI and TSC certification) on the sustainable development of their places of origin, based on the literature review, the central research question is as follows: *are place-linked agri-food products of differentiated quality recognised as the brand image of their territories of origin by the regional and local entities responsible for promoting and advertising tourism*?

Specifically, this research aims to: (1) determine the coverage of the websites and Facebook pages of local and regional entities responsible for promoting and advertising tourism; (2) identify and assess the importance and recognition of regional agri-food products, as brand images for their places of origin, by these same entities; and (3) identify and characterise the creative tourism initiatives developed in these places that have regional agri-food products as the main element or background.

The collective nature of an agri-food product, together with its capacity to include suitably the identity, qualities and culture of a territory, strengthens the forms of valorisation defined locally (Bérard & Marchenay, 2008). Therefore, promoting agri-food products has many more implications than the mere intention to sell them (Slee, 1993).

Quality agri-food products, and all the aspects involved in them, are the driver of a diversified range of rural development projects implemented in recent years in many areas that were previously considered marginal from the economic point of view (Brunori & Rossi, 2000; De Salvo et al., 2013). These projects have given new life to those regions, not only economically but also strengthening them socially (Armesto-López & Gómez-Martín, 2006).

The tourism sector has benefited greatly from these rural development projects. However, tourism is also fertile ground for valorisation of local food heritage (Espeitx Bernat, 2004). Here, the link between the agricultural and tourism sectors in predominantly rural areas has led to the development of bi-directional quality standards (Armesto-López & Gómez-Martín, 2006).

Given the great potential of the link between tourism and regional agri-food products in valorisation of rural territories and in their development, regional and local entities responsible for stimulating tourist activity can be expected to use this synergy to promote their localities. This leads to the first proposition:

Proposition 1 (P_1) Regional and local entities responsible for promoting tourist activity recognise place-linked agri-food products as tourist assets promoting their places of origin.

In recent years, new forms of tourism have emerged in geographical areas until then neglected by tourism agents, and therefore, by consumers (Cava-Jiménez et al., 2019). This aspect is particularly relevant if considering that the modern tourist industry pays more and more attention to the experiential dimension and intangible heritage of travelling. This leads to a transformation of tourism, ceasing to be only a service market and becoming an authentic market of emotions and feelings (De Salvo et al., 2013). In this context, the search for tourist activities with the potential for a sensory component requires re-organisation of the traditional form (Nocifora et al., 2011).

These experiences should ensure the use and valorisation of the most unique and attractive endogenous resources, in order to increase sustainability in the context of an integrated tourism supply (Saxena et al., 2007), and to raise unique positioning, based on independent and inimitable geographical, historical and/or cultural characteristics (Ritchie & Crouch, 2003). These strategies should produce the best destination-market combination possible, making optimal use of local resources and competences by projecting tourist experience opportunities targeted at those potentially most interested in this type of local experience (Kastenholz et al., 2016).

Therefore, since different places, territories, regions and even nations seek rapid ways to differentiate in the global economy (Turok, 2009), creative tourism initiatives based on regional agri-food products can be an important alternative in the tourist industry. Consequently, the second proposition is as follows:

Proposition 2 (P2) Agri-food products linked to places are the basis of creative tourism activities developed in their territories of origin.

The symbiotic, synergistic relation between place-linked agri-food products and tourism is a mechanism frequently used to promote the whole territory, through combined initiatives of conventional and creative tourism (Fig. 3).

3 Methododogy

3.1 Sample

In the empirical approach, the subject of analysis is a portfolio of place-linked, agri-food products, originating in two neighbouring border regions of Portugal and Spain: *Cereja do Fundão* (cherries) and *Queijo Serra da Estrela* (cheese) from the Central Region of Portugal and *Cereza del Jerte* (cherries) and *Torta del Casar* (cheese) from the province of Cáceresin Extremadura, Spain. The analysis made in

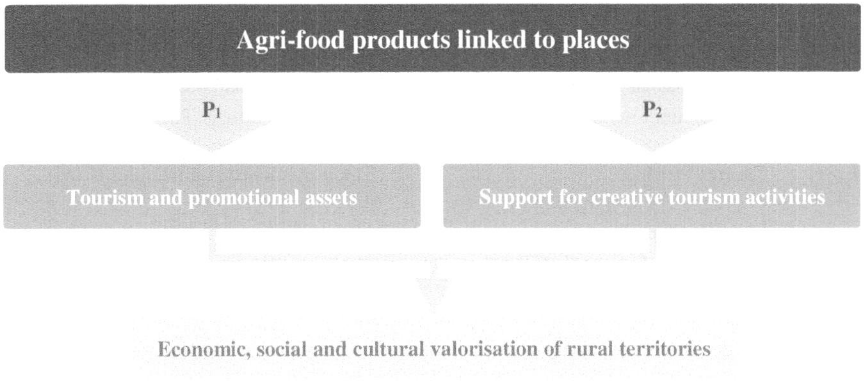

Fig. 3 Creative tourism initiatives based on territory and place-linked agri-food products. Source: Own elaboration

this study concerns the geographical areas of their production, dealing with the creative tourism initiatives promoted on the websites and Facebook pages of the local entities responsible for tourism promotion and advertising.

These entities were selected based on what is set out in Portuguese legislation. In this respect, the Portuguese government clearly states it is up to regional entities and local public entities to "*organise and divulge tourist information*" (sub-heading e, of number 2, of article 5, of Law n° 33/2013, p. 2912), considering these entities' important role in communicating the touristic and gastronomic identity of their respective regions.

Therefore, besides the national bodies promoting tourism in Spain and Portugal—selected for the sample due to being the institutional support for all tourism projects developed at the regional level—the same bodies but with regional and local responsibilities were selected, as well as producing and commercialising entities, since as Barroco and Augusto (2016) indicate, the promotion and valorisation of endogenous products should also be one of their responsibilities.

According to typology and geographical sphere of action, the selected entities are distributed in four groups: national entities (Level I); regional entities (Level II); municipal or supra-municipal entities, in Portugal, and municipal, local council or provincial entities, in Spain (Level III); and entities of a local, municipal, supra-municipal or regional nature in Portugal, or of a local, municipal, local council or provincial nature in Spain, with responsibilities for the promotion and valorisation of the agri-food products focused on in this study (Level IV) (Table 1).

The field of research comprised the websites and Facebook pages of regional and local entities responsible for promoting and advertising tourism in their territories. This domain was selected based on the success obtained in similar studies by authors using this methodology (see Cabeceiro et al., 2019; Jiménez Morales & de San Eugenio Vela, 2009; Lev-On & Steinfeld, 2015; Piñeiro-Naval et al., 2016; Piñeiro-Naval & Serra, 2018; Roque et al., 2012). The sample is formed of 85 entities, corresponding to a total of 83 websites and 81 Facebook pages (Table 2).

Table 1 Selected entities from the places of origin of agri-food products

	Spain	Portugal
Level I	Spanish Tourist Board	Portuguese Tourist Board
Level II	Extremadura Tourism	Centre of Portugal Tourism
Level III	Cáceres Province	Inter-municipal communities (CIM)
	Local Authorities	Local Authorities (Councils)
	Local Action Groups (LAG)	Local Action Groups (GAL)
Level IV	Regulating Councils of PDO	PDO and PGI Producer Groups
	Public entities promoting tourism at the local level	Public entities promoting tourism at the local level

Source: Own elaboration

Table 2 N° of websites and Facebook pages per regional agri-food product

Regional agri-food products	Entities	Websites	Facebook pages
Cereja do Fundão (cherries)	15	15	15
Queijo Serra da Estrela (cheese)	37	37	36
Cereza del Jerte (cherries)	17	15	14
Torta del Casar (cheese)	16	16	16
	85	**83**	**81**

Source: Own elaboration

3.2 Data Collection Method

At a first stage, to identify references to the regional agri-food products on the websites and Facebook pages, the exploratory method was used. In the second stage, aiming to characterise the creative tourism initiatives based on these products, one of the research methods most commonly used by researchers in the area of communication media (Berger, 2000) and tourism (Veal, 2006) was adopted, i.e., content analysis.

The search for references to regional agri-food products on the websites and Facebook pages of the selected entities followed a matrix of indicators based on the previous work by Bezerra and Correia (2019), Bonsón et al. (2015) and Hoffman and Fodor (2010). Application of this matrix of analysis reveals in which sections of the websites and Facebook pages there are references to agri-food products, besides identifying events, activities and initiatives (Table 3).

In the second stage, the identification, classification and characterisation of creative tourism events or initiatives were based on the work of Richards (2009, 2010), which is schematically represented in the decision tree below in Fig. 4.

Table 3 Types of references searched on the websites and Facebook pages per regional agri-food product

A. Websites
1. References to the product on the first page of the website
1.1. Tourist information
1.2. News
1.3. Coming events
1.4. Photographs
1.5. Links to specific pages on the product
1.6. Presence in the latest news/highlights
2. In the search box on the website, search for the name of the product related to the entity ("*name of the product*") and record the number and type of resulting entries
2.1. Information on the product
2.2. Tourist information
2.3. News
2.4. Activities and events:
(a) Creative tourism initiatives
(b) Other initiatives
B. Facebook pages (In the section on events, count the number of events held related to the agri-food product (or only promoted by the entity), during 2019 and 2020)
1. Events where the product is the central element
1.1. Creative tourism events and initiatives
1.2. Other type of events
2. Other type of events

Source: Own elaboration

4 Results

4.1 Place-Linked Agri-Food Products on Websites and Facebook Pages

The first proposition intends to determine to what extent regional and local entities responsible for tourism promotion and dynamics recognise regional agri-food products as touristic and promotional assets of their places of origin. In this context, given the relevance of websites in organisations' communication, local authorities and regional tourism bodies use them to attract potential visitors by compiling an appealing leisure and cultural offer.

Aiming to test empirically the validity of the above proposition, the analysis method described in the previous section was applied to the websites of regional and local entities responsible for tourism promotion and dynamics. This revealed that only 15.66%, of all the websites tested made explicit mention of place-linked agri-food products on their initial pages (Table 4).

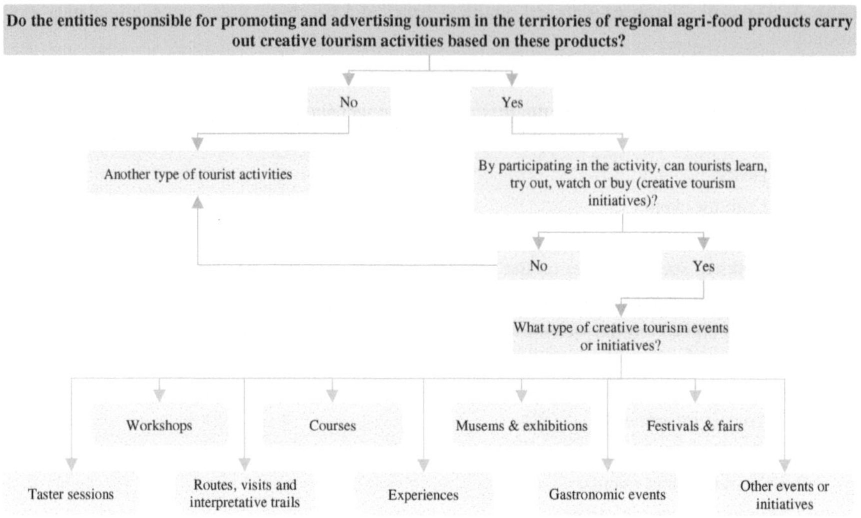

Fig. 4 Creative tourism initiatives based on the territory and agri-food products. Source: Own elaboration

Table 4 Number and type of references to regional agri-food products onwebsites

	Cereja do Fundão	Queijo Serra da Estrela	Cereza del Jerte	Torta del Casar	Total websites
Number of websites of the product's area of origin	15	37	15	16	83
1. References to the product on the first page	3	3	4	3	13
1.1. Tourist information	0	0	0	1	1
1.2. News	2	2	0	0	4
1.3. Events to be held	0	3	1	1	5
1.4. Photographs	2	3	3	3	11
1.5. Links to specific pages on the product	2	1	0	1	4
1.6. Presence in the latest news/ highlights	2	1	0	0	3
2. References to the product after placing its name in the search box on the website	4	18	6	6	34
2.1. Information about the product	1	4	1	3	9
2.2. Tourist information	1	10	2	2	15
2.3. News	3	16	3	2	24
2.4. Activities and events:	4	11	7	4	26
(a) Creative tourism initiatives	*3*	*9*	*5*	*3*	*20*
(b) Other initiatives	*1*	*2*	*2*	*1*	*6*

Source: Own elaboration

In relative terms, when grouped by area of origin of the regional agri-food product, the websites of entities in the territory of *Cereza del Jerte* refer most to the product, being recorded on 26.67% of the websites analysed, i.e., in four of the fifteen websites.

The type of reference most commonly found, i.e., in 13.25% of the websites, is through photographs of agri-food products, followed by promotion of related events.

Searching through entering its exact name in the search box on the websites revealed references to the agri-food products in 40.96% of cases, i.e., in 34 of the 83 websites analysed. In relative terms, when grouped by the area of the regional agri-food product's origin, the websites of entities for the territory of *Queijo Serra da Estrela* make most reference to the product, being recorded in 48.65% of the websites analysed, i.e., in 18 of the 37.

The most common type of reference, present in 26 websites, concerns activities and events, and among these, creative tourism initiatives are highlighted, present in 24.10% of the 83 websites analysed, and other types of activities and events, present in 7.23% of the 83.

In relative terms, the websites of entities from the territory of *Cereza del Jerte* show the greatest number of activities and events related to the agri-food product, and 33.33% of these are creative tourism initiatives.

As for the holding of events, and their promotion, an analysis was also made of entities' Facebook pages, since as highlighted by Cabeceiro et al. (2019), this social network is a communication channel much used by local bodies to promote tourism in their areas. This social network also reveals the events held over the years since the page was created. This is important, as the global restrictions since the beginning of 2020 caused by the COVID 19 pandemic have prevented events from being held, and so it was possible to see the events held and promoted on the various entities' Facebook pages in 2019, a year of full activity.

Besides revealing that only 81 of the 85 entities analysed have Facebook pages, the study showed that overall in 2019, only 58.02%, promoted events, and in 2020, this number was reduced to 43 pages corresponding to 53.09% (cf. Table 5), which is closely associated with the suspension of events due to the worldwide public health crisis.

Taking as a reference the data presented in Table 5, in relative terms, when grouped by the area of origin of the regional agri-food product, the Facebook pages of the entities of the territory of *Cereza del Jerte*, with 71.43% in 2019 and 57.14% in 2020, and those of *Queijo Serra da Estrela*, with 61.11% in 2019 and 55.56% in 2020, make most use of this channel to promote events related to agri-food products.

Of all the Facebook pages analysed, 12.35% in 2019 and 9.88% in 2020 promoted creative tourism events. Considering only the pages that promoted events, the percentage of pages promoting creative tourism events was 21.28%, in 2019, and 18.6%, in 2020.

It also stands out that despite the low number of entities promoting events through their Facebook pages, this social network was used very proficiently. Those Facebook pages promoted an average of 37.97 events in 2019 and 24.95 events in 2020. The lower average number of events in 2020 is due to the restrictions caused

Table 5 Number of Facebook pages promoting events

	Cereja do Fundão			QueijoSerra da Estrela			Cereza del Jerte			Torta del Casar			Total events		
Number of Facebook pages	15			36			14			16			81		
Years	2019	2020	Δ	2019	2020	Δ	2019	2020	Δ	2019	2020	Δ	2019	2020	Δ
(a) Entities promoting events on their Facebook pages	7	7	0	22	20	−2	10	8	−2	8	8	0	47	43	−4
(b) Facebook pages that promote creative tourism events related to the regional Agri-food product	2	2	0	2	1	−1	5	1	−4	1	4	3	10	8	−2
(c) Facebook pages that promote other events related to the regional Agri-food product	1	2	1	1	0	−1	1	2	−1	2	0	−2	5	4	−1
(d) Facebook pages that promote other types of events	7	6	−1	22	19	−3	10	8	−2	8	4	−4	47	37	−4

Source: Own elaboration

Table 6 Types of events promoted on Facebook pages

	Cereja do Fundão		Queijo Serra da Estrela		Cereza del Jerte		Torta del Casar		Total events	
Number of Facebook pages	15		36		14		16		81	
Years	2019	2020	2019	2020	2019	2020	2019	2020	2019	2020
Entities promoting events on their Facebook pages	7	7	22	20	10	8	8	8	47	43
1. Events on the Facebook page where the Agri-food product is the central element	14	10	3	1	8	5	12	4	37	20
1.1. Creative tourism events and initiatives	10	3	2	1	7	2	7	4	26	10
1.2. Other types of events	4	7	1	0	1	3	5	0	11	10
2. Other types of events	431	202	479	160	337	130	158	7	1405	499
Total events	445	212	482	161	345	135	170	11	1442	519

Source: Own elaboration

by the COVID 19 pandemic. Even so, in the first two months of 2020, the frequency of events promoted was normal. After the first semester of 2020, the promotion of online events began to appear on these entities' Facebook pages.

Once again, it is entities of the territory of *Cereza del Jerte* that make most use of their Facebook pages to promote the events held: 71.43% of the pages consulted promoted events in 2019 and 57.14% continued to do so in 2020 (cf. Table 6). In the two years analysed, it was also the Spanish entities that proportionately promoted a greater percentage of creative tourism events related to agri-food products in their territories. In 2019, 35.71% of entities of the territory of *Cereza del Jerte* that promoted events on their Facebook pages also promoted creative tourism events, and in 2020, 25.00% of entities of the territory of *Torta del Casar* promoting events on their Facebook pages also promoted creative tourism events.

Given the evidence obtained, it is not possible to confirm for all regional and local entities responsible for promoting tourist activity that all recognise their place-linked agri-food products as tourist assets promoting their places of origin.

4.2 *Innovative Creative Tourism Initiatives Based on Place-Linked Agri-Food Products*

This study also identified 49 creative tourism initiatives held in the places of origin of the four agri-food products. The products motivating mostly this type of initiative are *Torta del Casar* and *Queijo Serra da Estrela*, since, respectively, 34.69% and 26.53% of the tourism initiatives identified are based on these products.

It stands out that the creative tourism initiatives identified belong to the typologies and models of creative tourism proposed by Richards (2010, 2020). In these territories, except for "Taster sessions", "Courses" and "Workshops", the creative tourism initiatives fit in all the typologies proposed by this author (see Fig. 2), with more than 50.00% belonging to the typologies of "Festivals and Fairs" (26.53%) and "Experiences" (24.49%), as seen in Table 7.

The most prominent typologies of creative tourism initiatives on the websites and Facebook pages of entities in the place of origin of each agri-food product, for *Cereja do Fundão* and *Torta del Casar*, this is "Experiences", corresponding to 33.33% and 66.77% of all the creative tourism initiatives identified in each of these territories. Regarding the websites and Facebook pages of the territorial entities of *Queijo Serra da Estrela*, 46.15% of the initiatives identified fit the typology of "Routes, visits and interpretative trails", while for *Cereza del Jerte*, 63.64% correspond to "Festivals and Fairs".

The creative tourism initiatives identified on the websites and Facebook pages of the territorial entities of each agri-food product provide the tourist with different types of involvement. According to the description of each initiative identified, this involvement rarely belongs to just one of the involvement typologies identified by Richards (2010, 2020), these being "learning", "trying out", watching" or "buying".

Table 7 Typologies of creative tourism initiatives identified by territory

	Cereja do Fundão	Queijo Serra da Estrela	Cereza del Jerte	Torta del Casar	Total
Taster sessions	0	0	0	0	0
Courses	0	0	0	0	0
Gastronomic events	1	1	1	1	4
Experiences	4	0	0	8	12
Festivals and fairs	1	3	7	2	13
Museums and exhibitions	0	1	2	4	7
Workshops	0	0	0	0	0
Routes, visits, interpretative trails	1	6	1	2	10
Other events or initiatives	1	2	0	0	3
Total	8	13	11	17	49

Source: Own elaboration

These initiatives vary greatly in the form of tourist involvement. On the one hand, some initiatives imply more active forms of involvement, such as learning about a specific activity, for example in "*Little Shepherds, Little Craft people. Learn and have fun with your hands in the cheese in the open air*", held in the place of origin of *Torta del Casar*, or have a day fruit-picking, as in the activity "*It's cherry time—Pick Fundão cherries in the orchard*" in the place of origin of *Cereja do Fundão* or enter the life of the cooperatives, as in "*Guided tours of the Cooperative Group of Valle del Jerte*", in the territory of *Cereza del Jerte*. On the other hand, in some activities the tourist's involvement is less, for example, in visits to the "*Cheese Museum*" or the "*Cheese Manor*" in the territory of *Queijo da Serra da Estrela*, having a picnic when the cherry trees are in blossom through the initiative of "*Picnic Baskets—Cherry Blossom*" in the territory of *Cereja do Fundão* or appreciating the landscape in "*Rural Paths—Cherry trees in Blossom*" in the territory of *Cereza del Jerte*.

The evidence obtained overall does not allow complete confirmation of the second proposition, indicating that place-linked agri-food products are the basis of creative tourism activities organised in their territories of origin, despite finding a centrifugal force present in some initiatives of public–private co-creation, with the responsibilities and risks shared between local producers and municipal entities.

The following tables (cf. Tables 8, 9, 10, and 11) contain a description by type of initiative and typology of tourist involvement, of the 47 tourism initiatives identified in the course of this study, and the entities promoting or advertising them.

5 Conclusion

Creative tourism is an alternative sector in tourism which has seen strong, rapid growth in most parts of the world, but due to its newness does not yet have very well-defined limits. In recent years, Spain and Portugal have followed this global trend, with institutions linked to this type of tourism demonstrating efforts to achieve co-creation and have done so with the major involvement of local communities.

The main aim of the research presented in this chapter is to understand to what extent four place-linked agri-food products, originating in neighbouring border regions of Portugal and Spain, are considered tourist assets in their territories of origin, and the basis for developing creative tourist initiatives. Based on the literature of reference on the topic, two propositions were put forward and tested through methodology developed specially for this study and supported by the previous work by Jiménez Morales and de San Eugenio Vela (2009), Roque et al. (2012), Lev-On and Steinfeld (2015), Piñeiro-Naval et al. (2016), Bezerra and Correia (2019) and Cabeceiro et al. (2019), consisting of exploratory analysis of the websites and Facebook pages of regional and local entities responsible for promoting tourism initiatives in these agri-food products' territories of origin.

Concerning the first proposition, it is considered that regional and local entities responsible for promoting tourism activities recognise regional agri-food products as tourist assets able to promote their territories of origins.

Table 8 Creative tourism initiatives identified in the territory of origin of *Cereja do Fundão*

N°	Designation	Type of initiative	Description	Promoter	Typology of tourist involvement			
					Learn	Try out	Watch	Buy
1	Picnic Baskets—Cherry trees in Blossom	Experience	"Hiring" Picnic Baskets, intending to give everyone a unique experience in the midst of the protected area and get to know the tastes of Fundão.	Fundão Local Authority		✓	✓	✓
2	It's time to pick Fundão cherries	Experience	Visit to the Fundão cherry orchards with picking; visit to Fundão's historic centre; Free for lunch; Walk through the Slate Village of Janeiro de Cima with entry to the Weavers' House.	Fundão Local Authority	✓	✓	✓	
3	It's cherry time—Pick Fundão cherries in the orchard	Experience	Visit the Fundão Cherry orchards, with fruit-picking; visitFundão's historic centre; Free for lunch; Cherry Festival. The experience intended is picking and eating cherries directly from the tree.	Fundão Local Authority	✓	✓	✓	
4	Trips by hot air balloon	Experience	Activity aiming to give a different perspective of the cherry orchards, normally held when the orchards are in production.	Fundão Local Authority			✓	✓
5	Fundão, here you eat well: Cherry tastes	Gastronomic event	Gastronomic festival involving restaurant, cake-shops and bars in the territory, where the main ingredient is the Fundão cherry.	Fundão Local Authority		✓		✓
6	Cherry Festival in Alcongosta	Festivals & fairs	Event held in the mountain village of Alcongosta, with various activities related to the cherry and its sector.	Fundão Local Authority	✓	✓	✓	✓

7	Cherry trees in Blossom—Cherry tourist train	Other events and initiatives	Tourist train through the orchards in blossom in the Gardunhamountain range, Visit to Fundão's historic centre, Free for lunch, Visit the historic village of Castelo Novo	Fundão Local Authority			✓	
8	Cherry Trail	Routes, visits and interpretative trails	Walking tour on the north slope of the Gardunha mountain range, characterised by the abundance of cherry tree terraces and production of the famous Fundão cherry.	Fundão Local Authority			✓	

Source: Own elaboration

Table 9 Creative tourism initiatives identified in the territory of origin of *Queijo Serra da Estrela*

N°	Designation	Type of initiative	Description	Promoter	Typology of tourist involvement			
					Learn	Try out	Watch	Buy
1	Culinary Competition "With Serra da Estrela Cheese"	Gastronomic event	Activity promoting Serra da Estrela cheese which, besides presenting different sweet and savoury specialities made from Serra da Estrela cheese and its derivatives, seeks to encourage innovation and entrepreneurship for the possible creation of products of reference	Oliveira do Hospital Local Authority	✓	✓		
2	Festival of the Shepherd and Mountain Cheese	Festivals and fairs	An annual event held on the first Sunday of March at Mosteiro/Pena Verde, which includes a presentation and competition of bordaleira sheep, with around 40 local producers generally participating.	Aguiar da Beira Local Authority	✓	✓	✓	✓
3	Cheese Fair	Festivals and fairs	Event with many activities related to Serra da Estrela cheese, organised annually by Celorico da Beira Local Authority.	Celorico da Beira Local Authority	✓	✓	✓	✓
4	Serra da Estrela Cheese Fair	Festivals and fairs	Event in which shepherds, cheese-makers and producers are the main figures, with around 30 cheese producers selling their product. Visitors can taste and buy the best of the Local Authority's produce.	Gouveia Local Authority	✓	✓	✓	✓
5	Serra da Estrela Cheese Fair	Festivals and fairs	Event to promote the identity and rural life of shepherds, traditional cheese-makers and producers of PDO cheese. Other regional products are also present, such as olive oil, charcuterie, honey and regional sweets. Also involved are craftspeople, music and folk traditions, the Serra da Estrela dog and the Serra da Estrela sheep.	Fornos de Algodres Local Authority	✓	✓	✓	✓

6	Serra da Estrela Cheese Festival	Festivals and fairs	Event aiming mainly to promote Serra da Estrela cheese and other endogenous products of the region, including aspects of gastronomy, habits and customs, historical and archaeological heritage of Oliveira do Hospital local authority. There are also cookery demonstrations using Serra da Estrela cheese, seminars and exhibitions on the subject and cultural entertainment, and for the youngest, arts workshops at the "Cheese School" and dramatisation linked to the Serra da Estrela cheese and the area's history and culture.	Oliveira do Hospital Local Authority	✓	✓	✓	✓
7	Shepherd and Cheese Fair	Festivals and fairs	Event presenting the local authority's potential: Serra da Estrela cheese,"Dão de Penalva" wine, Bravo de Esmolfe apples and smoked ham, crafts, liquors and sweets.	Penalva do Castelo Local Authority	✓	✓	✓	✓
8	Serra da Estrela Cheese Fair	Festivals and fairs	Event aiming mainly to promote and raise the value of Serra da Estrela cheese and the whole sector in Seia Local Authority and the region.	Seia Local Authority	✓	✓	✓	✓
9	Cheese Museum	Museums and exhibitions	The Cheese Museum aims to spread knowledge about traditions, the sound of the fields being worked, the shepherd, flocks and transhumance. It provides a true journey of the senses around Serra da Estrela cheese and the plains of Cova da Beira, formerly known as the Valley of the Jews. It is also an excellent place to taste the delicious cheese produced in the region.	Covilhã Local Authority	✓	✓	✓	✓

(continued)

Table 9 (continued)

N°	Designation	Type of initiative	Description	Promoter	Typology of tourist involvement			
					Learn	Try out	Watch	Buy
10	Cheese Manor	Museums and exhibitions	Museum related to the activity of Serra da Estrela cheese. Exhibition of artefacts used in producing hand-made cheese, explaining all steps in the process of making Serra da Estrela cheese, from milking to the gourmet/consumer's table. Here there is also tasting of Serra da Estrela cheese and charcuterie for groups of visitors.	Celorico da Beira Local Authority	✓	✓	✓	✓
11	Serra da Estrela Cheese Trail	Routes, visits and interpretative trails	The Local Authority decided to reform the trail, and now besides the characteristics of this endogenous product, people can get to know the authorised producers and sellers of the cheese.	Oliveira do Hospital Local Authority			✓	

Source: Own elaboration

Table 10 Creative tourism initiatives identified in the territory of origin of *Cereza del Jerte*

N°	Designation	Type of initiative	Description	Promoter	Typology of tourist involvement			
					Learn	Try out	Watch	Buy
1	Valle del Jerte Cherry Fair	Festivals and fairs	The Valle del Jerte Cherry Fair takes place in Valle del Jerteas part of the cultural offer of "Cerecera" held in the 11 communities of the local authority. It includes a great many activities, such as cherry-tasting and classification, cherry market, street entertainment, games and guided tours, which can be enjoyed from the end of May to the beginning of August.	Soprodevaje LAG	✓	✓	✓	✓
2	Cherry Blossom Festival	Festivals and fairs	Festival of national tourist interest, running each year from the end of March to the beginning of April in Valle del Jerte, Extremadura, celebrating the cherry trees in blossom with various events, such as exhibitions, medieval markets, open-air parties, trails and tasting, in the villages in the valley.	Community of local authorities of Valle del Jerte	✓	✓	✓	✓
3	La Cerecera	Festivals and fairs	The harvesting of the famous Valle del Jerte cherries takes place in May, June and July. At this time the "Cerecera" has been celebrated for many years, a cultural, gastronomic and entertainment programme to promote discovery of the essence of the Valley in its most interesting, authentic period.	Soprodevaje LAG	✓	✓	✓	✓

(continued)

Table 10 (continued)

N°	Designation	Type of initiative	Description	Promoter	Typology of tourist involvement			
					Learn	Try out	Watch	Buy
4	Gastronomic Days of the Cherry	Gastronomic event	Event held in different restaurants in Valle del Jerte, including the tasting of dishes prepared with cherries.	Community of local authorities of Valle del Jerte		✓		✓
5	Cherry Museum	Museums and exhibitions	Housed in a traditional building in Valle del Jerte, showing traditional elements of traditional architecture. Enter the world of the Valle cherries, viewing the different displays over three floors. It uses various interactive resources to recreate original environments and objects and give the feeling of putting the "garabato" on one's shoulder, the basket that generations of locals have used to pick the Jerte cherries, with the instrument used, the "recolector", being very personal. The visitor can also get to know the indigenous varieties of cherries, especially the famous "picotas del Jerte", aspects related to caring for the health of the cherries and the fruit's production in terraces.	Valle del Jerte Tourism Association	✓		✓	
6	Open Days in the Cooperatives of Valle del Jerte	Routes, visits and interpretative trails	Activity including a guided tour of the premises of different cooperatives in Valle del Jerte and the respective cherry-tasting.	Valle del Jerte Tourism Association	✓		✓	

7	Cherry Blossom Mountain Bike trail	Routes, visits and interpretative trails	The prime mountain bike trail in the area in spring, following a route that cannot fail to impress those who love the sport.	Community of local authorities of Valle del Jerte, Casas del Castañar Town Hall and Valle del Jerte Cooperatives			✓	
8	Dog-Walking in the Cherry Blossom	Routes, visits and interpretative trails	Activity to appreciate the beauty of the countryside created each year by over a million cherry trees, the economic driver of this community. It is also an opportunity to take your dog along and get to know other animal lovers, and especially, a time to collaborate with the Animal Protection Shelter in Plasencia.	Community of local authorities of Valle del Jerte			✓	
9	Marked Trail—Cherry Blossom	Routes, visits and interpretative trails	The PR-CC 2 of the Cherry Blossom, marked out in white and yellow, is one of the most popular routes in Valle del Jerte with the numerous paths of Villa de Tornavacas. Over 22 kms, it follows the descent of the River Jerte, rising and falling among cherry orchards, oak and chestnut trees.	Community of local authorities of Valle del Jerte			✓	
10	Visits to cherry farms and leisure and adventure activities	Routes, visits and interpretative trails	Take a trip to Valle del Jerte, visit farms, pick cherries, learn the secrets of the Jerte cherries and appreciate beautiful landscape with the help of professional, local guides. Try out delicious, special menus highlighting the cherry.	Valle del Jerte Tourism Association		✓	✓	✓
11	Guided tours of the Valle del Jerte Group of Cooperatives	Routes, visits and interpretative trails	Visits to the premises of the Valle del Jerte Group of Cooperatives	Valle del Jerte Tourism Association	✓		✓	

(continued)

Table 10 (continued)

N°	Designation	Type of initiative	Description	Promoter	Typology of tourist involvement			
					Learn	Try out	Watch	Buy
12	Children's cooking. Junior workshops involving the Jerte cherry.	Other events and initiatives	At the workshop, children can learn by playing how the product is cultivated and how to prepare recipes using cherries with the help of a chef	Regulatory Council of the Protected Origin of the Cherry	✓	✓		
13	Cherry Market	Other events and initiatives	Cherry market with Valle del Jerte producers	Regulatory Council of the Protected Origin of the Cherry			✓	✓

Source: Own elaboration

Table 11 Creative tourism initiatives identified in the territory of origin of *Torta del Casar*

N°	Designation	Type of initiative	Description	Promoter	Typology of tourist involvement			
					Learn	Try out	Watch	Buy
1	The Secrets of the Pasture	Experience	Learn from an expert in gastronomy about the most interesting aspects of cheese, including Torta del Casar, accompanied by excellent wine.	Extremadura Cheese Trail: Navafríay country house La Bodega de Pérez (cellar).	✓	✓		✓
2	Magic Transformation. Discover the best-kept secret of Torta del Casar	Experience	Guided tour of the Quesos del Casar cheese-makers to discover the production process of the well-known Torta del Casar. You will learn how this delicious Torta del Casar is formed, seeing first-hand a transformation that is as surprising as it is magical, from the most traditional process to the modern, innovative production techniques in one of the most important cheese-makers.	Extremadura Cheese Trail: Quesos del Casar cheese-makers	✓		✓	✓
3	Cheese Safari	Experience	Visit the natural environment of Cáceres and surroundings in a 4 × 4 and learn about the workings of the pasture ecosystem as Extremadura's natural larder, visiting a farm raising sheep to produce the milk used in PDO Torta del Casar and tasting the famous and delicious cheese of that name.	Extremadura Cheese Trail: Extremadura Safari			✓	✓

(continued)

Table 11 (continued)

N°	Designation	Type of initiative	Description	Promoter	Typology of tourist involvement			
					Learn	Try out	Watch	Buy
4	Who has taken my cheese? From shepherd to seller in a few hours.	Experience	An entertaining and educational workshop making Extremaduran cheese after milking a sheep to obtain the raw ingredient to do so. The day includes a traditional shepherd's breakfast, a guided tour of a farm with transport from Cáceres included, and the milking of sheep and goats. The milk obtained will be used in a cheese workshop in Zamarilla Leisure Centre, from which the results can be taken home.	Extremadura Cheese Trail: Légola	✓	✓		✓
5	Little Shepherds, Little Crafts-people. Learn and have fun with your hands in the cheese in the open countryside.	Experience	Légola provides children with activities through an educational-experiential programme of fun and learning activities around the cheese-making cultura and nature. On the farm of "La Zamarrilla", you can become a shepherd milking goats and then make cheese from their milk. The day continues with a workshop of traditional sweets. There is also the chance to play various games or attend environmental education workshops.	Extremadura Cheese Trail: Légola	✓	✓	✓	✓

6	Traditional Tastes. Discover our villages, taste our products.	Experience	Accommodation in a country house retaining all the atmosphere of the old labouring houses, perfectly situated on the Silver Route less than an hour from the Heritage towns of Cáceres and Mérida. As a Cheese Trail customer, the country house welcomes you with tasting of the three PDO cheeses of Extremadura, accompanied by a glass of local wine. Demonstration and lessons in the use of some tools involved in making the famous PDO cheeses of Torta del Casar, presenting the history of its creation, as well as useful advice for its consumption.	Extremadura Cheese Trail: Silver Route, Country House	✓	✓		✓
7	Awaken your Senses through Cheese. Enjoy different tastes, smells and textures.	Experience	An educational trip through the factory showing the whole process of traditional cheese-making, with special emphasis on PDO Torta del Casar. You will then move to the tasting room where you will be able to try ten of the varieties produced. The unique characteristics of each will be explained. You will participate in making the cheese and realise the great difference between each of the 19 traditional cheese varieties produced.	Extremadura Cheese Trail: Doña Francisca cheese-maker	✓	✓	✓	✓

(continued)

Table 11 (continued)

N°	Designation	Type of initiative	Description	Promoter	Typology of tourist involvement			
					Learn	Try out	Watch	Buy
8	Combinations of Cheese and other Food. Tasting of traditional products in a natural setting.	Experience	The Pinotes House offers two types of tasting. Both take place in the Interpretation Centre of Cañada Real Soriana Occidental and shepherding culture of "Casa Pinotes". It is located in a rural setting, excellent for bird-watching as it is inside a ZEPA area, together with the surrounding nature. Torta del Casar combined with wine: this will be in the presence of cheese experts and wine specialists, who will share their knowledge about these wonderful gastronomic products in a combination *par excellence*.	Extremadura Cheese Trail: Interpretative Centre of Pasture Culture "Casa Pinotes"	✓	✓	✓	✓
9	Torta del Casar Tapa Trail	Gastronomic event	The Torta del Casartapa trail is an opportunity to taste the delicious Tortain the local bars and restaurants.	Torta del Casar PDO authority; Casar de Cáceres Town Hall; LAG TAGUS; Cáceres provincial authority;		✓		✓
10	Torta del Casar Week	Festivals and fairs	Held at the beginning of autumn every year since 1995, numerous activities aim to promote and advertise the culture of Torta del Casar to all village residents and visitors of all ages.	PDOTorta del Casar Cooperative Foundation	✓	✓	✓	✓

11	European Fair of Casar de Cáceres Cheese	Festivals and fairs	Event with the participation of cheese-makers from the territories belonging to the European Cheese Trail Association. The Fair is part of the 23rd Torta del Casar Week.	Casar de Cáceres Town Hall.		✓	✓	✓
12	Pinotes House. Interpretative Centre of Cañada Real Soriana Occidental	Museums and exhibitions	The Interpretative Centre of Cañada Real Soriana Occidental and shepherding culture "Pinotes House" has preserved its original structure. It was previously a working house in the area of stock raising, and now pays homage to transhumance and shepherding life through periodic and audio-visual exhibitions.	Casar de Cáceres Town Hall.	✓		✓	
13	Torta del Casar Interpretative Centre "Pastoralia"	Museums and exhibitions	Pastoralia is unique, with a wide variety of activities centred on Torta del Casar. You will be transported to the magical world of this magnificent, unique product and its related tasks and traditions. Everything is geared towards the visitor's enjoyment of a different and original environment.	Pastoralia	✓		✓	✓

(continued)

Table 11 (continued)

N°	Designation	Type of initiative	Description	Promoter	Typology of tourist involvement			
					Learn	Try out	Watch	Buy
14	Casar de Cáceres Cheese Museum	Museums and exhibitions	The Casar de Cáceres Cheese Museum presents the history and production process representing part of the identity of the local population: Torta del Casar. This museum shows how the locals have been able to maintain the characteristics of this singular product, preserving its essence through time.	Casar de Cáceres Town Hall.	✓		✓	
15	Tasting at the "Torta del Casar" Museum	Museums and exhibitions	Hourly guided tours and tasting of "Torta del Casar" combined withJerte cherries.	Casar de Cáceres Town Hall.	✓	✓	✓	
16	Transhumance route	Routes, visits and interpretative trails	Since 2006, the Transhumance Route has attracted more than a thousand people each year, including a visit to the Interpretative Centre of Shepherding Life and Cañada Real "Pinotes House" as well as having a traditional shepherd's breakfast.	Casar de Cáceres Town Hall.	✓	✓	✓	

17	Torta del Casar Trail	Routes, visits and interpretative trails	Paths and trails speak of the life of this land of transhumance and its most notable inhabitants, the shepherds and their flocks. Tanks for washing wool, mills, castles and ancient oaks mark those paths; and at the end of the hard day's work, the promise of a well-deserved rest in the ancient city of Cáceres, World Heritage Site, where history and gastronomy come together.	Extremadura Cheese Trail		✓	✓	✓

Source: Own elaboration

In the scope of this exploratory study, there is partial confirmation of the first proposition, in that the evidence obtained reveals that only some of the entities whose websites and Facebook pages were subject to analysis use place-linked agri-food products as brand images of the territory they represent. In these cases, the entities not only use photographs and references to agri-food products on their websites and Facebook pages, but also organise various initiatives to encourage economic and social development, as well as actively involving the local population.

It is also highlighted that as the territories of origin of these agri-food products are vast, the entities located closest to the centre of the industrial operations related to the product tend to be the ones resorting to them as a promotional tool. For example, in the case of the territory of origin of *Cereja do Fundão* or *Torta del Casar*, it is the entities located in Fundão, in the former, and Casar de Cáceres, in the latter, that use the cherry or cheese most frequently as a brand image.

As for the second proposition, it is considered that place-linked agri-food products are the basis of creative tourism activities developed in their territories of origin, and similarly to the first proposition, there is only evidence leading to partial confirmation of this second proposition.

Once again, the results obtained showed that only some of the entities in the territories of origin of the place-linked agri-food products studied here use them as a basis to promote a wide range of initiatives, among them creative tourism initiatives.

The field of analysis was restricted to public entities, or those of a similar nature, which promote creative tourism initiatives based on these products. This is a limitation to overcome in future studies. Currently, a number of local companies, for example, in rural accommodation, leisure activities or even producers of these products, as found by Duxbury and Richards (2019), seek to incorporate local specificities, i.e., those linked to places, for example, portfolios of regional agri-food products with a strong identity, in creative tourism provision aiming for differentiation.

Finally, this study has shown that creative tourism initiatives, through the wide range of types of experiences these can include, when supported by place-linked agri-food products can be a very important element for diversification of the local economy and to reinforce the social dynamics of their territories of origin. Therefore, in responding to the central research questions, it should be pointed out that local entities recognise the value of their territories' agri-food products. The emergence of economic and social development actions based on creative tourism are an expanding phenomenon, despite requiring more professional management of all associated activities and initiatives. These should be integrated in new versions of intelligent specialisation strategies that consider places' characteristics and specificities, in a new form of global thinking and acting in the place, that is, through place-linked products.

Recommended for future research is exploration of creative tourism as a viable alternative to traditional tourism, which requires more in-depth studies about the influence of place-linked agri-food products on initiatives of this type in other regions. Besides the obvious comparative effect, this will allow better definition of the limits of this type of tourism and contribute to strengthening its definition and characterisation, helping to draw up new policies for sustainable tourism.

References

Alencar de Farias, S., Aguiar, E. C., & Kovacs, M. H. (2011). Imagen de los destinos turísticos en los portales gobernamentales. Análisis de tres localidades brasileñas. *Estudios y Perspectivas En Turismo, 20*(6), 1345–1365.

Almeida, J. S. F. (2019). *Turismo Criativo: Uma nova visão do Turismo, uma nova geração de experiências e emoções*. Instituto Superior de Gestão. Retrieved from https://comum.rcaap.pt/handle/10400.26/30310

Armesto-López, X. A., & Gómez-Martín, B. (2006). Tourism and quality agro-food products: An opportunity for the Spanish countryside. *Tijdschrift voor Economische en Sociale Geografie, 97*(2), 166–177. https://doi.org/10.1111/j.1467-9663.2006.00510.x

Bakas, F. E., Duxbury, N., Silva, S., & De Castro, T. V. (2020). Connecting to place through creative tourism. In A. Gonçalves, J. Marques, M. Tavares, & S. Cabeça (Eds.), *Creative tourism dynamics: Connecting travellers, communities, cultures, and places* (pp. 1–15). Coimbra.

Barroco, C., & Augusto, L. (2016). Turismo Gastronómico em Portugal: Formas de comunicar os produtos endógenos da região Dão Lafões e alto Paiva. *Anais Brasileiros de Estudos Turísticos - ABET, 6*(2), 23–39.

Belletti, G., & Marescotti, A. (2011). Origin products, geographical indications and rural development. In E. Barham & B. Sylvander (Eds.), *Labels of origin for food: Local development, global recognition* (pp. 75–91). CAB International. https://doi.org/10.1079/9781845933524.0075

Belletti, G., Marescotti, A., & Touzard, J.-M. (2017). Geographical indications, public goods, and sustainable development: The roles of actors' strategies and public policies. *World Development, 98*, 45–57. https://doi.org/10.1016/j.worlddev.2015.05.004

Bérard, L., & Marchenay, P. (2008). *From localized products to geographical indications awareness and action*. Bourg-en-Bresse. Retrieved from www.ethno-terroirs.cnrs.fr

Berger, A. A. (2000). *Media and communication research methods. An introduction to qualitative and quantitative approaches*. SAGE Publications.

Bertella, G. (2011). Knowledge in food tourism: The case of Lofoten and Maremma Toscana. *Current Issues in Tourism, 14*(4), 355–371. https://doi.org/10.1080/13683500.2010.489638

Bessière, J. (1998). Local development and heritage: Traditional food and cuisine as tourist attractions in rural areas. *Sociologia Ruralis, 38*(1), 21–34. https://doi.org/10.1111/1467-9523.00061

Bezerra, R., & Correia, A. I. (2019). The potential of olive oil for creative tourism experiences in the northern region of Portugal. *Revista Portuguesa de Estudos Regionais, 51*, 55–72.

Binkhorst, E. (2008). Turismo de co-creación, valor añadido en escenarios turísticos. *Turismo de Co-Creación, Valor Añadido En Escenarios Turísticos, 1*(1), 40–51. Retrieved from http://revistes.ub.edu/index.php/ara/article/view/18968

Bonsón, E., Royo, S., & Ratkai, M. (2015). Citizens' engagement on local governments' facebook sites. An empirical analysis: The impact of different media and content types in Western Europe. *Government Information Quarterly, 32*(1), 52–62. https://doi.org/10.1016/j.giq.2014.11.001

Bonsón, E., Torres, L., Royo, S., & Flores, F. (2012). Local e-government 2.0: Social media and corporate transparency in municipalities. *Government Information Quarterly, 29*(2), 123–132. https://doi.org/10.1016/j.giq.2011.10.001

Brunori, G., & Rossi, A. (2000). Synergy and coherence through collective action: Some insights from wine routes in Tuscany. *Sociologia Ruralis, 40*(4), 409–423. https://doi.org/10.1111/1467-9523.00157

Buhalis, D., & Law, R. (2008). Progress in information technology and tourism management: 20 years on and 10 years after the internet—The state of eTourism research. *Tourism Management, 29*(4), 609–623. https://doi.org/10.1016/j.tourman.2008.01.005

Cabeça, S. M., Gonçalves, A. R., Marques, J. F., & Tavares, M. (2020). Creative tourism experiences in low-density territories: Valuing the Algarve's inland. In A. Gonçalves,

J. Marques, M. Tavares, & S. Cabeça (Eds.), *Creative tourism dynamics: Connecting travellers, communities, cultures, and places* (pp. 183–200). Grácio Editor.

Cabeceiro, D., Esteves, E., & Morais, E. P. (2019). Promotion of activities of tourist animation in the pages of Facebook of the municipalities of Terras de Trás-Os-Montes. *Iberian conference on information systems and technologies, CISTI*, 2019-June, 19–22. https://doi.org/10.23919/CISTI.2019.8760590

Cafiero, C., Palladino, M., Marcianò, C., & Romeo, G. (2020). Traditional agri-food products as a leverage to motivate tourists: A meta-analysis of tourism-information websites. *Journal of Place Management and Development, 13*(2), 195–214. https://doi.org/10.1108/JPMD-05-2019-0032

Cava-Jiménez, J. A., Millán-Vazquez de la Torre, M. G., & Hernández-Rojas, R. (2019). Analysis of the tourism demand for Iberian ham routes in Andalusia (Southern Spain): Tourist profile. *Sustainability, 11*(16), 4278. https://doi.org/10.3390/su11164278

Cei, L., Defrancesco, E., & Stefani, G. (2018). From geographical indications to rural development: A review of the economic effects of European Union policy. *Sustainability, 10*(10), 3745. https://doi.org/10.3390/su10103745

Cloke, P. (2007). Creativity and tourism in rural environments. In G. Richards & J. Wilson (Eds.), *Tourism, creativity and development* (pp. 37–47). Routledge.

Cohen, E., & Cohen, S. A. (2012). Current sociological theories and issues in tourism. *Annals of Tourism Research, 39*(4), 2177–2202. https://doi.org/10.1016/j.annals.2012.07.009

de Moragas Spà, M. (2015). Cruce de caminos. Tecnologías de la comunicación y convivencia en la ciudad global. *Telos: Cuadernos de Comunicación e Innovación, 100*, 70–80.

De Salvo, P., Hernández-Mogollón, J. M., Di Clemente, E., & Calzati, V. (2013). Territory, tourism and local products. The extra virgin oil's enhancement and promotion: A benchmarking Italy-Spain. *Tourism and Hospitality Management, 19*(1), 23–34.

Deakin, M. (2010). A review of city portals. In *Politics, democracy and E-government* (pp. 215–224). IGI Global. https://doi.org/10.4018/978-1-61520-933-0.ch013

Du Rand, G. E., Heath, E., & Alberts, N. (2003). The role of local and regional food in destination marketing. *Journal of Travel & Tourism Marketing, 14*(3–4), 97–112. https://doi.org/10.1300/J073v14n03_06

Duxbury, N., & Richards, G. (2019). Towards a research agenda for creative tourism: Developments, diversity, and dynamics. In E. N. Duxbury & G. Richards (Eds.), *A research agenda for creative tourism* (pp. 1–14). Edward Elgar Publishing Limited.

Espeitx Bernat, E. (2004). Patrimonio alimentario y turismo: una relación singular. *PASOS Revista de Turismo y Patrimonio Cultural, 2*(2), 193–213. https://doi.org/10.25145/j.pasos.2004.02.016

Eurostat. (2016). *Statistics on ICT use in tourism online article*. Retrieved from https://ec.europa.eu/eurostat/statistics-explained/index.php?title=Archive:Statistics_on_ICT_use_in_tourism#out_of_10_ Europeans_look_online_for_travel_related_inf

Ferreira, A. M., Alén González, E., & Liberato, D. (2018). Turismo criativo e sustentabilidade territorial. *Proceedings XX Congreso Internacional AECIT "Conocimiento, Creatividad, Innovación, Hacia El Turismo Del Futuro,"* 1–17. Retrieved from http://www.aecit.org/files/congress/20/papers/399.pdf

Florida, R. (2003). Cities and the creative class. *City & Community, 2*(1), 3–19. https://doi.org/10.1111/1540-6040.00034

Folgado-Fernández, J. A., Palos-Sánchez, P. R., Campón-Cerro, A. M., & Hernández-Mogollón, J. M. (2017). Productos gastronómicos con identidad y desarrollo del destino turístico. Un estudio sobre rutas del queso de España. *International Journal of Scientific Management and Tourism, 3*(1), 93–109. Retrieved from http://ijosmt.com/index.php/ijosmt/article/view/182

Gatti, S., & Incerti, F. (1997). The wine routes as an instrument for the valorisation of typical products and rural areas. *52nd EAAE Seminar: "Typical and traditional productions: Rural effect and agro-industrial problems,"* 213–224. Retrieved from https://pdfs.semanticscholar.org/91f0/4d1e5a8be022b1a02ecaf2543fc49d519f34.pdf

Gibson, C., & Gordon, A. (2018). Rural cultural resourcefulness: How community music enterprises sustain cultural vitality. *Journal of Rural Studies, 63*, 259–270. https://doi.org/10.1016/j.jrurstud.2016.11.001

Gu, H., & Ryan, C. (2008). Place attachment, identity and community impacts of tourism—The case of a Beijing hutong. *Tourism Management, 29*(4), 637–647. https://doi.org/10.1016/j.tourman.2007.06.006

Hall, C. M. (2012). Boosting food and tourism-related regional economic development. In *Food and the tourism experience. The OECD-Korea workshop* (pp. 49–61). OECD publishing.

Hildreth, P., & Bailey, D. (2014). Place-based economic development strategy in England: Filling the missing space. *Local Economy: The Journal of the Local Economy Policy Unit, 29*(4–5), 363–377. https://doi.org/10.1177/0269094214535712

Hjalager, A. M. (2002). A typology of gastronomy tourism. In A. Hjalager & G. Richards (Eds.), *Tourism and gastronomy*. Routledge.

Hoffman, D. L., & Fodor, M. (2010). Can you measure the ROI of your social media marketing? *MIT Sloan Management Review, 52*(1), 41–49.

Ilincic, M. (2013). Benefits of creative tourism - the tourist perspective. In G. Richards & A. P. Russo (Eds.), *Alternative and creative tourism* (pp. 99–113). Atlas.

Jarábková, J., & Hamada, M. (2012). Creativity and rural tourism. *Creative and Knowledge Society, 2*(2), 5–15.

Jensen, R. (1999). The dream society: How the coming shift from information to imagination will transform your business. .

Jiménez Morales, M., & de San Eugenio Vela, J. (2009). Identidad territorial y promoción turística: la organización de eventos como estrategia de creación, consolidación y difusión de la imagen de marca del territorio. *Zer: Revista de Estudios de Comunicación = Komunikazio Ikasketen Aldizkaria, 26*, 277–297. Retrieved from http://dspace.uvic.cat/handle/10854/1561

Kastenholz, E., Eusébio, C., & Carneiro, M. J. (2016). Purchase of local products within the rural tourist experience context. *Tourism Economics, 22*(4), 729–748. https://doi.org/10.1177/1354816616654245

Lee, A. H. J., Wall, G., & Kovacs, J. F. (2015). Creative food clusters and rural development through place branding: Culinary tourism initiatives in Stratford and Muskoka, Ontario, Canada. *Journal of Rural Studies, 39*, 133–144. https://doi.org/10.1016/j.jrurstud.2015.05.001

Lev-On, A., & Steinfeld, N. (2015). Local engagement online: Municipal Facebook pages as hubs of interaction. *Government Information Quarterly, 32*(3), 299–307. https://doi.org/10.1016/j.giq.2015.05.007

Long, P., & Morphet, N. D. (2016). *Tourism and the creative industries. Theories, policies and practices* (1th ed.). Routledge.

Lopes, C., Leitão, J., & Rengifo-Gallego, J. (2018). Place branding: Revealing the neglected role of agro food products. *International Review on Public and Nonprofit Marketing, 15*(3), 497–530. https://doi.org/10.1007/s12208-018-0211-9

Martin, D., & Woodside, A. G. (2008). Grounded theory of international tourism behavior. *Journal of Travel & Tourism Marketing, 24*(4), 245–258. https://doi.org/10.1080/10548400802156695

Millán-Vázquez de la Torre, G., Morales-Fernández, E. J., & Castro-Freire, M. S. (2013). Turismo Gastronómico y D.O.P.: Una Relación Simbiótica En Andalucía. In *Turismo e innovación: VI jornadas de investigación en turismo* (pp. 827–850). Universidad de Sevilla. Retrieved from https://idus.us.es/xmlui/bitstream/handle/11441/53111/vazquez-torre.pdf?sequence=1

Millán-Vazquez de la Torre, M. G., Arjona-Fuentes, J. M., & Amador Hidalgo, L. (2017). Olive oil tourism: Promoting rural development in Andalusia (Spain). *Tourism Management Perspectives, 21*, 100–108. https://doi.org/10.1016/j.tmp.2016.12.003

Moscardo, G. (1996). Mindful visitors. *Annals of Tourism Research, 23*(2), 376–397. https://doi.org/10.1016/0160-7383(95)00068-2

Mossberger, K., Wu, Y., & Crawford, J. (2013). Connecting citizens and local governments? Social media and interactivity in major U.S. cities. *Government Information Quarterly, 30*(4), 351–358. https://doi.org/10.1016/j.giq.2013.05.016

Nocifora, E., De Salvo, P., & Calzati, V. (2011). *Territori lenti e turismo di qualità. Prospettive innovative per lo sviluppo di un turismo sostenibile (1^a)*. Retrieved from http://books.google.it/books?id=pdEBtwAACAAJ

Norris, D. F., & Reddick, C. G. (2013). Local E-government in the United States: Transformation or incremental change? *Public Administration Review, 73*(1), 165–175. https://doi.org/10.1111/j.1540-6210.2012.02647.x

OECD. (2014). *Tourism and the creative economy. OECD studies on tourism*. Retrieved from https://doi.org/10.1787/9789264207875-en

Ohridska-Olson, R. V., & Ivanov, S. H. (2010). Creative tourism business model and its application in Bulgaria. *Proceedings of the Black Sea Tourism Forum' Cultural Tourism–The Future of Bulgaria*, 1–17. Retrieved from http://papers.ssrn.com/sol3/papers.cfm?abstract_id=1690425%0Ahttp://culturalrealms.typepad.com/files/creativetourismbm_article_1_lr.pdf%0Ahttp://www.culturalrealms.com/2010/07/the-creative-tourism-business-model.html

Oliveira, G. H. M., & Welch, E. W. (2013). Social media use in local government: Linkage of technology, task, and organizational context. *Government Information Quarterly, 30*(4), 397–405. https://doi.org/10.1016/j.giq.2013.05.019

Pacciani, A., Belletti, G., Marescotti, A., & Scaramuzzi, S. (2001). The role of typical products in fostering rural development and the efects of regulation (EEC) 2081/92. In *73rd Seminar of the European Association of Agricultural Economists*, 1–17. Ancona, Italy.

Pike, S., & Page, S. J. (2014). Destination marketing organizations and destination marketing: A narrative analysis of the literature. *Tourism Management, 41*, 202–227. https://doi.org/10.1016/j.tourman.2013.09.009

Pine, B. J., & Gilmore, J. H. (1999). *The experience economy: Work is theatre and every business a stage*. Harvard Business School Press.

Piñeiro-naval, P. O. R. V., José, J., & Perosanz, I. (2016). Un análisis de contenido de los Portales web municipales de España. *Revista TELOS, 102*, 1–16.

Piñeiro-Naval, V., & Serra, J. P. (2018). O potencial turístico da web local. Uma análise à comunicação digital das autarquias portuguesas. In V. Piñeiro-Naval & J. P. Serra (Eds.), *Cultura, Património e Turismo na Sociedade digital: Uma perspetiva ibérica* (pp. 157–184). LabCom.IFP.

Rachão, S., Breda, Z., Fernandes, C., & Joukes, V. (2019). Food tourism and regional development: A systematic literature review. *European Journal of Tourism Research, 21*, 33–49.

Richards, G. (2001). *Cultural attractions and European tourism*. CABI Publishing.

Richards, G. (2009). Creative tourism and local development. In R. Wurzburger, A. Pattakos, & S. Pratt (Eds.), *Creative tourism: A global conversation* (pp. 78–90). Sunstone Press.

Richards, G. (2010). Creative tourism and cultural events. In *2nd Forum on UNESCO creative cities network*, Icheon, Republic of Korea.

Richards, G. (2011). Creativity and tourism. The state of the art. *Annals of Tourism Research, 38*(4), 1225–1253. https://doi.org/10.1016/j.annals.2011.07.008

Richards, G. (2012a). An overview of food and tourism trends and policies. In *Food and the tourism experience*. OECD.

Richards, G. (2012b). Food and the tourism experience: Major findings and policy orientations. In D. Dodd (Ed.), *Food and the tourism experience* (pp. 13–46). OECD.

Richards, G. (2014). The role of gastronomy in tourism development. In *Fourth International Congress on Noble Houses: A Heritage for the Future*, Arcos de Valdevez. https://doi.org/10.21832/9781873150498-020.

Richards, G. (2016). The challenge of creative tourism. *Ethnologies, 38*(1–2), 31–45. https://doi.org/10.7202/1041585ar

Richards, G. (2020). Designing creative places: The role of creative tourism. *Annals of Tourism Research, 85*, 102922. https://doi.org/10.1016/j.annals.2020.102922

Richards, G., & Raymond, C. (2000). Creative trousim. *Atlas NEWS 23, 1*(23), 16–20.

Richards, G., & Wilson, J. (2006). Developing creativity in tourist experiences: A solution to the serial reproduction of culture? *Tourism Management, 27*(6), 1209–1223. https://doi.org/10.1016/j.tourman.2005.06.002

Ritchie, J. R. B., & Crouch, G. (2003). *The competitive destination: A sustainable tourism perspective*. CABI Publishing.

Roberts, E., & Townsend, L. (2016). The contribution of the creative economy to the resilience of rural communities: Exploring cultural and digital capital. *Sociologia Ruralis, 56*(2), 197–219. https://doi.org/10.1111/soru.12075

Rodrigo, I., Cristóvão, A., Tibério, M. L., Baptista, A., Maggione, L., & Pires, M. (2015). The Portuguese agrifood traditional products: Main constraints and challenges. *Revista de Economia e Sociologia Rural, 53*, S023–S032. https://doi.org/10.1590/1234-56781806-94790053s01002

Roque, V., Fernandes, G., & Raposo, R. (2012). Identificação dos media Sociais utilizados pelas organizações de gestão de destinos: O Caso de estudo do destino turístico Serra da Estrela. *Revista Turismo & Desenvolvimento, 1*(17), 311–320.

Saxena, G., Clark, G., Oliver, T., & Ilbery, B. (2007). Conceptualizing integrated rural tourism. *Tourism Geographies, 9*(4), 347–370. https://doi.org/10.1080/14616680701647527

Sidali, K. L. (2011). A sideways look at farm tourism in Germany and in Italy. In *Food, agriculture and tourism* (pp. 2–24). Springer Berlin Heidelberg. https://doi.org/10.1007/978-3-642-11361-1_1

Slee, B. (1993). Endogenous development: A concept in search of a theory. *Options Mediterraneenns, 54*, 43–54. Retrieved from https://www.researchgate.net/profile/Richard_Slee2/publication/242222244_ENDOGENOUS_DEVELOPMENT_A_CONCEPT_IN_SEARCH_OF_A_THEORY/links/00463539966329738000000.pdf

Spielmann, N., & Charters, S. (2013). The dimensions of authenticity in terroir products. *International Journal of Wine Business Research, 25*(4), 310–324. https://doi.org/10.1108/IJWBR-01-2013-0004

Szerszynski, B. (2003). Technology, performance and life itself: Hannah Arendt and the fate of nature. *The Sociological Review, 51*(2_suppl), 203–218. https://doi.org/10.1111/j.1467-954X.2004.00459.x

Tan, S. K., Kung, S. F., & Luh, D. B. (2013). A model of "creative experience" in creative tourism. *Annals of Tourism Research, 41*, 153–174. https://doi.org/10.1016/j.annals.2012.12.002

Tibério, M. (1998). Produtos tradicionais: importância sócio-económica na defesa do mundo rural. In *1.ª Jornadas de Queijos e Enchidos - Produtos Tradicionais*.

Torres, R., & Momsen, J. (2011). *Tourism and agriculture* (R. Torres & J. Momsen, Eds.). Routledge. https://doi.org/10.4324/9780203834404.

Tregear, A., Arfini, F., Belletti, G., & Marescotti, A. (2007). Regional foods and rural development: The role of product qualification. *Journal of Rural Studies, 23*(1), 12–22. https://doi.org/10.1016/j.jrurstud.2006.09.010

Túñez López, M., Altamirano, V., & Valarezo, K. (2016). *Comunicación turística colaborativa 2.0: promoción, difusión e interactividad en las webs gubernamentales de Iberoamérica*. Universidad de La Laguna. https://doi.org/10.4185/RLCS-2016-1094

Turok, I. (2009). The distinctive city: Pitfalls in the pursuit of differential advantage. *Environment and Planning A: Economy and Space, 41*(1), 13–30. https://doi.org/10.1068/a37379

UNESCO. (2006). *Towards sustainable strategies for creative tourism: Discussion report of the planning meeting for the 2008 international conference on creative tourism*. UNESCO.

Veal, A. J. (2006). *Research methods for leisure and tourism* (3rd ed.). Financial Times/Prentice Hall/Pearson Education.

World Trade Organisation. (1994). *The TRIPS agreement*. https://doi.org/10.1016/j.clsr.2005.12.010

The Rise of Half-Tourists and their Impact on the Tourism Strategies of Peripheral Territories

João Almeida and Fernando Belezas

Abstract Peripheral territories face significant challenges related to depopulation and lack of public and private investment. With the COVID-19 pandemic, lockdown and cancellation of all cultural activities that attract visitors, peripheral and rural territories, highly dependent on cultural activities and tourism, were deeply affected. Nevertheless, the COVID-19 pandemic has significantly pushed forward a trend that has increasingly been seen in recent years—remote work and digital nomadism. This *trend* boomed in the last year when most people went to work at home during the lockdown. Mostly due to the pandemic, the number of people searching for a more flexible work environment or for a more healthier and less crowded environment to live in has rapidly increased. Therefore, remote work and digital nomadism are an opportunity that peripheral territories have to embrace as a tool to attract an emerging target group—the half-tourists. This chapter analyses two projects—Digital Nomads Madeira (Portugal) and Nomad City Gran Canaria (Spain). By performing in-depth interviews with the people that implement these strategies, this chapter contributes to raising the awareness of the importance of this emerging target group and identifying the barriers and best practices to innovate in tourism strategies in peripheral places.

1 COVID-19, Tourism and Peripheral Territories

COVID-19 impacts are still challenging to measure and to understand to their full extent. Nonetheless, it is undoubtedly "a crisis like no other" as described by the World Economic Forum (2020), with an overwhelming impact on the global economy and an unequal impact on sectoral economies. In the case of the tourism and related sectors, those suffered "amongst the hardest hit" due to the closure of

J. Almeida (✉)
GOVCOPP, DCSPT, University of Aveiro, Aveiro, Portugal
e-mail: joaolopesalmeida@ua.pt

F. Belezas
DEGEIT, University of Aveiro, Aveiro, Portugal
e-mail: fernandobelezas@ua.pt

J. Leitão et al. (eds.), *Tourism Entrepreneurship in Portugal and Spain*, Tourism, Hospitality & Event Management, https://doi.org/10.1007/978-3-030-89232-6_9

international trips needed social distancing and lockdown in most countries of the world (Segal, 2020).

Estimates of the World Tourism Barometer pointed out an enormous economic impact—a loss of US$ 1.2 trillion in export revenues from tourism and 120 million direct tourism job cuts—seven times bigger than the impact of September 11th, the worst decline so far (UNWTO, 2020). The World Travel & Tourism Council also highlighted that tourism's contribution to the global economy fell 49.1% in 2020 due to the pandemic's devastating impact.

Hall et al. (2020) stated that the impact of COVID-19 will be uneven in space (and time). If this was already a "crisis like no other", the damage was even worse for countries and regions highly dependent on the tourism economy (Gössling et al., 2020). Many peripheral territories are highly dependent on tourism and its associated sub-sectors to attract young people, entrepreneurs, visitors and higher social and economic dynamism to their territory and its community. In the case of Portugal and Spain's outermost territories, this sector represents a quarter (or more) of their gross domestic product (GDP). In addition to this high dependence on the tourism sector, peripheral territories face significant structural challenges. Those suffer from severe population shrinking processes, ageing population, lack of public and private investment, lack of skilled labour caused by the increasing urbanisation and centralisation of infrastructures, services and opportunities to study and work in urban territories (Freitas & Kitson, 2018).

The high vulnerability of the tourism sector to external events associated with the high vulnerability of peripheral territories' economies (Kline et al., 2014) leads to even higher challenges for people and companies that want to thrive in these territories. With the lockdown and cancellation of all cultural activities that attract visitors, these territories were even highly affected. Silva (2021) showed that the COVID-19 pandemic had had a significant adverse impact on most tourist accommodation products in Portuguese rural territories.

Nevertheless, peripheral territories' characteristics and qualities might also provide opportunities for place differentiation (Credit et al., 2018), some of those boosted by the COVID-19 pandemic. These territories can now lead to people's increasing interests and concerns with sustainability, the environment, and less crowded spaces. Also, those can lead to social shifts through investments in renewable energy and circular economy. Emerging technologies and the increasing digital connectivity may help address the lack of human capital and unlock entrepreneurship and innovation in these territories' traditional sectors (OECD, 2018).

In the case of the tourism sector, Ianioglo and Rissanen (2020) highlight six major global trends for the *post-pandemic* world: digitalisation; responsible and sustainable tourism; domestic tourism; thriving wellness; changed views on mobility; changed business travel. Thus, *post-pandemic* tourism strategies need to be innovative and take into account these opportunities and the new types of tourists that search for new experiences and search for sustainability, wellness, and contribute to the local community of the destination.

2 Remote Work, Digital Nomadism and the Half-Tourists

In 1983, Margrethe Olson highlighted that "office automation technology permits many office workers to be potential telecommuters in that their work can be performed remotely with computer and communications support" (Olson, 1983, p. 182). The author explores almost 30 years ago the remote-work options, characteristics and needs, pointing that "remote work will only become successful when the concept is institutionalised" (Olson, 1983, p. 187).

Remote work is a work pattern in which workers can perform most of their working time and tasks at home or in coworking spaces. In turn, in digital nomadism, remote or self-employed workers travel indefinitely between different locations while continuously fulfilling their work obligations. In this case, the digital nomad is considered an "emerging class of highly mobile professionals, whose work is location independent working while travelling on (semi)permanent basis and vice versa, forming a new mobile lifestyle" (Olga, 2020).

Despite being discussed for over 30 years, the interest in remote work grew slowly. There was never (until 2020) a major change in the way people, companies, researchers and policymakers see new work trends and patterns. Apart from niche groups of freelancers, entrepreneurs and some employees of the technological field, few people thought that the work they perform could be done at home, at coworking space in any country of the world. Nonetheless, the emerging trends of remote work and digital nomadism were not set aside, even before the COVID-19 pandemic. In 2019, the WYSE Travel Confederation estimated that the number of digital nomads in 2035 would be more than 1 billion people.

In the last year, this trend *boomed* when most people went to work at home during the COVID-19 lockdown. Mostly due to the pandemic, the number of people searching for a more flexible work environment or for a more healthier and less crowded environment to live in has rapidly increased. Recent studies highlight that more than one third can be performed entirely at home (Dingel & Neiman, 2020). Besides, several companies are going to partial or full remote policies. For example, Twitter and Facebook will allow many of their employees to work from home (or from wherever they want) permanently. Mark Zuckerberg announced that half of Facebook employees would be working remotely within the next decade. This growing interest of companies is led by the increasing confirmation that a more flexible workplace can increase productivity, improve talent acquisition (that can be everywhere globally), and reduce costs related to real estate, for example.

Therefore, remote work and digital nomadism are an opportunity that rural and peripheral territories have to embrace as a tool to attract visitors or future residents. With the increasing digitisation and decentralisation of jobs, those territories can offer that healthier and less crowded environment that people are searching for.

In the light of this new trend, a new group of people with different characteristics from typical tourists is rapidly emerging—the half-tourists. The half-tourist is "someone who travels to a different city or country and spends part of the time working remotely while they are there" (Cambridge Dictionary, 2021). Their needs and desires are not the same as typical tourists. They search for other infrastructures and services, their programs and processes are different, they interact with other people (often, the *typical* tourist does not even want to interact with anyone).

Thus, it is essential to understand how territories, particularly peripheral and rural territories, and their stakeholders can innovate their tourism strategies and reconvert their infrastructures and services to attract and serve this emerging target group.

3 Methodology

To gain a deeper understanding of the topic under discussion—new trends of working and travelling and impact on peripheral territories—an in-depth case study methodology was conducted (Yin, 2012). This method allows to understand complex issues and is not limited to other types of evidence or data which contributes to triangulate the data and better answering the research question. Although a single-case study is more simple to implement, multiple case studies provide more convincing and reliable data. It also allows the investigation of broader topics than single-case studies (Yin, 2012). Thus, two case studies were selected in order to compare two peripheral territories with similar strategies. The cases selected were the *Digital Nomads Madeira* (Madeira, Portugal) and the *Nomad City Gran Canaria* (Gran Canaria, Spain). They were chosen for two reasons. Firstly, because this book aims to focus on Portugal and Spain and, secondly, because both cases are peripheral territories and present similar conditions—both are islands with a high focus on tourism.

Two semi-structured interviews were conducted with the project managers of Digital Nomads Madeira (**DN**—consultant and project manager) and Nomad City Gran Canaria (**NC**—founder and project manager). The interviewees are also specialists in the topic of remote work and digital nomadism. The interview script was divided into four main blocks (the strategy; the challenges of implementing in a peripheral territory; the social and economic impact; the impact of COVID-19 pandemic) with 13 questions. The interviews were conducted through an online communication platform in February 2021. The data gathered was transcript and analysed. The interview data was enriched with secondary data from various sources, including reports, websites and articles about the projects. These were essential sources of information to understand the context and to triangulate with interview data.

3.1 Case 1: Digital Nomads Madeira

Madeira is a Portuguese island and is the largest and most populous of the Madeira Archipelago. As of 2019, the island had a population of 249,052 (population density of 310.9/km^2) that constitutes approximately 98% of the archipelago population.

Tourism is an important sector in the region's economy, contributing 20% to its GDP.

The project Digital Nomads Madeira was launched in February of 2021, being promoted by Madeira's Regional Government and by Startup Madeira. The first initiative was to create a pilot project in Ponta do Sol, a seaside village with 8593 residents, and to turn this village into the first "Digital Nomad Village". See more about the project here—https://digitalnomads.startupmadeira.eu/ (Digital Nomads Madeira, 2021).

In two months, the project attracted the interest of about 5000 people from all over the world who signed up to come to work in Ponta do Sol and Madeira. Due to COVID-19 restrictions and fully-booked housing, most of them were not yet able to come. By March 2021, there were around 100 people at the Digital Nomad Village and around 400 remote workers on all the island.

3.2 Case 2: Nomad City Gran Canaria

Gran Canaria is the second-most populous island of the Canary Islands, a Spanish archipelago in the Atlantic ocean. As of 2019, the island had a population of 851,231 (population density of 545.6/km^2) that constitutes approximately 40% of the archipelago population.

The economy is based primarily on tourism, which makes up 32% of the GDP. The Canaries receive about 12 million tourists per year. And in the case of the island of Gran Canaria 4,267,385 (2019). Almost one-third of the visitors were residents in the United Kingdom, which will be an added challenge since Brexit and complex travelling between UK and Europe.

The project Nomad City Gran Canaria "emerged as the brand that gave shelter to an international conference organised in Gran Canaria for the last five years. Last year that conference was renamed Repeople Conference, and Nomad City has remained an information platform about the destination Gran Canaria for teleworkers and digital nomads" (NC). The project was born in 2015 to create a coworking space and organise a conference to answer an emerging movement of remote workers on the island.

"Las Palmas was already a natural destination for remote workers, although six years ago it was still a tiny niche and what we did was try to accelerate this process." (NC). See more about the project—https://www.nomadcity.org/.

Nowadays, there are 19 coworking spaces, according to their website (Repeople, 2021). Some of them are part of the private organisation that runs the project, but most of them are from other private organisations.

More recently, the regional government announced an investment of €500,000 in an international campaign to lure 30,000 Spanish and international remote workers to the archipelago for long-term stays (Vega, 2020).

4 Findings and Discussion

4.1 *Motivation and Goals*

The promotion of digital nomadism, remote work and the enhancement of the tourist destination are common motivations and objectives in both initiatives. However, the initiatives also had different motivations and goals.

Regarding motivations, while the DN initiative aimed to position Madeira in a new tourist market, by attracting "digital nomads, business professionals and entrepreneurs from all over the world, to come and work from Madeira, for periods between one and six months, making the island known as one of the best places to work remotely, where they can enjoy a specially created infrastructure, from work spaces, activities and community", the NC initiative aimed at "accelerating this process", since Las Palma de Gran Canaria was already a natural destination for remote workers. It is also curious that one of the motivations of the DN initiative was the experience of the participation of one of its promoters in the NC initiative.

These different motivations also lead to different goals. On the one hand, in the DN initiative, its promoters clearly define short- and long-term goals. In the short term, they intend to "create a community of digital nomads and remote workers with the primary objective of boosting the economy", while in the long term, they want digital nomadism to be the "first step to fixing expatriates who are very interesting, new expats who pay local taxes, but it also has a local economic impact". Because of these goals, measurable results of the initiative's impact on the territory are made public.

On the other hand, the NC initiative has no clearly defined goals. It focuses on remote workers and digital nomads and the growth of that community. In addition, it aims to be an information portal that provides "additional value to the entire community" that decides to choose Gran Canaria as its destination.

4.2 *Main Challenges*

The challenges that both initiatives faced are closely linked to the effects of the COVID-19 pandemic. If in the NC initiative, since it is prior to the pandemic, the challenges are related to the lack of knowledge about digital nomadism, in the DN initiative, due to the pandemic's impulse to remote work, the main challenges are related to the capacity of the infrastructures of support for the initiative.

Before the pandemic, Nomad City Gran Canaria was a project with a lack of recognition "because no one understood what we were talking about or its potential" (NC). This was one of the biggest challenges to involve other stakeholders (public government, the tourism sector and the local community) in this strategy and ecosystem of remote work/digital nomadism in Gran Canaria. With the COVID-19 pandemic, the increasing recognition of new work patterns and opportunities raise all stakeholders' interest in attracting this new target group. Something that has changed since the COVID-19 pandemic, with the increasing recognition of new work patterns and opportunities and with other destinations are following similar strategies. There is more awareness in the companies, public institutions and media of this emerging trend (NC).

Another challenge is that there are new needs and challenges for a new target group (in this case, the half-tourists). Thus, local governments and the private sector cannot answer this new target group with the same strategy, services and infrastructure to serve the traditional tourist.

In Madeira, for example, in the two months before the initiative started, local partners, such as restaurants and accommodation, were involved, preparing them to welcome this new type of tourist.

However, due to the success of the initiative, since the early days, the DN initiative faced growth problems, because in small villages, when a similar strategy is successful, it cannot grow quickly, or at least it is more difficult: "in a village if you are successful, there is no way to grow, you cannot build 50 or 100 beds or a thousand beds because there is limited space".

4.3 Governance and Stakeholders

In the case of Digital Nomads Madeira, the project was not created by the local government, but by a remote work consultant (our interviewee) and "the project was basically sold, and the Regional Government of Madeira contracted this project and I am the consultant for the project" (DN). Also, Nomad City Gran Canaria "has been created by private initiative" with no direct support from the government, only in specific events or initiatives throughout the last years.

Despite different approaches, both cases consider the local government as an accelerator of the strategy and its implementation. NC points that "the private sector will not have the capacity or the means to run a major campaign, because it will not have a direct return on that investment to justify it" and DN comments that "a big advantage (of having the local government involved) is that the project can move much faster". On another side, both interviewees argue that any exclusive strategy by the local government will tend to fail since "it is very important that the private environment is prepared and involved in this issue because otherwise, the effect is negative" (NC).

Besides the importance of the local/regional government in supporting these initiatives, other stakeholders should be involved. The local tourism business (accommodation, restaurants, cafes, tourism animation services, etc.) must be involved in the strategy since they are essential to serve this target group well. For example, DN commented that most restaurants did not have a vegetarian menu, and he had to talk and prepare them before the arrival of the first digital nomads since almost 30% of those are vegetarian. These simple details are essential for the success of these projects and emphasise the importance of evolving all the local stakeholders.

In the opposite way, both cases highlight that there were few or no partners from major tourism companies that are still not very aware of this emerging trend and target group. Nevertheless, both believe that those "will certainly be important partners in the coming years as well" (NC). Also, the national governments should be aware of these emerging trends and groups, despite not being directly involved in these local/regional projects. Both interviewees highlight the importance of policies to facilitate the attraction of this new type of tourists, for example, by providing a specific visa that goes beyond the typical tourist 3-month visa.

Finally, and one of the most important stakeholders—the *local leaders*. Those are essential to orchestrate all the interactions among the different stakeholders and to dynamise the community of remote workers of the destination. DN affirms that part of his job is "to find and educate local leaders who are living here (. . .) actually we already have two people identified (. . .) one of these people is going to start his own coworking space, create his community, do his business and above all foster the community which is by far the most important thing".

Thus, DN concludes that "the partnership between private and public is essential and the only way for things to happen", reinforcing the need for public-private-community cooperation in defining innovative tourism strategies in these territories.

4.4 The Half-Tourists and the Community

Most tourism ecosystems' frameworks include communities—the "individuals with a mutually recognised interest in the resources of a destination" (Roxas et al., 2020)—as important stakeholders of tourism strategies and initiatives. In this case, the community goes beyond this definition, and it is bidimensional. It includes not only the local residents, many of the owners/employees of small tourism-related businesses or simply participants of cultural activities, but also the community of remote workers/digital nomads that live in that place and have similar goals and interests.

DN highlights that "the digital nomads do not travel to places, they travel to communities". The community, in its bidimensional perspective, is probably the most important success factor highlighted by both interviewees.

With regard to the community of local residents, factors such as level of education, language skills and integration strategies may be essential for the successful implementation of these tourist strategies. In addition, local businesses must adapt to a new tourism opportunity. If, on the one hand, there has to be a perception that new target groups may require the provision of different services, different business models will also be needed to meet their needs. In the case of tourist accommodation, for example, in addition to the common overnight accommodation or long-term rental, there is a need to create a monthly medium- and short-term rental for half-tourists, whose price is "between 30 and 40% above long term rents".

While, with regard to the community of remote workers or digital nomads, the existence of a community is, in itself, a factor of success, as it is perceived by other half-tourists as support, welcoming and sharing community. However, the promotion of cultural activities, as well as entertainment infrastructures, is essential to foster social interaction between members of the community and even with the local resident community.

4.5 Social and Economic Impact

The social and economic impacts of these initiatives on the territories are similar, although there are some differences due to the fact that the DN initiative takes place in a small village, while the NC initiative takes place in a large city.

Regarding social impact, the interviewees pointed out that the cultural integration and cultural exchanges within the community of half-tourists and between those and the local community has great value and great social impact. Also, the knowledge dissemination can lead to new ideas and projects that can impact the lives of tourists, the local community or the territory.

In terms of economic impact, these types of projects and strategies lead to the "creation of new business and attraction of subsidiaries of big companies" (NC) and to the "creation of more businesses associated with these half-tourist communities" (DN) that need specific infrastructures and services. Also, both interviewees highlight that the economic impact of the half-tourists is more local and longer than the short term and many times not the local impact of a typical tourist. DN points that "(...) this is a completely different kind of tourist who spends much more money locally on local businesses". The half-tourist books directly in local housing, local restaurants, local activities and does not book an all-included plan to a hotel, many times through a tourism agency with no local presences (as most tourists in these locations do). Also, the economic impact is longer given that the half-tourist stays more time in a place (one to six months) and at any time of the year, rather than one/two weeks vacations and in specific seasons of the year, as the typical tourist does.

Apart from the positive impact, few negative risks were highlighted by the interviewees. DN stated that "there is always a risk of becoming too famous", which can lead to uncontrolled economic development, the rise of housing prices, more construction in the village and gentrification. Nevertheless, both agreed that these risks are far from relevant in the short or medium term.

5 Conclusions

In the midst of the COVID-19 pandemic, there were also opportunities and new perspectives of how people live, work and travel. While businesses were forced to adapt and leave the traditional office culture, a new *era* of remote working arrived.

The interest in remote working and digital nomadism will continue to grow due to the several perceived benefits for companies, employees and territories.

As remote working grows, the spatial relationship between employee and employer will be revamped. The understanding that physical proximity is not needed to maintain high cooperation and productivity levels is leading to an increasing decentralisation of employment and offices to suburban or more peripheral areas or to fully remote policies.

Peripheral territories, and particularly those highly dependent on tourism, were deeply affected by the COVID-19 pandemic and consequent lockdown and closure of international travel. Nonetheless, those have now the opportunity *to give people what they want* by creating infrastructures and services for this emerging group of half-tourists that is growing exponentially.

The cases analysed showed that this type of strategy and focus on this *new* target group can boost the recovery of peripheral territories after this pandemic crisis. However, several factors should be taken into account to create successful and innovative tourism strategies within this remote-work trend. The half-tourists do not have the same characteristics as the *typical* tourist. Those search for communities of people with the same (remote working) habits. Also, an increasing concern with wellness, sustainability and local impact are significant trends to take into account in post-pandemic tourism strategies.

Some of the factors needed in this new tourism strategy are the close cooperation between the local government, the private institutions and the community. One of those alone can develop a project/strategy but will face challenges, and it will take much longer to achieve results with impact. Also, local leaders are crucial to guarantee the dynamism of the ecosystem created around this community of *half-tourists*. Finally, it is crucial that the local businesses and community are prepared and engaged in this strategy, even before the *arrival* of the half-tourists, in order to avoid cultural shocks.

Future research should try to understand the "true" social and economic impact of this new target group in the territories. Existent frameworks (Roxas et al., 2020) in research should include these new trends and stakeholders of the hospitality and tourism ecosystem. Also, it is important to further clarify these strategies, their processes and important dimensions to help more and better the local policymakers of peripheral territories.

References

Cambridge Dictionary. (2021). Digital nomad definition. Retrieved March 15, 2021, from https://dictionaryblog.cambridge.org/2021/01/18/new-words-18-january-2021/

Credit, K., Mack, E. A., & Mayer, H. (2018). State of the field: Data and metrics for geographic analyses of entrepreneurial ecosystems. *Geography Compass, 12*(1), e12380. https://doi.org/10.1111/gec3.12380

Digital Nomads Madeira. (2021). Digital Nomads Madeira - HomePage. Retrieved March 15, 2021, from https://digitalnomads.startupmadeira.eu/

Dingel, J. I., & Neiman, B. (2020). How many jobs can be done at home? *Journal of Public Economics, 189*, 104235. https://doi.org/10.1016/j.jpubeco.2020.104235

Freitas, C., & Kitson, M. (2018). Perceptions of entrepreneurial ecosystems in remote islands and core regions. *Island Studies Journal, 13*(1), 267–284. https://doi.org/10.24043/isj.44

Gössling, S., Scott, D., & Hall, C. M. (2020). Pandemics, tourism and global change: A rapid assessment of COVID-19. *Journal of Sustainable Tourism, 29*, 1–20. https://doi.org/10.1080/09669582.2020.1758708

Hall, C. M., Scott, D., & Gössling, S. (2020). Pandemics, transformations and tourism: Be careful what you wish for. *Tourism Geographies, 22*(3), 577–598. https://doi.org/10.1080/14616688.2020.1759131

Ianioglo, A., & Rissanen, M. (2020). Global trends and tourism development in peripheral areas. *Scandinavian Journal of Hospitality and Tourism, 20*(5), 520–539. https://doi.org/10.1080/15022250.2020.1848620

Kline, C., Hao, H., Alderman, D., Kleckley, J. W., & Gray, S. (2014). A spatial analysis of tourism, entrepreneurship and the entrepreneurial ecosystem in North Carolina, USA. *Tourism Planning and Development, 11*(3), 305–316. https://doi.org/10.1080/21568316.2014.890127

OECD. (2018). Rural 3.0 - A framework for rural development.

Olga, H. (2020). In search of a digital nomad: Defining the phenomenon. *Information Technology and Tourism, 22*(3), 335–353. https://doi.org/10.1007/s40558-020-00177-z

Olson, M. H. (1983). Remote office work: Changing work patterns in space and time. *Communications of the ACM, 26*(3), 182–187.

Repeople. (2021). Repeople - HomePage. Retrieved March 15, 2021, from https://www.repeople.co/en/

Roxas, F. M. Y., Rivera, J. P. R., & Gutierrez, E. L. M. (2020). Mapping stakeholders' roles in governing sustainable tourism destinations. *Journal of Hospitality and Tourism Management, 45*, 387–398. https://doi.org/10.1016/j.jhtm.2020.09.005

Segal, S. (2020). The global impacts of COVID-19. Retrieved from https://www.csis.org/analysis/global-economic-impacts-covid-19

Silva, L. (2021). The impact of the COVID-19 pandemic on rural tourism: A case study from Portugal. *Anatolia*. https://doi.org/10.1080/13032917.2021.1875015

UNWTO. (2020). *UNWTO world tourism barometer*. World Tourism Organization.

Vega, G. (2020). Not just a tourist destination: Why Spain's Canary Islands are hoping to attract 30,000 remote workers. *El Pais*. Retrieved from https://english.elpais.com/economy_and_business/2020-11-10/not-just-a-tourist-destination-why-spains-canary-islands-are-hoping-to-attract-30000-remote-workers.html

World Economic Forum. (2020). COVID-19: What you need to know about the coronavirus pandemic on 4 April.

Yin, R. K. (2012). *Applications of case study research* (3rd ed.). Sage Publications.

Determinants of Profitability in the Tourism Sector in Portugal

Carlos Manuel Romão Travanca, Carlos Manuel Rodrigues Vieira, and Elisabete Gomes Santana Félix

Abstract This paper examines the impact on the profitability of firms in the Portuguese tourism sector of both internal (size, age, growth, debt) and external (real GDP growth rate, sovereign debt crisis, location) determinants. Panel data models have been employed on a large sample of small- and medium-sized firms, from 2009 to 2017, a period critically influenced by both the financial and the ensuing sovereign debt crisis in the southern European countries. Alternative profitability measures have been used and the results have been largely robust to the distinct models' specifications. The results suggested that profitability is positively affected by firm's size and age and by economic real growth rates, and negatively influenced by the firm's indebtedness ratio and the sovereign debt crisis. Firms in the main tourism regions are generally more profitable, and relatively less affected by the sovereign debt crisis, that was particularly harmful to micro-size firms.

1 Introduction

Portugal is currently one of the major tourism destinations in the world. The country's tourism industry has considerably developed since the 1990s, in a conscious internal effort to develop this strategic sector, partly aided by political instability in some more traditional touristic sites and also by numerous international awards that contributed to an increased visibility and attractiveness. The World Tourism Organization, a specialized agency of the United Nations, with

C. M. R. Travanca
Department of Management, University of Évora, Évora, Portugal

C. M. R. Vieira
Department of Economics, University of Évora and CEFAGE-UÉ, Évora, Portugal
e-mail: cvieira@uevora.pt

E. G. S. Félix (✉)
Department of Management, University of Évora and CEFAGE-UÉ, Évora, Portugal
e-mail: efelix@uevora.pt

J. Leitão et al. (eds.), *Tourism Entrepreneurship in Portugal and Spain*, Tourism, Hospitality & Event Management, https://doi.org/10.1007/978-3-030-89232-6_10

headquarters in Madrid, places Portugal in the 17th position of the world ranking of countries by number of incoming tourists in 2018.

Owing to its particular nature and type of clients, this sector is strongly affected by seasonal, conjunctural, and sometimes more structural fluctuations. The peak of sales of tourist services' providers in most locations is strongly related to the holidays' period, traditionally chosen in the summer months, when the climate is better suited for outdoor activities and the schools are closed. The profitability and overall performance of this economic sector is highly sensitive to climatic changes, the business cycle, political instability, and many other unexpected and disruptive circumstances with a major impact, such as the recent COVID pandemic.

In view of the extreme vulnerability of this economic sector to, mostly uncertain and uncontrolled, external circumstances, and also its growing importance for economic growth and employment rates in the Portuguese economy, this paper's main objective is to identify the determinants of firm's profitability in the tourism industry, focusing on a large panel of small- and medium-sized enterprises (SMEs). SMEs represent a very large proportion of all businesses in this sector, both in terms of volume of sales and employment. Uncovering the determinants of these firms' profitability is obviously important for a more efficient management, but is also important for national and regional authorities, interested in strengthening Portugal's position as a leading tourist destination.

The next section surveys the relevant theoretical and empirical literature, intending to identify the dependent and the explanatory variables previously proposed as determinants of a company's profitability in this particular industry. The following section describes the empirical model to be estimated and the dependent and independent variables chosen, bearing in mind the literature review, and the peculiarities of the country and time period considered. Afterwards we present and discuss the results obtained with the estimation of the empirical model. The final section reviews the main conclusions, limitations, and avenues for further research.

2 Literature Review and Hypotheses

In this section we will present the literature review associated with the hypotheses to be studied. This study considers a set of variables that can be aggregated according to: internal characteristics of the company; and, external constraints to the company.

Thus, internal characteristics of the company aspects such as size, age, growth of the company and the level of indebtedness of the company will be considered. The constraints external to the company that will be considered are GDP growth, the sovereign debt crisis, and the company's geographical location.

2.1 Internal Characteristics of the Company

2.1.1 Size

In the study of business profitability, it is important to take into account the company's size, mainly because large firms can take advantage of economies of scale in their activities, and this is therefore a differentiating factor compared to smaller companies. But, according to the Schumpeterian theory, it is the small and new enterprises (particularly in the area of technology) that are gradually growing and acquiring market share where large companies are already established (Papadogonas, 2007). This study analyzes tourism companies and, according to Dwyer et al. (2020), this is a sector in which economies of scale also seem to be present.

Moaveni (2014) shows us that a significant part of the financial performance of Turkish tourism companies, between 1998 and 2011, is affected by internal determinants, with size presenting a positive impact on profitability. More recently, Aissa and Goaied (2016) have drawn similar conclusions, referring that the size of a hotel has important implications for its profitability.

In the literature on hotel profitability, many studies have shown the positive relationship between size and profitability, such as Babayan et al. (2014), for hotels located in the Region of Alicante, and Dimitrić et al. (2019), for Spanish and Portuguese hotels. Even considering the impact that a financial crisis (considering pre, during, and post-crisis periods) may have on hotel profitability, Agiomirgianakis et al. (2012, 2013) and Magoutas et al. (2016) concluded that larger companies in the Greek tourism sector were more profitable, suggesting as justification the advantages that these companies have in price negotiation.

Based on the presented arguments, we formulated the following research hypothesis:

H1 There is a positive relationship between size and profitability in small- and medium-sized Portuguese hotels.

2.1.2 Age

The issue of age is important in the study of the economic performance of companies, including hotels. Wang et al. (2006) argued that the performance of the hotel increases with its age, to the extent that over time hotels gain organizational resources such as experience, reputation, and local and/or international brand. These resources positively influence strategic, operational, and innovation decisions.

Also, the research presented by Agiomirgianakis et al. (2012, 2013) and Magoutas et al. (2016), for the Greek tourism sector, show that as time goes by there is an impact of learning on organizations which becomes decisive for profitability.

The consideration of age allows to consider the life cycle of companies, in that in the first years of life of companies they aim to achieve a minimum of efficiency scale that allows them to survive in the markets in which they operate. In this way, it is normal that in the first years of life, companies present higher growth rates.

However, there is no consensus in the literature regarding the impact of the age of hotel profitability. Aissa and Goaied (2016), in the case of Tunisia, observed a negative relationship between the age and profitability of these hotels. Dimitrić et al. (2019) obtained different results depending on the country of the hotels, so that in Spain and Portugal, more recent hotels were more profitable, while for example for hotels in Croatia they could not obtain any statistically significant relationship.

Consequently, we formulate the following hypothesis:

H2 There is a positive relationship between age and profitability in small- and medium-sized Portuguese hotels.

2.1.3 Firm Growth

The company's pace of growth is also a variable with potential links to its profitability. On the one hand, growth shows the dynamism of the company, its willingness to take advantage of market opportunities to grow and increase its profitability. On the other hand, however, higher growth rates may also indicate that the company is still in the initial phase of the business, with higher investment requirements and therefore with higher costs. As demonstrated by Serrasqueiro and Nunes (2016), the issue of business growth has distinctive particularities in the case of the tourism sector and, within this, when we consider smaller companies, as is the case of the companies that are part of the sample used in the present study. It is a sector that is characterized by being very capital intensive, where the will for growth may require large investments.

However, there is little empirical evidence regarding the impact of a company's growth on its profitability. Note the work of Sandvik et al. (2014), who study the relationship between hotels' innovative actions and their profitability. Throughout the study the authors claim that sales growth has a positive impact on the business profitability.

Based on the arguments presented, we formulated the following research hypothesis:

H3 There is a positive relationship between hotel growth and profitability in small- and medium-sized Portuguese hotels.

2.1.4 Debt

In a simple way, it is possible to define the financing policy as being the set of decisions that aim to obtain the necessary funds for the growth and management of the company. The more financial resources companies have, the better their operation and results. The fewer resources, the greater the difficulties in achieving the goals. Management seeks maximum profitability of the resources needed to achieve the objectives.

According to Tsai et al. (2011), hotels prefer to fund their needs through debt, which has a lower cost than external equity. However, Gai et al. (2017) claim that an increase in leverage and, consequently, in debt levels, can have a positive impact (multiplier effect) on business profitability. But an increase in debt, particularly of a financial nature, carries a greater risk to the business, given its greater dependence on banks.

Leverage is negatively related to profitability in the tourism sector. High interest rates (and therefore costs), due to the company's increased risk, associated with high levels of indebtedness probably result in lower profits (Agiomirgianakis et al., 2012, 2013; Magoutas et al., 2016). In addition, when a large part of the revenue is intended to cover interest payments, fewer funds are available for reinvestment, thereby reducing the company's growth opportunities.

Also, Aissa and Goaied (2016) claim that the level of indebtedness is an influential factor in profitability but has a negative impact on the company.

Thus, we have formulated the following research hypothesis:

H4 There is a negative relationship between leverage and profitability in small- and medium-sized Portuguese hotels.

2.2 External Constraints to the Company

2.2.1 Economic Growth

The real GDP change rate is an indicator of economic growth with an expected positive impact on profitability. In the event of a decrease in GDP growth, there will be a negative impact on the consumer's purchasing power, which may reduce demand for products, thereby reducing the company's profitability (Tandelilin, 2010).

Oh (2005) looked at the relationship between GDP and tourism growth in South Korea and concluded that economic expansion causes a short-term increase in the tourism sector. Moaveni (2014) tested the effect of this variable on profitability obtaining a positive impact. According to this author, it is expected that as GDP increases over time, the number of tourist arrivals increases, which means that hotels can achieve higher profitability levels. As hotels are important tourism-related service industries, it can be said that "real GDP growth" has a positive impact on their profitability.

So, our hypothesis is:

H5 There is a positive relationship between economic growth and profitability in small- and medium-sized Portuguese hotels.

2.2.2 Sovereign Debt Crisis (2011–2014)

The sovereign debt crisis, that had significant impact in Portugal, predominantly from 2011 to 2014, began as part, and a consequence, of the global financial crisis that began in 2007–2008, developing in the context of the eurozone public debt crisis, which mainly affected southern European countries and also Ireland.

The ensuing economic crisis can affect the tourism sector due chiefly to the reduction in disposable income. In this way, this reduction in income is expected to adversely affect traditional tourist destinations, especially in the luxury segment (Stabler et al., 2009).

Agiomirgianakis et al. (2013), when analyzing the impact of the economic crisis on Greek tourism companies, conclude that it had a negative impact on the performance (profitability) of those companies.

A similar result was obtained by Babayan et al. (2014), who conclude that in times of crisis, hotel companies do not seem to be able to cover their costs, namely financial costs. The authors even argue that, as a general rule, all hotels in a crisis situation obtain relevant losses as a result of their financial results.

Based on the arguments presented, we formulated the following research hypothesis:

H6 In times of sovereign debt crisis, the profitability of small- and medium-sized Portuguese hotels is lower.

2.2.3 Location

The ideal location for a hotel establishment presupposes certain levels of occupancy, seasonality, and competitive intensity, among others. All these aspects directly impact the profitability of the establishments. In addition, decisions about the location of the business are one of its main success factors (Lundberg et al., 1995), because in addition to establishing the main forces that act in the profitability of the hotel, these are decisions that cannot be modified in the short term, and therefore have a marked strategic nature (Martínez-López & Vargas-Sánchez, 2013).

Location is a particularly relevant issue in the case of the sector under review. Lado-Sestayo et al. (2016) analyzed the effect of location through the market structure on the profitability of the Spanish hotel sector, concluding that profitability depends largely on the structure of the market.

In the case of this study, focused on Portugal, the Algarve region proves to be the main destination of tourists, both national and international. A mirror of this is the fact that in 2019 it was, for the second consecutive year, the NUTS II region with the highest GDP growth, with more than 50% of SMEs operating in the tourism sector. Thus, it is extremely important to study the particularities of the profitability of companies in this region.

Thus, our research hypothesis with respect to location is as follows:

H7 There is a positive relationship between the location of hotels in the Algarve and the profitability of small- and medium-sized Portuguese hotels.

3 Methodology

3.1 Sample

The data used in this study were obtained through the *Amadeus* database as supplied by Bureau van Dijk. The companies comprising the sample meet the SME conditions set out in the Commission Recommendation 2003/361/EC of 6 May 2003. These are companies with: (1) a number of employees below 250 and (2) turnover less than or equal to EUR 50 million or total annual balance sheet less than or equal to EUR 43 million. Companies belonging to the category hotels and similar accommodation were also elected, which corresponds to code 551 of CAE—Rev. 2 (*Portuguese Classification of Economic Activities*).

This study is therefore based on financial accounting data from a final sample composed of 2508 SMEs from the tourism sector of Portugal, for the period 2009 to 2017, and it is a non-balanced sample.

3.2 Variables

Profitability ratios relate the company's profits to its assets. Although there are multiple profitability ratios with different variations and specific interpretations, the principle is always the same: compare the profit that the company was able to generate in a given period with data relating to the size of the company (be it the amount invested, the value of the asset, or the net value of the company). Thus, as dependent variables we used the ratios: Return on Investment; Return on Equity; and Return on Assets.

Based on the results of the relevant literature, the variables presented in Table 1 were thus defined and computed.

3.3 Estimation Method

In this study, the estimation models with panel data are adopted. These have numerous advantages over sectional data or time data models, combining them and leading to more accurate statistics and increased power of statistical tests. They allow to perceive the relationship between variables under study over time and the level of individual heterogeneity. Thus, the panel data provide a greater

Table 1 Definition of variables

Variables	Description	Acronym	
Dependent variables			
Return on investment (%)	$\frac{\text{Net Income}}{\text{Total Assets}} \times 100$	ROI	
Return on equity (%)	$\frac{\text{Net Income}}{\text{Equity}} \times 100$	ROE	
Return on assets (%)	$\frac{\text{EBIT}}{\text{Total Assets}} \times 100$	ROA	
Independent variables			Expected effect
Size (log)	Turnover logarithm	size	+
Age (log)	Logarithm of: Data year—year of constitution	age	+
Firm growth (%)	$\frac{\text{Total Assets}_t - \text{Total Assets}_{t-1}}{\text{Total Assets}_{t-1}} \times 100$	firmgrowth	+
Leverage ratio	$\frac{\text{Total Liabilities}}{\text{Total Assets}}$	debt	–
Economic growth (%)	$\frac{\text{GDP at constant prices}_t - \text{GDP at constant prices}_{t-1}}{\text{GDP at constant prices}_{t-1}} \times 100$	GDPgrowth	+
Dummy sovereign crisis	*Dummy* variable, with 1 from 2011 to 2014 and 0 for the remaining years	dsdcrisis	–
Dummy location (Algarve region)	*Dummy* variable, with 1 in Algarve region and 0 in the remaining regions	dalgarve	+
Dummy location (Lisbon region)	*Dummy* variable, with 1 in Lisboa region and 0 in the remaining regions	dlisboa	+
Dummy location (Porto region)	*Dummy* variable, with 1 in Porto region and 0 in the remaining regions	dporto	+
Dummy micro firms	*Dummy* variable, with 1 for micro firms and 0 in the remaining dimensions	dmicro	+
Dummy small firms	Dummy variable, with 1 for small firms and 0 in the remaining dimensions	dsmall	+

amount of information, greater variability of the data, lower linearity between the variables, greater number of degrees of freedom, and greater efficiency in estimation.

The main equation is:

$$Y = f\ (\text{size}, \text{age}, \text{firmgrowth}, \text{debt}, \text{GDP growth}, \text{dsdcrisis}, \text{local}, \text{dmicro}) \tag{1}$$

being Y given by ROI, ROE, and ROA. Thus, the three equations under study will be (for firm i in year t, with μ being the error term):

$$\begin{aligned} \text{ROI}_{it} = {} & \beta 0_{it} + \beta 1\ \text{size}_{it} + \beta 2\ \text{age}_{it} \\ & + \beta 3\ \text{firmgrowth}_{it} + \beta 4\ \text{debt}_{it} \\ & + \beta 5\ \text{GDPgrowth}_{it} + \beta 6\ \text{dsdcrisis}_{it} \\ & + \beta 7\ \text{local}_{it} + \beta 8\ \text{dmicro}_{it} + \mu_{it} \end{aligned} \tag{2}$$

$$\begin{aligned} \text{ROE}_{it} = {} & \beta 0_{it} + \beta 1\ \text{size}_{it} + \beta 2\ \text{age}_{it} \\ & + \beta 3\ \text{firmgrowth}_{it} + \beta 4\ \text{debt}_{it} \\ & + \beta 5\ \text{GDPgrowth}_{it} \\ & + \beta 6\ \text{dsdcrisis}_{it} + \beta 7\ \text{local}_{it} \\ & + \beta 8\ \text{dmicro}_{it} + \mu_{it} \end{aligned} \tag{3}$$

$$\begin{aligned} \text{ROA}_{it} = {} & \beta 0_{it} + \beta 1\ \text{size}_{it} + \beta 2\ \text{age}_{it} \\ & + \beta 3\ \text{firmgrowth}_{it} + \beta 4\ \text{debt}_{it} \\ & + \beta 5\ \text{GDPgrowth}_{it} \\ & + \beta 6\ \text{dsdcrisis}_{it} + \beta 7\ \text{local}_{it} \\ & + \beta 8\ \text{dmicro}_{it} + \mu_{it} \end{aligned} \tag{4}$$

Additionally, the models have also been estimated including alternative interaction terms between some of the variables, that may help interpret the obtained results and attain a better fit.

The chosen estimation method is validated through several tests: R-Square; *White* and Modified *Wald* (tests to heteroscedasticity); *Hausman* (fixed and random effects test); Test F; and Test F for fixed effects.

4 Empirical Results and Discussion

The appendix includes Tables 2 and 3 with, respectively, descriptive statistics and the correlation matrix of the variables employed, and Tables 4, 5, and 6, which present the results obtained for the various estimated models.

Let us start by analyzing the results of Table 2, with regard to dependent variables, which reflect the profitability of hotels measured in three different ways. As we may observe, ROI, ROE, and ROA present negative average values (3.71%, 5.40%, and 1.27%, respectively), thus indicating that, on average, the activities carried out by the companies generated losses in the period under analysis. Note that our sample comprises the period when economic activity was negatively affected by the global financial crisis and the European sovereign debt crisis.

As regards the hotels included in the sample under study, it is noted that: their average size, measured by turnover, is EUR 36.45 thousand; have an average age of 12.95 years; grew on average by about 0.84% (again, note that this low growth rate may be explained by the financial and sovereign debt crisis in Portugal); the average leverage ratio is 1.44, a high value but that may be due to the fact that hotels need a high initial investment to start their business. It is also observed that 73% of the hotels are microenterprises, 23% are small and 4% medium sized. With regard to the location of the hotels, 15% are located in the Algarve region, 27% in the greater Lisbon area, 7% in the Porto region, and 51% are located in the rest of the country.

The average change in GDP for the period under review is very close to zero. This scenario is mainly due to the figures recorded in 2009, 2011, 2012, and 2013, where there was a negative variation in GDP caused by the effects of the financial crisis.

Analyzing Table 3, which presents the correlation coefficients between the variables used in this study, we can observe that the correlations between the independent variables are less than 30%. Therefore, according to Gujarati and Porter (2010), the problem of multicollinearity between these variables is not relevant.

Next, we present and discuss the results obtained with the models, which are presented in Tables 4, 5, and 6. Initially, the results of the base model are analyzed in more detail and then the impact of the inclusion of variables related to the sovereign debt crisis and the location of hotels. Afterwards, we examine the impact of introducing interaction terms, combining the sovereign debt crisis and financial leverage with the location and size of enterprises (micro-enterprises).

4.1 Base Model

Let us start by analyzing the base model (relative to the above eqs. 1, 2, 3, and 4) whose results are evidenced in Table 4, in appendix. The Hausman *test* showed statistical evidence that the choice between the two model specifications should favor the fixed effects, since the *P*-value associated with the test statistics of the three models (ROI, ROE, and ROA) is lower than the usual level of significance (5%). However, as observed in Table 4, the results do not differ qualitatively between the two models, except for the variable *size,* which is statistically significant with the random effects specification, but, conversely, is not statistically significant when the fixed effects specification is applied. Thus, for the purposes of analysis of the study, we selected the random effects, since the fixed effects model admits that the intersecting coefficient is different between the companies, allowing to detect differences that do not change with time and homogeneity in the slope coefficient. Although the fixed effects model allows capturing differences not observed between companies, the inclusion of *dummy* variables reduces the degrees of freedom, which can make the model estimates less efficient than in the random effects model.

All models present strong overall significance, since the F and Wald tests for joint significance of all covariates allow us to clearly reject the null hypothesis that all coefficients are equal to zero.

Analyzing the impact of each explanatory variable, we observe that the *size* variable is statistically significant at 1% in all models, except for the random effects model 3.1 where it is statistically significant at 5%, showing a positive impact on all models. Thus, we can confirm hypothesis 1 in which we suggested a positive relationship between *size* and *profitability* in small- and medium-sized Portuguese hotels. If we analyze the impacts of this same variable in Tables 5 and 6, we observe the same results, which indicates that the result is consistent. This result confirms those of authors as Agiomirgianakis et al. (2012, 2013), Moaveni (2014), Aissa and Goaied (2016), Magoutas et al. (2016), Dimitropoulos (2018), and Dimitrić et al. (2019). Our results thus show that the larger the size of the hotel, the higher the profitability it achieves, which seems to be consistent with the impact of scale economies ensuring a lower average cost (even in the case of tourism companies).

The result is also an indicator that larger companies can better control resources, but also have an easier and cheaper access to funding, which allows them to achieve higher levels of profitability.

The variable *age* has a positive and statistically significant impact at 1% in all models of Tables 4, 5, and 6 (with the exception of random effects models 3.3, 4.6, 5.1, and 5.3, in which it is statistically significant at 5%). It is therefore possible to confirm our hypothesis 2, there is a positive relationship between *age* and profitability in small- and medium-sized Portuguese hotels. These results suggest that older companies can achieve greater profitability, thus confirming previous results by Agiomirgianakis et al. (2012, 2013), Aissa and Goaied (2016), Magoutas et al. (2016, 2018). This result reflects the impact of accumulated "learning by doing" or "incumbent" effect, indicating that experience and reputation is an advantage for the portuguese hotels. However, when we consider ROE as the dependent variable, we observe that the impact of *age* on profitability is negative and statistically significant (between 1% and 5%). In fact, Dimitrić et al. (2019) in a recent paper on the determinants of profitability in some Mediterranean countries (Croatia, Greece, Spain, and Portugal) for the period 2007 to 2015, found a similar result for the case of Spain and Portugal (although considering ROA as the measure of profitability). These authors present as a possible justification the fact that the newest hotels in these countries are more modern, prone to new technologies and services and can more easily adapt to customers. Indeed, in the recent past there has been a high number of renovations of existing hotels, as well as the emergence of a large number of hotels. According to data from PORDATA, from 2012 to 2013 there was a rate of variation of 10.1% of beds in total tourist accommodations (such high peaks had only occurred in 1973, 1979, and 1986). For Spain, this peak was observed in 2003, in the order of 10.3% compared to 2002, and later in 2005 and 2011 but of a smaller relative dimension (annual growth rates of 3% and 2.7%, respectivley) (PORDATA, 2020).

In what the *growth of the company* is concerned, we have not found any statistically significant relationship with profitability, so we do not confirm the third hypothesis.

With regard to the impact of the amount of *debt* on profitability we found, in all models, an expected negative impact with a statistical significance of 1%, so our fourth hypothesis is confirmed. Our result is in line with the literature and, in particular, the specific literature on the tourism sector, for example Agiomirgianakis et al. (2012, 2013), Murugesu (2013), Dimitropoulos (2018) and Magoutas et al. (2018). Highly leverage firms are required to pay higher interest amounts, which will increase their expenses and lower their profits. They also have to show higher returns to potencial investors because of the amount of leverage they have. On the other hand, higher leverage leads to higher levels of risk, leading to greater difficulties on the part of these companies in obtaining financing. This ends up having an adverse effect on the company's future investment opportunities which in turn can lead to a negative impact on its long-term operating performance and solvency (Dimitropoulos, 2018; Singh & Faircloth, 2005; Shin-Ping & Tsung-Hsien, 2009).

In terms of external determinants, it is observed that the impact of economic growth on hotel profitability is always positive and statistically significant at 1% (with the exception of models 4.1, 4.2, and 5.1, in which it is equal to 5%), allowing to confirm hypothesis 5. In periods of higher economic growth, there is an increase in consumer purchasing power, which may be reflected in the increased demand for hotel stays (whether national or international). Indeed, tourism is a key industry worldwide, having spread rapidly in recent decades, it is an activity with increasing importance in economic development and vice versa. These conclusions are in line with those of Moaveni (2014).

Finally, two dummies were included that intended to analyze the impact of the fact that the hotels are micro and small. As we may observe, these do not have statistical significance, which seems to indicate that being classified in a specific dimension does not favor the profitability of the Portuguese hotels in our sample.

4.2 Sovereign Debt Crisis and Location

Table 5 presents the results obtained when incorporating some other variables in the base model described above. Note that the results for the variables from the base model remain qualitatively similar, both in terms of sign and statistical significance of the coefficients. The country's real GDP growth ratio and the firm's dimension and age (in this case, except when ROE is the dependent variable) positively affect profitability. The firm's leverage ratio has a negative effect on profits. Notably, these results are robust across all model specifications and with distinct profitability ratios.

The effect of the *sovereign debt crisis* is clearly negative, as expected, and strongly significant in statistical terms, whatever the dependent variable considered, thus confirming our initial research hypotheses (H6). The sovereign debt crisis, a by-product of the financial crisis had a large impact in Portugal, extending the economic recession and forcing harsh austerity policies during a large part of the sample period considered in this study. The result was a sharp fall in disposable income and a large increase in the unemployment rate and in business bankruptcies. Overall demand fell significantly, particularly for non-essential goods and services, such as tourism.

This sovereign debt crisis had also a strong negative impact on the economies of other European countries, thereby diminishing external demand for tourism services in Portugal, particularly from Spain, one of the major sources of tourists in Portugal. Furthermore, tourists from Spain are typically not concentrated in the summer months, alleviating the seasonal characteristics of this economic sector.

These results confirm previous findings by Agiomirgianakis et al. (2013) and by Babayan et al. (2014), who have also uncovered a statistically significant negative effect of the financial crisis on firm's profitability in the tourism sector. Both studies have also focused on Southern European countries, the most negatively affected. While the former study the Greek tourism sector, the latter examine the impact on hotels in Spain.

Table 5 also reports the results of estimating the models with dummy variables for the three major *tourism regions* in the country: the Algarve, the capital, Lisbon, and the second largest city, Oporto. Although the results are not uniform across all estimated models, location in one of these three regions positively affects profitability relatively to firms located elsewhere. The exception is when the dependent variable is return on investment (ROI), in which case only firms in the Algarve show a positive effect, albeit with a low statistical significance. These results confirm our research hypothesis H7 and may possibly be explained by higher occupancy rates across the year and the advantages of possible network effects in the major tourism centers.

These results compare with those obtained by Lado Sestayo et al. (2016), who have studied the effects of location on profitability in a large sample of Spanish hotels, from 2005 until 2011, thereby including the effects of the financial crisis but not yet the combined impact of the following sovereign debt crisis. Employing dynamic panel data models, the authors conclude that profitability depends largely on the market structure and the level of demand of the tourist destination.

This model presented in Table 5, with dummy variables for the sovereign debt crisis and for the main tourist destinations in Portugal, has been further extended to allow for possible interaction effects between some of the variables (see Table 6). Note that the previous results are largely robust to this extended model. Profitability in this economic sector is conditioned by a number of internal firm characteristics, such as its *size*, *age*, and *leverage ratio*, but also by external determinants, like the *real growth rate of GDP*, *location*, and the *sovereign debt crisis*.

As expected, the results from these interaction terms are less uniform across the different models presented in Table 6, both in terms of sign and statistical significance. Overall, there is a suggestion, with a weak statistical significance, that hotels' profitability in the Algarve region has been less severely affected by the debt crisis, possibly because a large part of the hotels' clients come from northern European countries, less affected by this crisis.

A more convincing result is the particularly harsh effect of the sovereign debt crisis on the smaller firms. Businesses in the tourism sector ranked as micro firms, a very large part (73%) of the sample employed in this paper, have experienced a particularly negative effect of the crisis on profitability. This result corroborates the conclusions from a paper by Ferreira and Saridakis (2017), focused on Portuguese firms' survival rates across the business cycle and in particular during to the two recent financial and sovereign debt crises. The authors found that smaller firms were more likely to shut down than larger firms and were more affected by the sovereign than by the financial crisis.

Location and firm size have also been interacted with leverage, trying to uncover a significant relationship between these determinants. The results for the interaction term between leverage and location in the Algarve region are mixed, not allowing convincing conclusions. The results for the interaction term between leverage and the firms' rank as micro companies are more robust across the distinct dependent proxies of profitability. The statistically significant negative effect of a firm's leverage ratio on profitability, consistently found in all estimated models, is lower for micro firms. Given its very small size these firms may be more credit constrained, and therefore less affected when credit conditions, and the economic cycle, unexpectedly worsen.

5 Conclusion

The main goal of this paper was to identify the factors determining the profitability of Portuguese companies in the tourism sector, including internal variables (size, age, firm growth, leverage, and a dummy variable categorizing as a micro or small firm, since these represent, respectively, 73% and 23% of the companies in the sample) and also some relevant external determinants (economic growth, a dummy variable for the financial and sovereign debt crises that caused a major disruption in the Portuguese economy and geographical location).

Our results confirmed hypotheses 1, 2, 4, 5, 6, and 7, confirming, respectively, that the profitability of companies in the Portuguese tourism sector is positively influenced by the size and age of the company, negatively influenced by the amount of debt, positively influenced by GDP growth, has suffered a negative impact with the sovereign debt crisis and, finally, is positively influenced by being located in the Algarve region.

With regard to the variables that were used to measure the profitability of companies, ROI, ROE, and ROA, all three demonstrated to fit well with a strong overall significance, and the ROA seems to be the one with the best explanatory power, with an R-square of around 30%, a value very close to the coefficients of determination obtained by Dimitropoulos (2018). We believe that it was important to use these three profitability measures because they measure the relationship between results and alternative dimensions of the firm, particularly in the case of ROI and ROA, with the former taking into account the weight of financial expenses in the company.

This study reveals that a significant part of the financial performance of tourism companies is affected by internal determinants. It is demonstrated that the profitability of these companies increases with their size and age, evoking the impact of possible economies of scale as well as the increase in the level of reputation and experience that occurs with age. As entirely predictable, the impact of the level of leverage in hotels is negative. This type of companies requires high levels of financing (in particular when entering the market), but funding has high expenses and increases the level of risk, which ultimately has a negative effect on access to new financing, with adverse effects on opportunities for future growth.

GDP real growth rates exhibit the largest coefficient in the model, displaying a strong positive and highly statistically significant effect on these firms' profitability, suggesting that in periods of stronger economic expansion, consumers affect a larger share of their income to the demand for hotels' services. This result shows that economic growth is of high importance for the countries' tourism activities and vice versa.

When analyzing the impact of the sovereign debt crisis, we observe a significant negative impact on the profitability of companies in the Portuguese tourism sector. It was also observed that the impact of this crisis is significantly enhanced for microenterprises, but lower in companies located in the Algarve.

The present study has implications at various levels. At the level of the literature on this research area, its impact is threefold: it studies the influence of various determinants of profitability on alternative measures, as ROI, ROE, and ROA, in companies in the tourism sector of Portugal; studies the influence of the sovereign debt crisis that deeply affected the Portuguese economy at the first half of the decade; and it studies the impact on profitability of different geographical locations.

At the level of the companies in this particular economic sector, it sets out the main external and internal determinants of their profitability. It warns these companies for the need of a balanced and efficient management of their external financing, given the strong negative impact of financing costs on profitability. On the other hand, it draws attention to companies for taking advantage of periods of economic expansion, usually characterized by a greater influx of tourists. In these periods, these companies should focus on a good management of the internal capital they generate, for amortization of their debts of external capital, and to implement investments that would allow them to take advantage of opportunities for future sustained growth.

Finally, the results obtained with the variables of leverage, sovereign debt crisis and dimension, should serve as a warning to the need for government policies that presenting more appropriate forms of financing for companies in this sector. It also suggests the need for the development of public policies that address the impact that possible crises have on profitability of the tourism sector businesses in the face of the weight that this type of sector already represents in GDP. In Portugal, the sectors of trade and repair of vehicles, accommodation and catering accounted for 19.4% of Gross Domestic Product in 2018 (PORDATA, 2021).

This study has a limitation related to the specific nature of small and medium hotels in Portugal, that is, many companies do not present correct or complete data, thus limiting the size of the sample to be collected. In the future, as this limitation relaxes, it would be interesting to include the use of more comprehensive and recent data, along with additional internal variables such as marketing expenses and/or innovative activities, as well as the weight of fixed costs. It would also be interesting to extend the study of profitability to other countries, at the European Union level, for example, considering the difference between hotels located in Nordic countries and those located in the Mediterranean countries, so that international analyses can be made. Finally, while we are still living in a pandemic situation, it would be important (after extending the time series) to analyze the effect it had on the profitability of hotels in Portugal, but also in other countries, particularly those where this industry represents a large fraction of the economy.

Acknowledgments This paper is financed by National Funds of the FCT—Portuguese Foundation for Science and Technology within the project "UIDB/04007/2020".

Appendices

Table 2 Descriptive statistics of the variables under study

Variables	Observations	Mean	Standard deviation	Minimum	Maximum
ROI	19,512	−3.71	198.27	−6675	25,559
ROE	14,948	−5.40	80.77	−997.81	881.87
ROA	18,737	−1.27	16.38	−99.60	98.49
Size	9151	36.45	10.58	−3	3
Age	20,342	12.95	2.62	0	2.03
Firmgrowth	19,274	0.84	20.31	−1	1493.42
Debt	12,227	1.44	3.30	−0.05	98.07
GDPgrowth	22,410	0.00	0.02	−0.04	0.03
Dsdcrisis	22,410	0.44	0.50	0	1
Dalgarve	22,410	0.15	0.36	0	1
Dlisboa	22,410	0.27	0.44	0	1
Dporto	22,410	0.07	0.25	0	1
Dmicro	22,410	0.73	0.44	0	1
Dsmall	22,410	0.23	0.42	0	1

Table 3 Correlation matrix

Variable	ROI	ROE	ROA	Size	Age	Firmgrowth	Debt	GDPgrowth
ROI	1							
ROE	0.5180	1						
ROA	0.9995	0.5179	1					
Size	0.1627	0.0392	0.1626	1				
Age	−0.0212	−0.0001	−0.0212	0.0589	1			
Firmgrowth	0.0243	−0.0067	0.0238	−0.0246	−0.1087	1		
Debt	−0.0932	−0.1052	−0.0930	−0.1189	−0.2431	0.0458	1	
GDPgrowth	0.2076	0.1229	0.2078	0.0778	0.0187	0.0128	−0.0239	1

Table 4 Regressions of the base model of profitability determinants

	Dependent variables					
	ROI		ROE		ROA	
	3.1		3.2		3.3	
Independent variables	FE	RE	FE	RE	FE	RE
Internal determinants:						
Size	0.302	1.215**	12.064***	7.562***	3.22***	2.194***
	(1.025)	(0.565)	(2.94)	(1.548)	(0.395)	(0.245)
Age	21.205***	4.325***	−24.09**	−12.78***	10.317***	1.795**
	(3.857)	(1.595)	(11.65)	(4.53)	(1.48)	(0.7086)
Firmgrowth	0.061	0.059	−0.244	−0.114	0.017	0.023
	(0.059)	(0.05)	(0.15)	(0.125)	(0.022)	(0.0197)
Debt	−8.796***	−7.085***	−20.195***	−15.84***	−2.669***	−3.283***
	(0.327)	(0.211)	(3.25)	(2.425)	(0.187)	(0.144)
External determinants:						
GDPgrowth	65.827***	96.478***	283.575***	329.501***	53.302***	78.063***
	(16.87)	(14.69)	(47.357)	(40.507)	(6.445)	(5.748)
Dmicro	2.601	0.252	−9.47	−2.705	0.595	0.56
	(3.2)	(2.048)	(8.955)	(5.55)	(1.217)	(0.872)
Dsmall	0.834	−0.605	−6.607	−6.692	−0.536	−0.357
	(2.705)	(1.912)	(7.481)	(5.138)	(1.027)	(0.799)
_cons	−19.49***	−1.409	25.92	13.183	−15.792***	−3.215**
	(5.402)	(2.867)	(16.58)	(8.513)	(2.075)	(1.255)
R-square	0.182	0.204	0.026	0.047	0.093	0.233
Wald		1205.44***		152.11***		849.27***
F test	111.88***		15.65***		73.31***	
F test (FE)	2.6***		2.31***		4.34***	
Hausman test	70.70***		29.60***		115.16***	
Observations	6440	6440	5218	5218	6363	6363
Firms	1327	1327	1153	1153	1319	1319

***$p < 0.01$; **$p < 0.05$; *$p < 0.1$; Standard errors in parentheses

Table 5 Regressions of the profitability determinants model (dummy sovereign debt crisis and dummy of companies located in the Algarve region)

	Dependent variables					
	ROI		ROE		ROA	
Independent variables	4.1	4.2	4.3	4.4	4.5	4.6
Internal determinants:						
Size	1.195**	1.311**	7.526***	7.997***	2.183***	2.225***
	(0.564)	(0.565)	(1.55)	(1.541)	−0.244	−0.243
Age	4.616***	4.389***	−12.07***	−13.931***	2.046***	1.758**
	(1.588)	(1.59)	(4.534)	(4.506)	−0.708	−0.704
Firmgrowth	0.06	0.06	−0.111	−0.12	0.025	0.024
	(0.05)	(0.05)	(0.125)	(0.125)	−0.02	−0.02
Debt	−7.083***	−7.07***	−15.518***	−15.683***	−3.255***	−3.275***
	(0.21)	(0.21)	(2.425)	(2.414)	−0.144	−0.143
External determinants:						
GDPgrowth	39.491**	39.889**	230.615***	233.844***	48.551***	49.35***
	(17.52)	(17.53)	(48.275)	(48.21)	−6.834	−6.83
Dsdcrisis	−4.696***	−4.686***	−8.217***	−8.136***	−2.411***	−2.4***
	(0.792)	(0.792)	(2.191)	(2.189)	−0.307	−0.307
Dalgarve		3.204*		23.159***		3.737***
		(1.903)		(5.082)		−0.877
Dlisboa		0.87		13.142***		2.011**
		(1.704)		(4.58)		−0.787
Dporto		−3.52		17.776**		4.626***
		(2.87)		(7.723)		−1.333
Dmicro	0.79	1.048	−1.693	0.362	0.877	1.167
	(2.046)	(2.059)	(5.554)	(5.541)	−0.871	−0.871
Dsmall	−0.33	−0.161	−6.188	−4.5799	−0.184	0.02
	(1.909)	(1.912)	(5.137)	(5.114)	−0.797	−0.795

(continued)

Table 5 (continued)

	Dependent variables					
	ROI		ROE		ROA	
Independent variables	4.1	4.2	4.3	4.4	4.5	4.6
_cons	−0.025	−0.611	15.001*	7.916	−2.693**	−3.85***
	(2.87)	(2.936)	(8.52)	(8.584)	−1.253	−1.276
R-square	0.206	0.21	0.047	0.071	0.232	0.252
Wald test	1245.97***	1253.44***	166.51***	194.29***	916.62***	954.63***
Observations	6440	6440	5218	5218	6363	6363
Firms	1327	1327	1153	1153	1319	1319

***$p < 0.01$; **$p < 0.05$; *$p < 0.1$; Standard errors in parentheses

Table 6 Regressions of the profitability determinants model (final model)

	Dependent variables		
	ROI	ROE	ROA
Independent variables	5.1	5.2	5.3
Internal determinants:			
Size	1.226**	8.243***	2.247***
	(0.56)	(1.535)	(0.243)
Age	3.56**	−15.566***	1.55**
	(1.576)	(4.495)	(0.703)
Firmgrowth	0.058	−0.124	0.022
	(0.049)	(0.124)	(0.019)
Debt	−9.55***	−37.107***	−3.557***
	(0.511)	(4.611)	(0.253)
External determinants:			
GDPgrow	38.722**	231.689***	49.049***
	(17.49)	´(48.098)	(6.8)
Dsdcrisis	−2.343**	−10.122***	−1.96***
	(1.006)	(2.7)	(0.386)
Dalgarve	2.731	0.932	7.819***
	(2.184)	(8.926)	(1.141)
Dlisboa	1.003	13.06***	2.021**
	(1.679)	(4.555)	(0.786)
Dporto	−2.694	17.356**	4.735***
	(2.828)	(7.681)	(1.333)
Dmicro	−0.303	−22.122***	0.664
	(2.208)	(7.073)	(0.951)
Dsmall	−0.77	−4.458	−0.256
	(1.898)	(5.103)	(0.794)
Dsdcrisis × Dalgarve	2.44	9.103*	1.122
	(1.88)	(5.108)	(0.722)
Dsdcrisis × Dmicro	−6.003***	1.865	−1.304**
	(1.345)	(3.777)	(0.521)
Debt × Dalgarve	−0.377	22.293***	−4.13***
	(0.630)	(8.641)	(0.622)
Debt × Dmicro	2.953***	26.336***	0.72**
	(0.55)	(5.141)	(0.294)
_cons	2.55	27.295***	−3.357**
	(2.98)	(9.2)	(1.297)
R-square	0.226	0.081	0.248
Wald test	1321.18***	231.54***	1022.19***
Observations	6440	5218	6363
Firms	1327	1153	1319

***$p < 0.01$; **$p < 0.05$; *$p < 0.1$; Standard errors in parentheses

References

Agiomirgianakis, G. M., Magoutas, A. I., & Sfakianakis, G. (2012). Determinants of profitability and the decision-making process of firms in the tourism sector: The case of Greece. *International Journal of Decision Sciences, Risk and Management, 4*(3–4), 294–299.

Agiomirgianakis, G. M., Magoutas, A. I., & Sfakianakis, G. (2013). Determinants of profitability in the Greek tourism sector revisited: The impact of the economic crisis. *Journal of Tourism and Hospitality Management, 1*(1), 12–17.

Aissa, S. B., & Goaied, M. (2016). Determinants of Tunisian hotel profitability: The role of managerial efficiency. *Tourism Management, 52*, 478–487.

Babayan, N., Darós, L. C., & Mascarell, C. S. (2014). Financial performance of Spanish hotels during the crisis. *Working Papers on Operations Management, 5*, 15–30.

Dimitrić, M., Žiković, I. T., & Blecich, A. A. (2019). Profitability determinants of hotel companies in selected Mediterranean countries. *Economic research-Ekonomska istraživanja, 32*(1), 1977–1993.

Dimitropoulos, P. E. (2018). Profitability determinants of the Greek hospitality industry: The crisis effect. In *Innovative approaches to tourism and leisure* (pp. 405–416). Springer.

Dwyer, L., Forsyth, P., & Dwyer, W. (2020). *Tourism economics and policy*. Channel View Publications.

Ferreira, P., & Saridakis, G. (2017). Firm shutdown during the financial and the sovereign debt crises: Empirical evidence from Portugal. *International Journal of the Economics of Business, 24*, 153–179.

Gai, L., Ielasi, F., & Ventisette, E. (2017). Overcoming borders: A financial analysis of tourism sector in Italy. *Journal of Economics and Development Studies, 5*(4), 17–30.

Gujarati, D., & Porter, D. (2010). *Essentials of econometrics*. McGraw Hill International.

Lado-Sestayo, R., Otero-Gonzalez, L., Vivel-Bua, M., & Martorell-Cunill, O. (2016). Impact of location on profitability in the Spanish hotel sector. *Tourism Management, 52*, 405–415.

Lundberg, D. E., Krishnamoorthy, M., & Stavenga, M. H. (1995). *Tourism economics*. John Wiley and sons.

Magoutas, A., Papadoudis, G., & Sfakianakis, G. (2016). The financial performance of Greek hotels before and during the crisis. *Journal of Investment and Management, 5*(6), 189.

Magoutas, A., Papadoudis, G., & Sfakianakis, G. (2018). Determinants of profitability in the Greek tourism sector–assessing the effect of the crisis. *International Journal of Tourism Policy, 8*(1), 65–72.

Martínez-López, A. M., & Vargas-Sánchez, A. (2013). Factores con un especial impacto en el nivel de innovación del nsector hotelero español. *Tourism & Management Studies, 9*(2), 7–12.

Moaveni, G. (2014). *Determinants of profitability in tourism industry: Evidence from Turkey*. Doctoral dissertation, Eastern Mediterranean University (EMU)-Doğu Akdeniz Üniversitesi (DAÜ).

Murugesu, T. (2013). Effect of debt on corporate profitability (listed hotel companies Sri Lanka). *European Journal of Business and Management, 5*(30), 13–18.

Oh, C. O. (2005). The contribution of tourism development to economic growth in the Korean economy. *Tourism Management, 26*(1), 39–44.

Papadogonas, T. A. (2007). The financial performance of large and small firms: Evidence from Greece. *International Journal of Financial Services Management, 2*(1–2), 14–20.

PORDATA. (2020). Retrieved March 2021, from https://www.pordata.pt/Subtema/Portugal/Alojamentos+Tur%C3%ADsticos-345

PORDATA. (2021). Retrieved March 2021, from https://www.pordata.pt/DB/Portugal/Ambiente+de+Consulta/Tabela

Sandvik, I. L., Duhan, D. F., & Sandvik, K. (2014). Innovativeness and profitability: An empirical investigation in the Norwegian hotel industry. *Cornell Hospitality Quarterly, 55*(2), 165–185.

Serrasqueiro, Z., & Nunes, P. M. (2016). Determinants of growth in Portuguese small and medium-sized hotels: Empirical evidence using panel data models. *Tourism Economics, 22*(2), 375–396.

Shin-Ping, L., & Tsung-Hsien, C. (2009). The determinants of corporate performance: A viewpoint from insider ownership and institutional ownership. *Managerial Auditing Journal, 24*(3), 233–247.
Singh, M., & Faircloth, S. (2005). The impact of corporate debt on long term investment and firm performance. *Applied Economics, 37*(8), 875–883.
Stabler, M. J., Papatheodorou, A., & Sinclair, M. T. (2009). *The economics of tourism*. Routledge.
Tandelilin, E. (2010). *Portfolio and investment theory and application*. Kanisius.
Tsai, H., Pan, S., & Lee, J. (2011). Recent research in hospitality financial management. *International Journal of Contemporary Hospitality Management, 23*(7), 941–971.
Wang, F. C., Hung, W. T., & Shang, J. K. (2006). Measuring the cost efficiency of international tourist hotels in Taiwan. *Tourism Economics, 12*(1), 65–85.

Good Practices Within the European Project *Smart Rural*

Antonio Martínez-Puche, Xavier Amat-Montesinos, Carlos Cortes Samper, Jose Antonio Larrosa Rocamora, Alberto Lorente Saiz, Samuel Ortiz Pérez, and Isabel Sancho Carbonell

Abstract Since November 2019, the University of Alicante, through the Human Geography research group, has participated as a partner in the European *Smart rural* project. Its objective is to identify the existing gaps and best practices in the EU countries to develop a training program in smart rural tourism aimed at students, authorities, and local communities (*stakeholders*). In total there are seven universities involved. The purpose of the communication is to illustrate the process followed to date, determine the content of the training program, the topics and the methodology for its development, which must then be tested to obtain learning that improves the final product. The University of Alicante contributes the value of geography in the development of these training programs linked to tourism, reinforcing the importance of an integrated rural and territorial approach, with the local population as a fundamental pillar. It also provides an analysis of the good practices of each of the partners and a methodological guide.

1 Introduction

The European project SMART RURAL (2019–2021),[1] dedicated to smart rural tourism, aims to develop a MOOC with applied content and that collects good practices and case studies that make it easier to understand and disseminate. For this reason, it contemplates improving training in the field of smart rural tourism, as well as creating a common training program between seven European universities. Hence the need to undertake, from training, the possibility of training and sensitizing local populations and rural entrepreneurs, of the possibility of applying innovations

[1] 2019–1- TRO1-KA203–073781. "Towards an intelligent development of rural tourism in Europe" (K203-Erasmus+).

A. Martínez-Puche (✉) · X. Amat-Montesinos · C. Cortes Samper · J. A. Larrosa Rocamora · A. Lorente Saiz · S. Ortiz Pérez · I. Sancho Carbonell
Universidad de Alicante, Alicante, Spain
e-mail: antonio.martinez@ua.es; xavier.amat@ua.es; carlos.cortes@ua.es; ja.larrosa@ua.es; alberto@crea360.es; samuel.ortiz@ua.es

J. Leitão et al. (eds.), *Tourism Entrepreneurship in Portugal and Spain*, Tourism, Hospitality & Event Management, https://doi.org/10.1007/978-3-030-89232-6_11

in rural areas. It is not easy, and we have verified it in the current project. The digital divide, but also the cultural one between the countryside and the city, is notable, and the approaches that prevail in this matter (from the partner Universities of the SMART RURAL project) are excessively technological and theoretical (IoT), and not so much territorial and applied (differentiated rural contexts).

Traditionally, innovations have been generated in places where resources and knowledge were concentrated in physical proximity. Large cities or business clusters ensured sufficient physical density to generate the processes of economic growth and territorial development. However, the digital transformation has prioritized virtual density over physical density (Gallardo, 2016), which implies greater independence regarding the location of the various groups of people or companies, providing an extraordinary opportunity for rural areas. It is in this context that the notion of territorial intelligence associated with research, innovation, and knowledge emerges as a means to achieve economic growth supported by sustainable development (Naldi et al., 2015), with increasingly notorious implications in rural regions.

Smart rural development is considered as part of the strategies introduced in response to the low rates of growth and productivity in rural areas (Naldi et al., 2015), but also in the face of the dichotomy caused by the digital divide between urban and the rural (Gallardo, 2016) and even as a stimulus to satisfy the aspirations of rural youth in marginal territories (Fennell et al., 2018). But, in addition to all this, the intelligence paradigm incorporates technology as a component for the transformation of economic activities in rural areas, and some productive sectors such as tourism are proving to be especially prone to the rapid expansion of the smart phenomenon. Smart tourism or smart tourist destinations thus constitute another step in the evolution of smart cities (Buhalis & Amaranggana, 2013). In rural areas, with the application of new technologies and the optimization of the cultural, environmental, and human resources of the territory (Fig. 1).

2 Theoretical Framework

The rise of rural tourism is a complex social construction (Fig. 2). A territory is a space where the economic, the social, the cultural, the environmental and the political are present. In these cases, a destination image can spread even further through various channels, such as residents, tourists, advertising brochures, tourist guides, novels, travel writing, literary works, the media, and image campaigns (Kim & Oh, 2009). Technology (IT) is also heavily involved in imaging tourist destinations (Choi et al., 2007). Smart tourism is recognized as a term that describes the convergence of IT with the tourism experience. It includes forms of information exchange in tourism based on mobile technology and Internet connectivity. As a result, the tourism public culture is no longer totally dependent on information provided in printed or even official form and commercial websites. Visitors now exchange user-created content via social media in the form of storytelling, tips, and shared comments. At present concepts and terms such as "e-tourism," "connected"

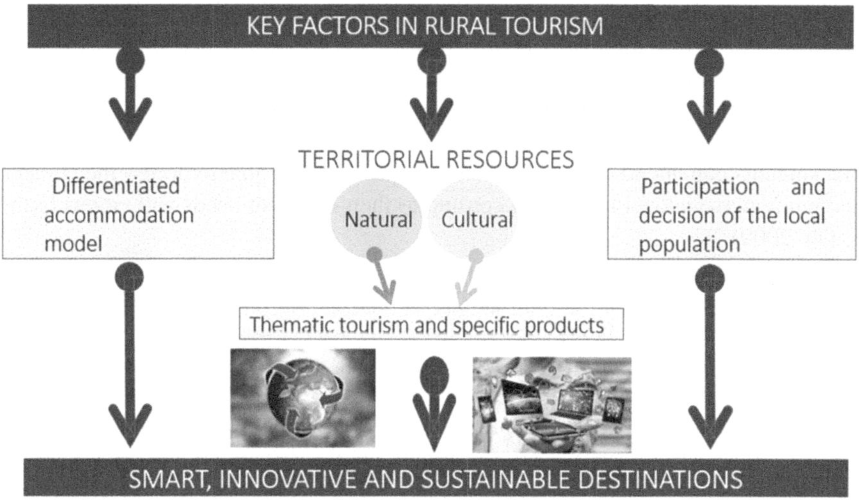

Fig. 1 New trends in rural tourism. Source: Own elaboration

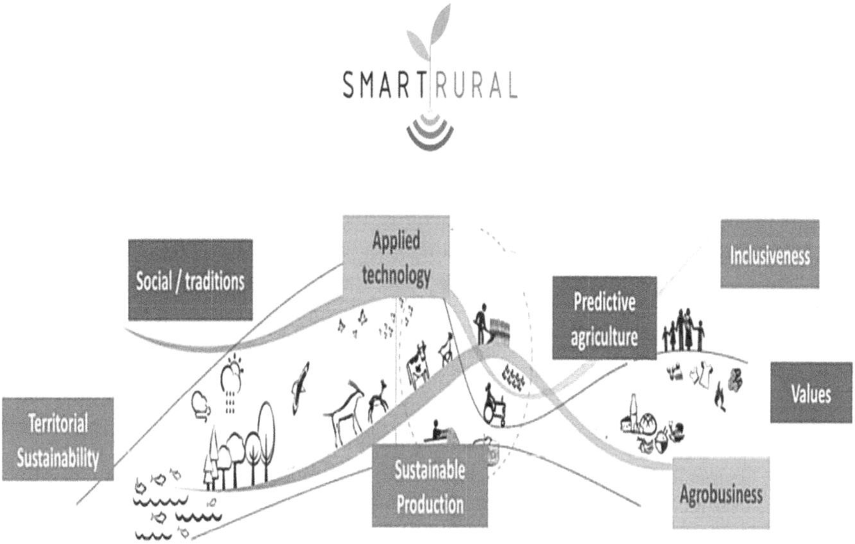

Fig. 2 Smart rural project elements. Source: Own elaboration

tourist, "social" tourist, "prosumer," "ewom," "electronic commerce," "tourist applications," "geographical location" or all those related concepts with intelligence or "smart:" "smart cities," "smart destinations," "wired cities," "smart destinations" have become benchmarks in the study, planning, and management of destinations, as well as in the business tourism (Beltrán, 2018).

In this context, the development of new technologies also allows students to learn more effectively, efficiently, flexibly, and comfortably. Students use smart devices to access digital resources over the wireless network and to immerse themselves in personalized and continuous learning. Smart education, a concept that describes learning in the digital age, has gained increased attention. In general, the intelligent learning environment is effective, efficient, and attractive (Merrill, 2013). And the objective is to provide self-learning, self-motivation, and personalized services so that students can attend courses at their own pace and be able to access the content and enjoy personalized learning according to their different needs and targets (Kim & Oh, 2009).

The concept of smart tourism implies the technological integration in all the organizations and entities that make up a tourist destination, with the aim of exploiting the synergies between technologies and users to enrich tourist experiences (Buhalis & Amaranggana, 2013). This perspective is based on guaranteeing maximum connectivity in destinations through new technological applications (Ballina et al., 2019) and is crucial for the competitiveness of companies in the tourism sector (Polo & Frías, 2010). In this sense, the academic literature agrees that smart tourism must inevitably address the needs of travelers before, during, and after their trip, with the tourist experience being the central nucleus in the development of smart tourism destinations (Ballina, 2019; Buhalis & Amaranggana, 2013), which only in this way can be competitive (Ballina et al., 2019; Luque et al., 2015).

The conceptualization of a destination as smart has undergone an accelerated transformation in recent years, maturing from a limited approach to tourism promotion through web pages and online management of reservations (San Martín & Herrero, 2012) to another much more complex and diverse approach, which refers to the integral management of tourist destinations and incorporates multiple and disparate categories. Thus, for example, important elements are revealed, such as political and governance conditions, available infrastructures, the way of analyzing the data generated by tourists or the way in which sustainability variables are incorporated (Shaifee et al., 2019), taking into account that the implementation of the smart paradigm in tourist destinations implies an inescapable commitment to their habitat and the quality of life of its inhabitants (Sigalat-Signes et al., 2020). Thus, the new approach to smart destinations requires a global project that ideally incorporates five areas of action: governance, sustainability, connectivity, information systems and innovation, its adaptation to each territorial environment being unavoidable (Ivars et al., 2016).

However, when referring to rural areas, the notion of intelligence is most often reduced to connectivity and communication options that, in the best of cases, can have a catalytic impact on the lives of rural inhabitants when they are integrated adequately with other rural development initiatives (Fennell et al., 2018; Holmes, 2017). But the use of technology can also facilitate the participation of rural communities in the processes of socio-economic development and, from the tourist point of view, can promote the conservation and appreciation of rural heritage through the stimulation of activities such as environmental interpretation and education (Martini et al., 2017).

In rural settings, experience has shown that the application of new technologies is related to the creation of new tourism products, favors the dissemination of information and ultimately contributes to the promotion of rural destinations. Thus, for example, the potential offered by new technologies in the creation of products has been identified through the incorporation of GPS, QR codes, mobile applications or augmented reality. However, in most cases, the use of technology is limited to information functions for visitors, while the rest of the stages of tourism consumption still show an incipient development in the digital tourism environment (Coma et al., 2016).

In this context, development strategies through smart tools aspire for the moment to complement the existing tourism offer by adding new products and services to the generic offer of rural tourism (Romão & Neuts, 2017). However, smart tourism specialization seems to show more and more relevant impacts in rural areas where tourism already plays a prominent role, incorporating improvements in aspects such as mobility, access to facilities and services, or the supply of local resources. In addition, new technologies could also be improving relations between rural tourism and other productive sectors by exploring the synergies that can be established in the tourist appreciation of goods and services produced in the same region (Meyn, 2017).

In short, the notion of intelligence is still diffuse and incorporates disparate characteristics, variables, and dimensions. The conditions to affirm a territory or a productive sector as intelligent are not exempt from academic and institutional controversy (Desdemoustier et al., 2019) and, in the absence of an adequate conceptualization, many villages, tourist destinations or rural regions are pushed to define themselves as intelligent without a basis to justify such consideration. Consequently, the academic literature and applied research through projects such as Smart rural face the challenge of conceptualizing through rigorous analysis the notion of intelligence and its application to tourist destinations in rural areas, but starting from education and training.

3 *Smart Rural* European Project: Content and Objectives

It must be said that smart rural tourism is the basis of the SMART RURAL project. A program—ERASMUS+, understood as a key cooperation action for innovation and the exchange of good practices and framed in a strategic alliance in higher education. The project began in October 2019 and will run until December 2021. It is made up of seven partners, led by Eskisher Osmangazi University, Faculty of Tourism (Turkey). The rest of the partners are:

1. VDU Faculty of Bioeconomics Development (Lithuania)
2. University of Warmia and Mazuryin Olsztyn (Poland)
3. Liubliana University (Slovenia)
4. Department of Humain Geography of Alicante University (Spain)
5. Unipegaso University (Italy)
6. University of Economics (Czech Republic)

Through an in-depth analysis of existing smart rural tourism training programs in the EU and beyond, our project identifies existing gaps and best practices in EU countries, developing a new smart rural tourism development program for students, graduates, entrepreneurs, and local authorities, as well as rural tourism communities.

Table 1 Modules/subjects and responsible partners in the SMART RURAL project

Modules/subjects	Responsible university
1. Smart rural tourism: Introduction to the concept of smart rural tourism, identification of problems related to the rural tourism community/village	VDU, Lithuania
2. Design and co-creation of smart rural tourism experiences (the importance of territorial resources and the local community in the rural world. Diagnostics and instruments to "co-create" between public and private initiatives. Experiences and studies of case	U. Alicante, Spain
3. Models and theories of gamification/ICT and internet of things in rural tourism. Connectivity and digitalisation	U. Liubliana, Slovenia
4. Sustainable rural tourism	University of Economía, Czech Republic
5. Accessible rural tourism. Physical accessibility to the rural destination and inclusiveness	University Eskisher Osmangazi, Turkey
6. Generate and develop smart ideas for rural tourism destinations. Creativity	University of Warmia and Mazury, Olsztyn, Poland
7. Smart business models/scenarios/strategies of rural tourism (entrepreneurship and entrepreneurship)	UNIPEGASO, Italy

Source: Own elaboration

It is planned to test the training program in each country and rely on the input of the participants. These will be students who will test the training program, with 30 students from each of the seven partner countries. In addition, it will also be tested by stakeholders in rural communities. The intention is to improve and adjust, as best as possible, an open online course (MOOC). The objective is to propose a useful tool to analyze, train, communicate, and optimize rural spaces, with modules distributed among the different partners (Table 1).

So far, the production of 16 tutorials, of up to 15 min, which will be in English, for each of the seven modules, with materials (videos and complementary readings) has been completed. Each module will last 4 h. Adding a total of 30 h (2 final hours for evaluation). The course, which will be certified by the participating universities, has been evaluated by groups of students and professionals in the field of rural tourism. The results will be edited for their MOOC format on the UNIPEGASO platform (Italy). In addition, a practical guide has been produced for operators in smart rural tourism that aims to act as a tool for the transfer of good practices.

We can already advance some achievements, despite the fact that the project has not finished. As a result of the heterogeneity of the different partners of the project, complementarity has been achieved in the materials produced. Adding modules with a more technological or business bias, with others that are more social or territorial. In this sense, the University of Alicante, from its geographical profile, has focused on overcoming the "urban centric" perspective of some of the proposals and has contemplated the rural environment from a holistic approach, needing not only investments and finalist grants, with a marked welfare character. Overcoming the perception that the territory is only limited to hosting technology-based activities and infrastructures (smart). This may require competitive, pedagogical, and territorial strategic plans and actions, led by the administrations (governance), but accompanied by the rest of the stakeholders in rural áreas (Martínez-Puche, 2020). Hence, the module proposal, made by the University of Alicante, and taking into account our profile of geographers, has contemplated innovation, from a social and territorial perspective, with an integrated approach, where consultation between public and private entities is necessary (cooperatives, agritourism, etc.). Also, putting in value what has been done and known by entities of our country in the field of rural tourism and innovation. Thus, among others, Wine Routes of Spain and the characteristics of the province of Alicante, INVAT.TUR (Valencian Institute of Tourism Technologies) of the Generalitat Valenciana, SEGITTUR, (Spanish State Society, dedicated to the management of innovation and tourism technologies) and the agroecological cooperative from Alicante La CAMPEROLA,[2] which has been chosen by the National Rural Network of the Ministry of Agriculture, Fisheries, and Food, to be included as a good entrepreneurial practice in rural areas.

Thus, in a rural context, a tourist destination can become intelligent when it makes use of technological infrastructures (connectivity through 5G) in order to improve the tourist experience and strengthen the decision-making process. Although without neglecting aspects related to sustainability, empowerment of the local population, economic diversification of rural areas, quality of life and improvement of services of all kinds. That is what we are working on from our geographical perspective in the different meetings held to validate proposals and training actions.

It is a shared and community learning, which is part of a process, which we will measure when we apply it. And fundamentally, from an attempt to share tools and knowledge in relation to training in smart rural tourism. And all this, in a European context, which also introduces participatory action-research (PAR), to continue advancing in a field where concepts must be nurtured by experiences and needs, and in which the territorial and integrated approach is really important to achieve optimal and useful results for rural communities.

The innovative elements it provides are (Martínez-Puche, 2020):

- There is no similar smart rural tourism development program at the EU level.
- It is counting on the participation of students (of tourism degree), academics, rural tourism professionals and institutions, for the joint creation of the course content.
- The content of the training is being enriched, with the skills necessary for its development. That is applied knowledge on rural tourism, sustainability, ICTs, citizen participation and public and private initiatives (co-creation).
- Use of best practices, evaluating their strengths and weaknesses, to propose a high-quality training program.
- Application of new teaching approaches. That is, discussion and analysis of real case studies of various rural tourism destinations; personalized tutorials, visits to well-known rural tourism destinations, guest speakers, gamification, storytelling, and design thinking).

[2] http://www.lacamperola.org/

There will also be pedagogical tools (infographics, educational videos, and other support material).

- Use the media to attract attention, raise awareness and develop more positive and sustainable perceptions of rural tourism in the minds of young people (students).

  The main target groups for this training program will be:

- Students and teachers.
- Rural tourism entrepreneurs.
- Local authorities, rural communities and stakeholders related to rural tourism.

4 Good Practices in the Framework of the *Smart Rural* Project

The development of rural communities depends on the enhancement of their resources and their adaptation to new market conditions and consumer habits. The emergence of the Internet and ICTs has brought about a change in tourist destinations, including in rural areas, which have become a digital tourism scenario (Jacobsen & Munar, 2012). For their part, tourists have also changed, as they are now hyper-connected and multi-channel people closely linked to smart phones and the use of information and activity management apps (Buhalis & Foerste, 2014; Xiang et al., 2015). Tourists make use of technology to obtain personalized services according to their needs and interests. The new generation of tourists highly values being able to digitally integrate with the destination before, during, and after the visit (Fernández et al., 2017). This leads to the empowerment of tourists thanks to the democratization of ICT, which allows them to make decisions throughout the tourist trip, thus being the protagonist of the creation of their own experiences (Celdrán-Bernabeu et al., 2018). Hence the importance, in the Smart rural project, of identifying, making visible and transferring good practices.

4.1 Why a Manual of Good Practices? What for?

The Good Practices in Intelligent Rural Tourism is a response to the needs and concerns expressed by the partners of this ERAMUS+ project regarding the promotion and articulation of intelligent rural tourism. Understood as a useful instrument to achieve collective benefits for rural communities and renewed entrepreneurs in European rural areas. All this in order to collect and capture those experiences or initiatives that respond to the needs of the territory, such as:

- Promoting economic activity and diversification through small enterprises.
- Promoting entrepreneurship and new business models through the association of tourism activity with ICTs.
- Making environmental conservation compatible with tourism development (sustainability).
- Creating an offer of accommodation and complementary activities, not concentrated and small-scale, but coordinated with the local environment.
- Favouring cooperation and co-governance networks between stakeholders in the territory.
- Developing more advanced tourism management based on knowledge and a new framework of relations between tourism agents.
- Facilitating connectivity and accessibility to rural destinations, both physically and digitally.
- Improving the proposals in the offer of activities based on creativity through Design Thinking.
- Differentiating the destination and positioning it in the market.
- Favouring the retention of socio-demographic staff and attracting new settlers.
- Improving the efficiency of destination management in all areas (marketing, environmental management, mobility, etc.).

The experiences collected in the manual that we have developed in the project are examples of good practices that stand out for their innovation, reinventing new actions to promote smart rural tourism. Not only for applying new technologies (connectivity) but also for being more sustainable, cultural, environmental, creative, and making destinations more accessible (digitalization). It also covers how to undertake and design new tourism projects (activities). Hence, this manual of good practices seeks transferability, i.e., the transferability to other geographical areas of the rural environment, by having a demonstrable impact (demonstrative example). They are experiences that have been evaluated and have a positive impact on the community and the territory. For this, good practices must be well planned, have an integrated approach, and be efficient in the use of local resources. Furthermore, they must be transformative and integrated actions, beyond the project proposed, and with a long-term perspective (durability and permanence over time).

Phase 1—Definition of Intelligent Rural Tourism Best Practice. In order to establish a methodology for the selection of Best Practices, the first step to follow is to define the concept. A Smart Rural Tourism Best Practice is understood as an experience or initiative that responds in an innovative and satisfactory way to the objectives set out above.

Phase 2—Drawing up a summary sheet. Brief sheet for the collection of information by the partners participating in the project. The purpose of this sheet is to collect the name of the initiative, as well as to summarize what the activity consists of, its location, objectives, the attributes of being a good practice, the difficulties and problems detected and the contact details of the person/s or bodies directly related to it.

Table 2 Location of partners and list of good practices

Partner	Good practice 1	Good practice 2
Lithuania	Adaptation of Zypliai manor ensemble for cultural tourism	Homestead "BARONO VILA"
Spain	CAMPEROLA TOURS—RURAL COMMUNITY TOURISMCooperativa Agroecològica La Camperola	Clúster Alicante wine route
Slovenia	Alpe-Adria-Trail	LOCALS FROM ZERO. Slovenia—Worldwide (online platform)
Czech Republic	Leading quality trails best of Europe: Lužnice Valley Hiking Trail. Czech Republic, South Bohemia, TOULAVA DESTINATION	Regional product brand. Czech Republic, 27 regions
Turkey	Monteverde, Italy. Accessible Village	REMOURBAN PROJECT—regeneration model for accelerating the smart URBAN transformation. TEPEBASI DISTRICT MUNICIPALITY—ESKISEHIR TURKEY
Poland	Warmia and Mazury thematic villages partnership. "PLACES WITH A SOUL." Poland, Vistula lagoon, Elbląg upland and the area of the Elbląg Canal	Cittaslow week. Poland, Warmia, and Mazury region. (Cittaslow towns associated in Cittaslow organization)
Italy	TURI.S.MO. (TURIsmo Sostenibile a Montepulciano—Sustainable tourism in Montepulciano	The Bufala tour. (Buffalo tour) RIS BUFALA association—Location of members offering the Bufala tour connected to the presence of certified cattle rearing acivities complying with the relative dairy DOP brands

Source: Own elaboration

Phase 3—Compilation and validation of experiences, which have been collected by theme, two for each of the partners (Table 2).

4.2 Methodological Design of the Guide

In each of the good practices that have been considered, and which are structured by thematic modules: Introduction, co-creation, digitalization, sustainability, accessibility, creativity, and entrepreneurship, you will find a series of headings that will help you to identify the key issues that you can consider in your project to help you (Martínez-Puche, 2019):

- *General Data About the Project: Name, Location*, etc.

For information purposes, the general data of the project considered as a good practice are given below.

- *Introduction and Context.*

In the introduction, you will find some brief basic introductory information about the project.

- *Why Is it a Good Practice?*

This section specifies the reasons why the project has been considered a good practice. In short, it explains why it has been classified as such.

- *Participants in the Project.*

In the following section, the participants in the project must be listed. This is essential to consider the economic and social agents that have participated directly or indirectly in the project developed in order to assess the degree of impact of the project from a socio-economic point of view.

- *Lessons Learnt.*

This section of the fiche indicates the lessons learnt after the development of the project. Here you can find the key issues of each project, and that can surely be useful to be considered in other projects.

- *Developed Activities.*

The developed activities section specifies the actions that have been carried out in each of the partners' projects. The list of these activities can help other territories to identify some of them that can also be considered for the development of similar proposals and that work (benchmarking).

- Results

In this section of the sheet, we have considered the effects and achievements of the development projects that have been considered as a good practice. You will be able to see the results obtained after the implementation of the project. A brief explanation of why the good practice has had a social and territorial impact has been included here. In addition, the impact has been measured, and each partner has rated from 1 to 5, according to the achievements obtained (Table 4).

5 Conclusions

The concept of smart rural tourism implies the technological integration of all the entities that make up a tourist destination. The objective is to exploit the synergies between the territory and the visitors to enrich the tourist experience. This perspective is required to achieve maximum connectivity in destinations through new technological applications. In addition, it is decisive to achieve the competitiveness of the tourism sector (Table 3).

However, the concept of good practice in smart rural tourism has evolved in recent years, from a limited approach to the partial use of new technologies, towards

Table 3 Traditional rural tourism *versus* smart rural

Traditional destination	Smart destination
It is all about commercial relations	What matters are the interactions between all components of the tourism system (supply and demand)
Creation of standardized products and unidirectional promotion	Co-creation (based on conversation) and product customization
Stable destination configuration. Defined products and predictable tourist flows	Dynamic configuration of the destination. Continuous adaptation and self-production of products and flows
Relatively simple behavior of tourist flows with a tendency to overcrowding	More complex behavior and very limited predictability
Limited (and delayed) access to information	Abundance of information (in real time)

Source: Own elaboration

a much more complex and diverse approach. Thus, smart rural tourism incorporates elements such as political and governance conditions, infrastructures, analysis of data generated by visitors, the way in which sustainability variables are included, in addition to the unavoidable commitment to the quality of life of the local inhabitants.

In summary, the definition of good practice in smart rural tourism implies a global strategy that ideally incorporates seven areas of action: (1) Introduction to the smart rural tourism, (2) design and co-creation of the tourism experience, (3) digitization, (4) sustainability, (5) accessibility, (6) creativity and (7) entrepreneurship.

1. *Introduction to the Smart Rural Tourism*

The basic potential elements to facilitate smart rural tourism are identified. There are capacities to promote a process of smart rural tourism, generating the first jobs along with the fulfillment of these principles, putting in value the basic resources that make up a conceptualization of rural tourism. Some services offered throughout the territory are identified, and it is seen whether networking relationships exist between the different actors.

2. *Smart Rural Tourism Experience, Design, and Co-creation*

The use of technology can facilitate community participation in the design and co-creation processes of new economic activities in rural areas, such as tourism. The smart rural tourism experience is based on the links between public and private entities of a territory, capable of generating networks and rich and diversified governance structures. Collaboration, communication, and active participation of the different entities (city councils, companies, associations, cooperatives, etc.) in the process of design and co-creation of the smart rural strategy will determine its economic, social, and territorial impact.

3. *Use of ICT and Internet of Things for Rural Tourism. Digitalization*

In rural areas, digitization facilitates the creation of tourist products, helps to disseminate information and favors the promotion of destinations. A good practice in

Table 4 Characteristics and contents of the levels of good practice by topic. Smart rural Project

LEVEL		INTRODUCTION TO THE SMART RURAL TOURISM	SMART RURAL TOURISM EXPERIENCE, DESIGN AND CO-CREATION	USE OF ICT AND INTERNET OF THINGS FOR RURAL TOURISM	SUSTAINABLE RURAL TOURISM	ACCESSIBLE RURAL TOURISM	CREATIVITY AND SMART IDEAS FOR RURAL TOURISM DESTINATIONS	STRATEGIES AND BUSINESS MODELS
Low	1	Potential elements for smart rural tourism are identified	Individual and autonomous work relationships are established, without cooperation or collective integration	Digital tools have individual, residual and unidirectional use. Limited and delayed access to information by tourists (web 1.0)	Small entrepreneurs offer local products available to tourists	Is detected a need for improve accessibility to the territory and universal accessibility for people with other abilities to tourist resources and services	The tourist resources are identified. Tourists come to the destination, but without generating sufficient interaction with the place or the local population	Some certified smart solutions are identified that a minority of local public and private stakeholders want. Initial formation of a territorial brand image
Medium	2	There are capacities to promote a smart rural tourism process, generating the first jobs in conjunction with compliance with these principles	There are private companies that works with public administrations to improve tourist activities but without a strategy, plan or tourist product defined yet	Digital tools are used collectively (common digital platforms) and two-way communication mechanisms are incorporated between tourism offer and demand (social networks)	Tourists visit the region attracted by the territory's own tourist resources. The region offers typical products, but also tourist services, thanks to local entrepreneurs and suppliers. The services satisfy the basic needs of the tourist (guide, accommodation, restaurants, etc.)	The degree of connectivity with the transport nodes that connect rural areas with other cities or municipalities has been analyzed. Access to resources and tourist services in the rural area has been evaluated	There is tourism linked to thematic, cultural, creative and experiential resources, but there is not certification or the distinctive label. There are tourism companies but no relationship between them	An increase in the number of sustainable smart solutions adopted in the business environment is established. Implementation of a coordinated elementary strategic plan to promote the territory's global identity brand. Initial use of specific public funds and an attraction of private investment
	3	Territorial coordination work is carried out between the different social, economic, and political agents	The organization of the groups is regulated through internal operating regulations, specifying tasks, activities, and responsibilities. There is a strategy formally established with the partners and stakeholders of the territory	Digital tools provide dynamic and personalized information for tourists. Digital ICTs facilitate the co-creation of the tourist experience (web 2.0 and social networks)	Local communities set up a local coordinating body. Tourism agents cooperate with the objective of offering tourists an integrated (tourism) product, to join the system of regional label	A strategy or plan for territorial and universal accessibility has been coordinated among the different agents of the rural territory. Access to information on tourist resources and services has been initially improved	There are business associations for creative tourism and a network. They have the distinctive label, or some type of thematic tourism certificate. The recognition has been obtained through an official body	The majority of stakeholders are aware of the territorial brand identity and implement business plans to upgrade their offer accordingly. Structural public funding and incentives are available
High	4	There is an integration and consolidation of the services offered throughout the territory and there are networking relationships between the different stakeholders	There is a definition of tourist products. Services are offered, and there is a strategy which has been adapted to the characteristics and singularities of the territory and its resources (cultural and natural). There is institutional support and multilevel governance	Digital tools are used bi-directionally and are an essential component of the tourism experience. In addition to consuming, the tourist informs and creates information about the destination through online resources (web 3.0 and social networks)	Development of marketing tools – websites, social networks, promotional materials. The tourist product is marketed and promoted through a brand recognized in the market for its certified guarantee of quality.	Some of the activities defined in the strategy were implemented. Stakeholders collaborate as network. Initially they have been adapted some tourist resources and services in demand of better territorial and universal accessibility. Adapted activities have already been started and tourist information has been adapted for people with other abilities.	There are specific tourist products linked to the territory and its socioeconomic development. It is a widely consolidated offer, with high levels of quality and standards required by the corresponding regulations. They are part of a strategy formally established with the partners and actors of the territory (national, regional, local)	The integration of social, economic and political agents is established in a coordinated plan. Funding is consolidated, becomes rotating and always available from both public and private sources
	5	New projects and territorial strategies are valued in common and from the beginning. The strategies have achieved work generation, reduce environmental impacts, improve accessibility, creativity and digitization	Existence of political, business and institutional support around a certified tourist product. Establishment of networks within and outside the rural areas, such as universities, financial and cultural institutions, etc.	The digital experience is functional, interactive and fully satisfying. An abundance of information and interaction in real time (incorporation of "advanced" tools: WIFI networks, QR, beacons, augmented reality ...)	A place branding is developed. There is a coordinating body which manages the system of regional product certification. Regional products are promoted regarding sustainable development of the region and tourism.	In the rural area is a tourist offer fully adapted to universal accessibility, and jobs have been created for people with other abilities. Accessibility to the rural environment stands out for its complete connectivity with other territories that may be the origin of visitors and tourists	There is an established partnership with institutional support and a regulatory framework. There is a diversified offer of activities for tourists, where there is a clear definition of tourism products. The network offers creative activities, experiential products and a clear program of activities (festivals and events) in the territory. There is international networking with other smart destinations	Consolidation and expansion of an entrepreneurial model that achieves an attractive and shared territorial image (place branding), good living conditions, with resources and job creation

Source: Own elaboration

smart rural tourism must be properly integrated in a digital environment. On the one hand, new technologies provide dynamic and personalized information for visitors, who are capable of being involved in the tourist experience. On the other hand, the ability to collect and analyze data generated by tourists allows destinations to learn about behavior patterns and adapt the offer in real time to the needs of visitors. The abundance of information and the capacity for interaction make the digital tourism experience more functional, interactive, and satisfying.

4. *Sustainable Rural Tourism*

The definition of sustainability for a rural tourist destination implies the adequate use of natural resources (natural heritage and biodiversity), respect for the social and cultural identity of the host communities and the consolidation of a viable and lasting economic model (stable employment, poverty eradication, etc.). The smart tourism strategy is based on the reasonable and weighted use of a territory's own resources, with decisive support for goods and services produced on a local or regional scale according to the parameters of sustainability. Digital tools can become the support of tourism marketing, helping to spread the values of sustainability and raise awareness among local inhabitants and visitors.

5. *Accessible Rural Tourism*

The smart strategy must have relevant impacts in rural areas. It implies incorporating improvements in aspects such as mobility, access to facilities and services, or the inclusion of people with specific needs. Accessibility is a multidimensional concept that has an impact on people's daily lives in many areas (building, transport, social activities, communication, etc.) and on the tourist experience. Accessibility refers, therefore, to the design of products and environments accessible to as many people as possible. Technology can improve territorial and universal access to tourism resources and services, so good practice requires the existence of plans to ensure accessibility (physical and virtual) to rural destinations.

6. *Creativity and Smart Ideas for Rural Tourism Destinations*

Smart ideas in the field of tourism allow the creation of innovative products and services based on an understanding of the problems and needs of visitors. Therefore, a good practice must be adapted to the needs of the demand, be technologically feasible and be economically justified. The entities that make up the tourist destination carry out a continuous search for innovation and develop original solutions based on demand trends. In this way, the offer consists of creative activities and experiential products based on its own territorial resources, supported and promoted from the institutional and business spheres.

7. *Strategies and Business Models*

Rural spaces are no longer exclusively associated with agricultural production but are seen as multifunctional places for the stimulation of new socio-economic activities that often incorporate tourism. In this context, development strategies through smart tools allow to expand and complement the tourist offer by adding new

Image 1 Trainer of trainer's seminar organized by the University of Alicante with all partners. Evaluation of subjects and good practices

products and services based on technology. But in addition, digitization also improves relations between rural tourism and other productive sectors by generating synergies in the tourist valorization of goods and services produced within the same territory (online marketing, marketing platforms, etc.).

It is necessary to establish levels that measure the degree of materialization of this good practice, distributed from 1 to 5, and organized by basic, medium and high level. The characteristics and levels are summarized and reflected in the following Table 3. With these standards, we believe that the partners themselves can measure the degree of development and impact of their good practice, with the action being valued more than the intention. It is important that all the themes show a degree of development above level three (medium) in order to be considered good practice in this project (Table 4). This aspect was agreed with the rest of the partners, and we consider that the results are optimal. We hope to finish the materials by the end of the year and have the meeting next September to finalize the assignments and launch the different intellectual projects and subjects through the MOOC (Image 1).

References

Ballina, F. J. (2019). Smart tourism destination, experienciaphygital y turismo rural. *International Journal of Information Systems and Tourism, 4*(1), 41–52.

Ballina, F. J., Pélaez, L. V., & del Valle Tuero, E. (2019). Tourism destination: Urban *versus* rural technological behaviours. *Revista Internacional de Turismo, Empresa y Territorio, 3*(2), 16–37.

Beltrán, G. (2018). *Turismo rural online. Páginas web y redes sociales* (p. 168). UOC.

Buhalis, D., & Amaranggana, A. (2013). Smart tourism destinations. In Z. Xiang & I. Tussyadiah (Eds.), *Information and communication technologies in tourism* (pp. 553–564). Springer International.

Buhalis, D., & Foerste, M. (2014). SoCoMo marketing for travel and tourism: Empowering co-creation of value. *Journal of Destination Marketing and Management, 4*(3), 151–161.

Celdrán-Bernabeu, M. A., Mazón, J. N., Ivars-Baidal, J. A., & Vera-Rebollo, J. F. (2018). Smart tourism. Un estudio de mapeo sistemático. *Cuadernos de Turismo, 41*, 107–138.

Choi, S., Lehto, X. Y., & Morrison, A. M. (2007). Destination image representation on the web: Content analysis of Macau travel-related websites. *Tourism Management, 28*, 118–129.

Coma, J., Sanz, B. E., & Torres-Delgado, A. (2016). La incidencia de las TIC en destinos turísticos de la montaña española. Un análisis de casos. *Ara, 6*(2), 75–86.

Desdemoustier, J., Crutzen, N., & Giffinger, R. (2019). Municipalities' understanding of the smart city concept: An exploratory analysis in Belgium. *Technological Forecasting and Social Change, 142*, 129–141.

Fennell, S., Kaur, P., Jhunjhunwala, A., Narayanan, D., Loyola, C., Bedi, J., & Singh, Y. (2018). Examining linkages between smart villages and smart cities: Learning from rural youth accessing the internet in India. *Telecommunications Policy, 42*, 810–823.

Fernández, A., López, J. M., Moreno, L., Perles, J., Ramón, A., & Such, M. J. (2017). Innovación y destinos inteligentes: oportunidad para el Know How turístico español. *ICE, Revista de Economía, 894*, 137–150.

Gallardo, R. (2016). *Responsive countryside: The digital age and rural communities* (p. 174). University Extension Center for Technology Outreach.

Holmes, J. (2017). *The smart villages initiative: Findings 2014-2017* (p. 134). CMEDT-Smart Villages Initiative.

Hortelano, L. A. (2019). El turismo rural como estratègia de la lucha contra la despoblación del interior de España. In F. García-Moreno (Ed.), *La despoblación del mundo rural* (pp. 503–533). Thomson Reuters Aranzadi.

Ivars, J. A., Monzonís, F., & Sánchez, D. G. (2016). Gestión turística y tecnologías de la información y la comunicación (TIC): El nuevo enfoque de los destinos inteligentes. *Documents d'Anàlisi Geogràfica, 62*(2), 327–346.

Jacobsen, J. K. S., & Munar, A. M. (2012). Tourist information search and destination choice in a digital age. *Tourism Management Perspectives, 1*(1), 39–47.

Kim, H. K., & Oh, S. J. (2009). A Q methodological approach to tourism motivation and preference typology. *Journal of Tourism and Leisure Research, 21*(2), 69–88.

Luque, A. M., Zayas Fernández, B., & Caro Herrero, J. L. (2015). Los Destinos Turísticos Inteligentes en el marco de la Inteligencia Territorial: conflictos y oportunidades. *Investigaciones Turísticas, 10*, 1–25.

Martínez-Puche, A. (2019). Territorios inteligentes y desarrollo local. Del concepto al precepto. In T. Cantó (Ed.), *Los territorios rurales inteligentes: administración e integración social* (pp. 21–56). Thomson Reuters Aranzadi.

Martínez-Puche, A. (2020). Turismo rural inteligente: Capacitación aplicada en el marco del proyecto europeo smart rural. In G. X. Pons Buades, A. Blanco-Romero, R. Navalón, L. Troitiño, & M. Salom (Eds.), *Sostenibilidad turística: "overtourism vs. undertourism"* (pp. 359–368). Mon. Soc. Hist. Nat. Balears.

Martini, U., Buffa, F., & Notaro, S. (2017). Community participation natural resource management and the creation of innovative tourism products: Evidence from Italian networks of reserves in the Alps. *Sustainability, 9*(12), 1–16.

Merrill, M. D. (2013). *First principles of instruction: Identifying and designing effective, efficient and engaging instruction* (p. 479). Wiley.

Meyn, M. (2017). *Digitalization: Status quo and future trends – A new impact on life in rural areas? Artículoocasional* (p. 17). American Council on Germany.

Naldi, L., Nilsson, P., Westlund, H., & Wixe, S. (2015). What is smart rural development? *Journal of Rural Studies, 40*, 90–101.

Polo, A. I., & Frías, D. M. (2010). The relationship between business characteristics and ICT deployment in the rural tourism sector. The case of Spain. *The International Journal of Tourism Research, 12*(1), 34–48.

Romão, J., & Neuts, B. (2017). Territorial capital, smart tourism specialization and sustainable regional development: Experiences from Europe. *Habitat International, 68*, 64–74.
San Martín, H., & Herrero, A. (2012). Influence of the user's psychological factors on the online purchase intention in rural tourism: Integrating innovativeness to the UTAUT framework. *Tourism Management, 33*, 341–350.
Shaifee, S., Ghatari, A. R., Hasanzadeh, A., & Jahanyan, S. (2019). Developing a model for sustainable smart tourism destinations: A systematic review. *Tourism Management Perspectives, 31*, 287–300.
Sigalat-Signes, E., Calvo-Palomares, R., Roig-Merino, B., & García-Adán, I. (2020). Transition towards a tourist innovation model: The smart tourism destination, reality or territorial marketing? *Journal of Innovation & Knowledge, 5*, 96–104.
Xiang, Z., Magnini, V. P., & Fesenmaier, D. R. (2015). Information technology and consumer behavior in travel and tourism: Insights from travel planning using the internet. *Journal of Retailing and Consumer Services, 22*, 244–249.

Tourism Events: Bringing Innovation to the North of Portugal

Susana Silva, Cândida Silva, and João Alves

Abstract Tourist destinations are places or regions that offer a set of tourism products, which provide experiences to the consumer according to their itinerary, culture, travel motivation, education, or life experience. Events usually provide the consumer with an opportunity for leisure and social interaction that goes beyond the experience of everyday life and by the desire of people to live an original and exclusive experience. Therefore, an event can be used to promote a destination and to make known the traditions, cultures, and experiences of local communities. In this way, it was carried out a study to understand the importance of events for the promotion of a tourist destination, identifying the themes of the events used to that promotion, and the role of events in the promotion of a touristic destination. This study was conducted in the Municipalities of the North of Portugal, following a qualitative methodology to collect and analyze data. The municipalities participating in this study hold entertainment, cultural, sports, and educational events, and the results showed that those events are a great attraction for visitors creating a good economic impact in the region, contributing to the increase of consumption and bigger dissemination of their products facilitating potential new contacts for business and partnerships.

S. Silva (✉)
Institute Polytechnic of Porto, School of Hospitality and Tourism, CEOS.PP and CITUR, Vila do Conde, Portugal
e-mail: susanasilva@esht.ipp.pt

C. Silva
Institute Polytechnic of Porto, School of Hospitality and Tourism, CITUR, Vila do Conde, Portugal

Algoritmi Research Center, University of Minho, Braga, Portugal
e-mail: candidasilva@esht.ipp.pt

J. Alves
Institute Polytechnic of Porto, School of Hospitality and Tourism, Vila do Conde, Portugal

J. Leitão et al. (eds.), *Tourism Entrepreneurship in Portugal and Spain*, Tourism, Hospitality & Event Management, https://doi.org/10.1007/978-3-030-89232-6_12

1 Introduction

Tourist destinations are a set of tourism products that offer experiences to the consumer. However, it is increasingly that destinations can have a more subjective concept, which can be interpreted by consumers taking into account their itinerary, culture, travel motivation, education, or life experience (Buhalis, 2000). In turn, the World Tourism Organization (WTO) defines tourist destinations as a physical space in which a visitor stays at least one night. This space has to include several products, such as ancillary services and tourist attractions, and resources that can be visited in less than a day's travel. The organization of a specific event is, many times, also developed to be a tourist attraction.

Over the years, numerous events have been held with the most diverse themes, from fairs, festivals, exhibitions, sports, culture, business, among others (Mair & Whitford, 2013). Although they have existed for a long time, events are phenomena that have been growing worldwide both in quantity and popularity (Gursoy et al., 2004), as they provide a way to mark important personal or public occasions and personal or public occasions and celebrations of important milestones throughout life (Mair & Whitford, 2013).

Given that the essence of tourism is linked to human activity and the will of each person to achieve certain goals or satisfy their needs related to the act of traveling (Ratkowski & Ratkowska, 2018), these will give rise to a desire in people to participate in events in other parts of the world, such as events. These can be used as attractions to enhance the tourism growth of a region (Saayman & Rossouw, 2010).

Therefore, it can be noticed that the degree of the importance of events in the quality of the tourism offer has been increasing. This is because tourists' interest in travel has increased to attend a specific event, which makes attending an event one of the most important motivations for a tourist to travel (Panfiluk, 2015).

Thus, considering this possible effect of events on the increase of tourism, this study aimed to understand the importance of events in the promotion of a tourist destination in a smaller region, in this case, in the North of Portugal. Thus, the main goals were to characterize the importance of events for the promotion of a tourist destination, identify the theme of events in the promotion of a tourist destination, and identify the role of events in the promotion of a touristic destination.

This paper is structured as follows: the next section makes a literature review on the main concepts related to the study, namely, events, and their types, events in tourism, and the impacts of events. Section 3 will describe all the methodology used to carry out the study; Sect. 4 presents and discuss the main results of the study, and the last section is dedicated to conclusions and final considerations.

2 Literature Review

Events are commonly characterized by a one-time or infrequent occurrence, by a limited time that provides the consumer with an opportunity for leisure and social interaction that goes beyond the experience of everyday life, and by the desire of people to live an original and exclusive experience (Hernández-Mogollón et al., 2014) that only events can offer, and thus differentiate them from other permanent attractions.

Events are considered important activities and even a factor that socioculturally unites communities or even nations. In this sense, more and more governments are investing invest in this area due to the growing impact they can have on society (Jamieson, 2014).

Events make it possible to create different kinds of experiences with a very wide diversity, from fun and entertainment to carnival and festivals, to the spirituality of religious pilgrimages and rituals of celebration. However, the purposes associated with the event can be in the promotion of commerce or learning, that is, using the event to enhance a transformation of behaviors, beliefs, values, and attitudes (Getz, 2008).

Events were, in the beginning, mostly an area of individual and community initiative. Nowadays, with the growth of the interest in events, it has become a professional and entrepreneurial activity and typified in four main categories according to its nature (Getz & Page, 2016).

2.1 Types of Events

Getz and Page (2016) classify events into four categories, namely business events, sporting events, festivals, and other cultural celebrations, and entertainment events. The categories are summarized in Fig. 1 as follows.

Business events are of great importance, as they can bring together professionals from one or more industries in one place, providing great opportunities for business, learning communication, and market research (Bathelt & Schuldt, 2008).

Trade fairs are a type of business event that have been growing in importance increasing in importance because of the experiences that visitors can get visitors can get from their participation in these events. This type of event was firstly more used for marketing and sales purposes. More recently, a greater impact has been felt in the use of these fairs with innovative spaces that allow and facilitate the sharing of knowledge from different companies or market sectors (Zhong & Luo, 2018).

Sporting events have been growing in the tourism industry and are considered to have an economic, socio-cultural, and political impact on the places where they are organized (Maussier, 2017). Nowadays, people have different motivations for participating in these events, whether for health reasons, pleasure, or passion for the sport itself, for the contact with the local community, or the authenticity of the event.

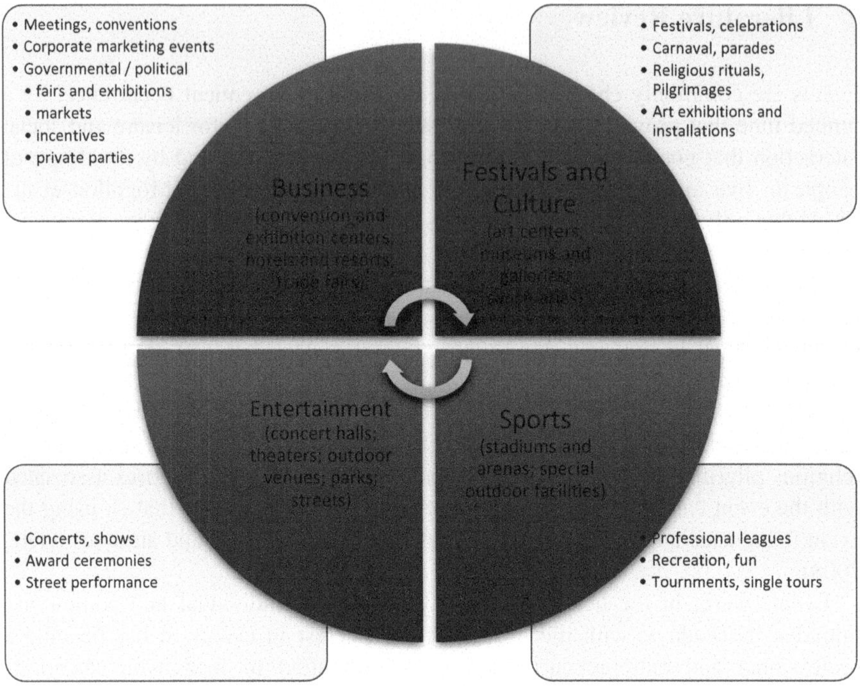

Fig. 1 Event types (adapted from Getz & Page, 2016)

These events are usually able to generate benefits for the city, region, and the host country. For example, sometimes these events are used as a means of promotion of the host city in the national or international market, and thus they can differentiate themselves from the competition (Hemmonsbey et al., 2018).

The link between sports and tourism is recognized in different areas. In the area of health, leisure, leisure time management, cognitive values, and their economic effect, and this same connection leads to the organization of sporting events with tourism aspects (Ratkowski & Ratkowska, 2018). For this reason, tourism is one of the industries that greatly benefit from sporting events.

Festivals and other cultural celebrations events encompass several types of events, from cultural events, historical commemorations to festivals.

Cultural events are a source of creative experiences and have a strong economic component for regions because they enhancement the existing infrastructures, or the creation of new ones, and drivers of creative expression and local animation (Liu, 2014).

As for historical commemoration events, these can be incorporated into event development strategies for tourism, i.e., they should be considered as tools that are valuable and with great potential to be exploited at the level of the organizing destination's tourism development. Nevertheless, to be used as tourism events, it will be necessary to overcome the domestic orientation of the celebration, converting them to an international level so that tourists do not feel like intruders (Viol et al., 2018). This way of utilizing the culture and traditions specific to each region increases the ability of destinations to attract visitors and increase the competitiveness of the different regions (Liu, 2014).

Entertainment events are difficult to define, as they can encompass a large number of events and diverse activities, such as sports, celebrations, or firework displays. These activities have entertainment value, meaning that anything that creates fun or joy can be considered entertainment (Getz & Page, 2016).

These types of events coupled with tourism activity can be used to revitalize cities. This revitalization includes the construction of various infrastructures appropriate for different types of entertainment, from atriums, convention centers, rehabilitation of historic centers, among others. This change ends up diversifying the existing offer and attracting tourists and professionals appropriate to the different types of attractions (Che, 2008). The theatre, concerts, shows, and spectacles are a good sample of these types of events, and their realization is normally provided by the private sector (Getz & Page, 2016).

2.2 Events in Tourism

The degree of importance of events in the quality of the tourism offer has been increasing. This is due to the fact that tourists' interest in travel has increased for the purpose of attending a specific event, which makes attending an event one of the most important motivations for a tourist to travel (Panfiluk, 2015).

Liu (2014) mentions that for tourism development, events are seen as a solution to the problem of seasonality and the issue of differentiation of the offer that each region can offer and consequently increase the competitiveness of the tourism market. Additionally, Getz and Page (2016) refer that their great boost in tourism is due to the strong economic impact that events have on the regions where they are organized.

However, although there is a great demand in the tourism sector, combating seasonality remains one of the biggest challenges for the supply of tourist destinations (Connell et al., 2015).

Another important factor for the development of tourism is the fact that events can promote the image of a destination both nationally and internationally. As a less positive point, we can say that events are very associated with the image of the destination. If the event does not go as planned or serious problems occur, it will end up being very negative for the destination, due to the great exposure it will have in the media (Liu, 2014). Given the association between events and destination image, care must be taken to choose the events that are best suited to project a genuine image of destinations (Hemmonsbey et al., 2018).

To take advantage of the fact that events promote a destination's image, many countries use mega-events to gain prestige by attracting attention to their achievements, for trade promotion, tourism, or to help promote their countries' image in the world (Getz, 2008).

As a general rule, smaller-scale events, i.e., more local and regional events, are created by people who have an interest in a particular activity and who use events as a means of celebrating that activity and emphasizing their identity. These events do not have tourism as an objective, but when they end up having a more touristic aspect, they can lose their value and authenticity (Kelly & Fairley, 2018).

From a tourism industry perspective, through the vision of Destination Management Organization and event organizing agencies, events are very valuable as attractions, as promoting agents, animators, and creators of a positive image of the destination (Getz, 2008).

Getz and Page (2016) refer that events are animators of a tourist destination, making it more attractive. Thus, they are the key to marketing in the promotion of a destination. Given the great increase in global competitiveness, they also serve to attract and encourage tourists to spend more money during their stays. The authors also mention that events are used to develop unique selling propositions that differentiate the destination from other competing markets.

In this sense, several effects caused by events in tourist destinations are identified (Getz & Page, 2016):

- Great attraction of tourists, sponsors, and media, among others, that without the events might not get to visit a particular region.
- Leverage to maximize the value of destinations and a key factor in combating seasonal demand, extending tourism geographically, and contributing to urban and economic development.
- Their value can be used to strongly influence the market, especially if several market segments are involved in the process.
- They can create a very positive destination image, help in the promotion of region-specific brands, reposition the different destinations in the market, be part of marketing strategies.
- They make cities more dynamic, being able to animate cities, resorts, parks, urban spaces, among others.
- They make destinations more attractive to visit and offer more reasons for tourists to return to the already visited places, increasing the efficiency of the places.

2.3 *Impacts of Events*

The events produce positive and negative impacts on the regions and communities where they are held. These impacts can be divided into three types: economic, environmental, and personal, social, cultural, and political (Getz & Page, 2016).

2.3.1 Economic Impact

The events organisation in tourism change the consumption and income patterns of local inhabitants. The taxes that arise due to the increase of tourism activity during events ultimately benefit government organizations (Getz & Page, 2016). Saayman and Rossouw (2010) reinforce this idea with the results obtained in their study of music events that reinforce the idea that not only government organizations benefit economically but also the private sector, local businesses, and the tourism industry in general. It is noteworthy that not only do these events help in creating more jobs and attracting more investors, more investors, but they also help in the economic growth of the region. Since this financial return exists, along with your entity's exposure during the event, public and private entities will continue to financially support the events, even though the economic crises in the world have been decreasing the amount of funding these organizations invest. This funding is essential for the economic sustainability of the event (van Wyk et al., 2015). Financial support is not only important for organizing events but also to support the high costs of the revitalization of some cities that are beginning to focus on tourism (Che, 2008).

If the events take place during peak demand season, there may be an opportunity for hotels to raise their prices even higher, since they would already be in high demand because it is high season. Then they can further inflate their prices due to the extra tourists that want to attend the event. Likewise, this can be seen as a negative point (Andersson & Lundberg, 2013).

2.3.2 Environmental Impact

Events, by their nature, have an impact on the environment where they are organized, since they are large consumers of energy and other resources, i.e., they generate a high ecological bill to be paid. However, sustainable venues are beginning to be created to reduce the pollution caused by events and restore the balance of the environment (Getz & Page, 2016) in addition, and there are beginning to be changed in the management and disposal of waste that arises from the events and thus minimize the negative impacts on the environment (Dickson & Arcodia, 2010).

The construction of permanent infrastructure needed to good accessibility to the events and to the actual the event itself end up modifying the landscapes of the regions where they are organized, also causing an increase in costs and costs and demand for resources for their construction. However, there are already practices and rules that try to educate and encourage the protection and conservation of nature (Getz & Page, 2016).

Dickson and Arcodia (2010) note that there is great complexity in creating the most environmentally friendly industry. Nevertheless, the events industry is developing research into the relationship between events and the environment and acting on the results. The authors also emphasize that researchers, professionals, and governments need to work together to help predict, plan for and if possible overcome challenges as a matter of urgency. Some of these challenges they mention are the fact that events depend on consistent and stable weather patterns or depend on environmental conditions for the event to take place, such as ice fishing competitions. They also highlight how the impact of climate change continues to be felt; events have to adapt to changes in the environment to maintain their continuity.

2.3.3 Personal, Social, Cultural, and Political Impact

Getz and Page (2016) state that there is difficulty in distinguishing impacts on a personal social, and cultural level, and can have political effects on destinations. Some of the social and cultural impacts are due to the fact that the change of traditional culture events begins to have a more touristic aspect and consequently lose authenticity. Gursoy et al. (2004) state that the growth of events can cause problems of overpopulation and inflations in local commerce, especially if they are organized only for the purpose of generating income. Overpopulation can also generate problems of insecurity, new conflicts, excess traffic, and pollution.

From another point of view, events can help preserve traditions, promote civism, national pride, and support culture and the arts (Getz & Page, 2016).

The social aspect that is disseminated by the media eventually affects the perceptions and attitudes of those who attend through demonstrations of disgust, protests, or antisocial behavior (Getz & Page, 2016).

The organization of events can create social and cultural capital by increasing cooperation networks between employees and the companies involved, such as volunteering, to promote personal development (Getz & Page, 2016). As mentioned earlier, events increase cooperation, unity, and social cohesion.

Gursoy et al. (2004) state that understanding the perceptions of event organizers about the impacts of festivals and special events on the local community is vital to the success of events.

The fact that events have both positive and negative impacts leads many people to be either for or against, depending on the benefits each person derives from the events (Getz & Page, 2016).

3 Methodology

3.1 Study Design

A qualitative methodology was used to answer the research question "How tourism events promote a tourist destination?" The main aims were to characterize the importance of events for the promotion of a tourist destination, identify the theme of events in the promotion of a tourist destination, and identify the role of events in the promotion of a touristic destination.

3.2 *Instrument*

It was used a semi-structured interview to reach the study objectives. This interview had two sections: the first section had seven sociodemographic questions such as gender, age, marital status, graduation, and professional experience; the second section had 16 questions about events. The section about events had questions about the infrastructure used, the objectives of the events, how the events are publicized, what results are obtained with the events, what is the link between events and tourism to questions related to the possible future development of the region. The questions were constructed more openly to give some freedom of response and to give some freedom of response and thus possibly obtain more information.

3.3 *Participants*

The participants in this study were responsible for the organization of events in several departments in the Municipalities of North of Portugal. The choice of these participants was made taking into account that they were the informants who best know the reality under study; therefore a theoretical sample was used.

Regarding the sociodemographic characteristics of the participants, 11 were female, and 4 were male, with a mean age of 45 years (SD = 7.51), 10 participants were married. Most of the participants were from the cultural department ($n = 8$) from several hierarchical levels and had a high education level. The participants were in the current function, on average, for 15.03 years (SD = 4.5).

3.4 *Procedure and Data Analysis*

The study dissemination started with an inviting e-mail to several municipalities in the North of Portugal. To the municipalities that answered the e-mail, a second e-mail was sent scheduling the interview. The interviews occurred during September and November 2019. All the interviews were record and verbatim transcripted. All ethical procedures were taken.

Data were analyzed according to grounded theory procedures (Strauss & Corbin, 1990). Each interview was analyzed line by line to extract its meaning, coding it initially in open codes and later organizing it in central themes, from where rise the axial and selective codes. This phase ends with the formalization of the concepts that emerged from the information collected.

At the end of this process, it was obtained a set of concepts and relationships between them, based on the information analyzed and confirmed by the literature.

To ensure the validity of the analysis and the coding process, two researchers were consulted as auditors (independent researchers that discuss and validate the categories) throughout the entire data analysis process to assist the primary author by challenging ideas and assisting in the construction of the categories. The main results are presented in the next section.

4 Results

This section presents the main results of the grounded theory analysis performed. Table 1 presents the structure of the codes created.

This analysis has one central theme, Events, emerging from Type of Events, Purposes, Planning, Infrastructure, Local Community, Dissemination, Disadvantages, Evaluation Methods, Results, Organization, Projects, and Future.

Regarding the main category, Events, the interviewees state that events are increasingly important for the Municipalities; all of them assume this importance and are often associated with the region logos. These logos are used to transmit the image of the Municipalities, its traditions, culture, and gastronomy to the world, as can be verified in the following quote "I think that this executive had the great vision to promote what is genuine. And this is what we are moving forward and this is what is that is managing to put the Municipality on the map." (Interview 2).

The category Type of Events is related to the several kinds of events, themes that are organized by the Municipalities, namely cultural events, entertainment events, sports events, and educational events. Illustrating this category in interview 9, it is mentioned, "There are different types of events, in the environmental area, in the cultural area, in the sports area, in the education area, and the social action area."

As for the category Purpose, given the different types of events, there are several purposes. The purposes may range from promoting the identity of the region to the practice of sports, to the creation of personal and professional skills in young people, or to combat school failure. It should be noted that the infrastructures that Municipalities have for the practice of outdoors are a strong point in this field for the growth of sporting events, as can be seen in the following quote "We can train young people competently to prepare them for the job market, but we also want schools to prepare people for life, to live in society, with values, with a taste for what are the roots of their land, to realize their potential, and somehow this also contribute to their personal development, we want educational success to also allow us to have habits of

Table 1 Codes structure

Events											
Type of Events	Purpose	Planning	Infrastructures	Local Community	Disclosure	Disadvantage	Evaluation methods	Results	Organisation	Projects	Future

healthy living, healthy eating healthy life habits, a healthy diet, sports, in the end, the goal is all of this goal is all this, educational success in all dimensions" (Interview 6).

The category Planning encompasses the factors most mentioned by respondents in the planning stage, assuming extreme importance in the organization of events, not losing sight of the purposes of the events and the culture, heritage of the region, and its recognition as a territory, as mentioned in interview 10, "So the most important events are those that in some way exalt our cultural, historical, and natural roots, are the brands, the sierras, the bread, the cookie, the toys, our religious heritage, the slate. And I think that somehow a series of events have been held and developed that have these same identity marks have been praised."

In the category Infrastructures, there are no infrastructures built from scratch for events, for this reason, several existing infrastructures are used that have other purposes such as Museums, auditorium, Forum, or other infrastructures, or outdoor spaces, such as streets and squares of the cities in the Municipalities, as shown in interview 2: "The infrastructures can be varied according to the nature of each event."

The category Local Community is a factor to be taken into account when organizing events where it is held. More and more the involvement and willingness to participate of the local population in the events. An important factor is a fact that the population can revive the old traditions of the region, which helps in the approval of the region to the events held, as shown in interview 3, "We notice that we are getting more adhesion from the public, people are recognizing these events that are being held periodically, so they are looking for participation in the events."

The category Disclosure refers to the way how dissemination is done to reach a greater number of people, as mentioned in the quote: "We have various outreach mediums including the posters, billboards, we have the digital channels, the municipality's website, Facebook" (Interview 11). This disclosure is performed through Digital media, such as social media, website, or television, and Non-Digital Media, such as posters, flyers, and billboards.

In the category Disadvantages, these refer to the events and are because roads have to be cut for the events to take place, which makes it difficult to access certain places. The noise that is originated by the events ends up bothering some of the local population revealing itself as another disadvantage. As mentioned in interview 2: "But there is always a person who complains because they cut a street or they don't have access to a restaurant, the access to the gas stations is or because a lot of noise was made in the street and the baby wanted to sleep, and could not sleep at 10 pm, he only got to sleep at midnight."

The category Evaluation Methods represents the mentioned ways of how the entities, evaluate the performance and the results of the events. It is a sector of the events that the interviewees state that there is a need for improvement. Sometimes they state that they only perceive the results by some kind of feedback from the public and not something structured such as surveys, as presented in interview 3: "We end up understanding how the events are going through the feedback from the public and the participants of the activities themselves."

The Results category reveals that the results obtained from events are diverse. You can get different results depending on the type of event you are organizing. But most of the events end up being a strategy for the Municipalities to show themselves to the world and present all their cultural heritage and all their added value, as the quote reveals: "By attending events, they get to know the region and realize that there are more activities they can do throughout the year, not only cultural ones but also others" (Interview 1). These results are related to the number of visitors, the amount of consumption, the economic impact, and the exposure of the Municipality.

In the subcategory Organization, some aspects that influence the organization of events are identified. In this case, there are some limitations for those who are organizing the events and some aspects that should be improved in the organization, cited in interview 10: "For now we are still planning with the outside in mind, but we are still planning a lot inward, we have a minimally defined strategy, but we are still planning a lot to educate internally and with the perspective of attracting externally, but for now we are responding to the influence of tourism is null" (Interview 10). Limitations are related to some aspects that end up limiting the organization of events were identified, namely the reduced number of workers and the event's costs, which ends up hindering the growth of the events or hindering the work of those who organize them, as can be seen in the quote: "Of the closed spaces that we needed, for example, a large multipurpose pavilion for certain activities which we also don't have so this is a disadvantage for us when we are preparing an event" (Interview 3).

The category Projects has new infrastructure under development in the Municipalities. These projects aim at the growth of the Municipalities, in having more attractions and creating new reasons to visit, as exemplified in interview 4, "we have some projects in agenda that will help to develop the Municipality, to do more events, and to attract more visitor. . ."

In the category Future, the same is expected to be promising. The Municipalities have been developing strategies for the region with the creation of logos that represent the region's greatest assets, with all the events organized, with the various infrastructures that have stood out in the outdoor sports level of outdoor sports and the new strategies that are being developed, as previously mentioned, to grow the region and the growth of the region and thus improve prospects, as mentioned in interview 2: "We are a municipality rich in history, rich in tradition, and a very good path is beginning to be taken towards that promotion. I hope that the future will enlighten even more those who are in this area and that in the future they will be able to help transform the Municipality into a special land."

5 Final Considerations

The organization of events is an important strategy for the promotion of a tourist destination, and it is advisable to focus on different themes (Getz & Page, 2016). In our results, this strategy is reinforced to the extent that municipalities hold entertainment, cultural, sports, and educational events. The first three types of events mentioned belong to three of the four main event typologies elaborated by Getz and Page (2016).

One of the purposes of events is to promote the recognition of the region at regional and national levels. In this regard, Hemmonsbey et al. (2018) note that it is necessary to carefully select events that are best suited to project a genuine image of destinations. The choice of events that give greater visibility to the Municipalities seems to be related to the region's logos that aim to preserve the culture, traditions, and recognition of the region's image. In this way, the events become an important reason to visit the region, as Getz (2008) states that events with their increasing growth allied to tourism are included in the development and marketing strategies of a tourist destination.

In this case, and in line with the literature (Getz & Page, 2016), events are a great attraction for visitors creating a great economic impact in the region, contributing to the increase of consumption on the days of the events and greater dissemination of their products facilitating potential new contacts for business and partnerships.

The existence of events throughout the year seems to have a great financial impact on the economy of a region due to a greater consumption caused by an increase in the number of visitors. Thus, events are a strategy to combat seasonality, making cities more dynamic and lively, and making the region a more attractive place to visit, as stated by Getz and Page (2016).

Events are a reason for tourists to travel and contribute to the increase in the number of visits to a region (Panfiluk, 2015). Additionally, events promote the dissemination of the characteristic products of the region, thus being a differentiating reason to attract visitors to the region. This is why they are increasingly used in the development and marketing strategies of a region (Getz, 2008).

On the other hand, events contribute to the valorization of existing infrastructures in the region. By taking place in the infrastructures themselves or their immediate surroundings, events attract people to visit spaces that they would hardly get to know otherwise.

Although there is a growth in the number of events, some aspects need to be improved for the growth to be even greater and the quality of the events to be improved. The lack of evaluation of the events' results is one of the aspects to be improved. For that to happen, it is imperative to know the opinion of visitors and partners about the opportunities for improvement as well as to improve the way events are promoted.

One of the limitations of this study, it is important to mention that we only have the perspective of one actor in the organization of events—the Municipalities. In future studies, it is suggested to replicate this study with visitors, but also with event partners as well as with local communities and companies to get a more global view about the impact of events in the promotion of a tourist destination.

References

Andersson, T. D., & Lundberg, E. (2013). Commensurability and sustainability: Triple impact assessments of a tourism event. *Tourism Management, 37*, 99–109.

Bathelt, H., & Schuldt, N. (2008). Between luminaires and meat grinders: International trade fairs as temporary clusters. *Regional Studies, 42*(6), 853–868.

Buhalis, D. (2000). Marketing the competitive destination of the future. *Tourism Management, 21* (1), 97–116.

Che, D. (2008). Sports, music, entertainment and the destination branding of post-Fordist Detroit. *Tourism Recreation Research, 33*(2), 195–206.

Connell, J., Page, S. J., & Meyer, D. (2015). Visitor attractions and events: Responding to seasonality. *Tourism Management, 46*, 283–298.

Dickson, C., & Arcodia, C. (2010). Environmentally sustainable events: A critical review of the literature. In *Global events congress IV: Festivals and events research: State of art* (Vol. 14, p. 16). Leeds Metropolitan University.

Getz, D. (2008). Event tourism: Definition, evolution, and research. *Tourism Management, 29*(3), 403–428. https://doi.org/10.1016/j.tourman.2007.07.017

Getz, D., & Page, S. J. (2016). Progress and prospects for event tourism research. *Tourism Management, 52*, 593–631. https://doi.org/10.1016/j.tourman.2015.03.007

Gursoy, D., Kim, K., & Uysal, M. (2004). Perceived impacts offestivals and special events by organizers: An extension and validation. *Tourism Management, 25*(2), 171–181.

Hemmonsbey, J., Tichaawa, T. M., & Knott, B. (2018). Conceptual framework for strategic destination branding through leveraging home-grown sport events. *South African Journal for Research in Sport, Physical Education and Recreation, 40*(2), 81–98.

Hernández-Mogollón, M., Folgado-Fernández, J., & Duarte, P. (2014). Event tourism analysis and state of the art. *European Journal of Tourism, Hospitality and Recreation, 5*, 83–102.

Jamieson, N. (2014). Sport tourism events as community builders – How social capital helps the "locals" cope. *Journal of Convention & Event Tourism, 15*(1), 57–68.

Kelly, D. M., & Fairley, S. (2018). What about the event? How do tourism leveraging strategies affect small-scale events? *Tourism Management, 64*, 335–345.

Liu, Y.-D. (2014). Cultural events and cultural tourism development: Lessons from the European capitals of culture. *European Planning Studies, 22*(3), 498–514.

Mair, J., & Whitford, M. (2013). An exploration of events research: Event topics, themes and emerging trends. *International Journal of Event and Festival Management, 4*(1), 6–30.

Maussier, B. (2017). The new ethical dimension of sports events: A reflection on the evolution from the ancient Greek Olympic sports festival to postmodern sports events. *Cultura, Ciencia, Deporte, 12*(34), 15–25.

Panfiluk, E. (2015). Impact of a tourist event of a regional range on the development of tourism. *Procedia – Social and Behavioral Sciences, 213*, 1020–1027. https://doi.org/10.1016/j.sbspro.2015.11.520

Ratkowski, W., & Ratkowska, J. (2018). Sports events as a determinant of sport tourism. *Baltic Journal of Health and Physical Activity, 10*(1), 86–94.

Saayman, M., & Rossouw, R. (2010). The Cape Town international jazz festival: More than just jazz. *Development Southern Africa, 27*(2), 255–272.

Strauss, A., & Corbin, J. (1990). *Basic of grounded theory methods*. Sage.

van Wyk, L., Saayman, M., Rossouw, R., & Saayman, A. (2015). Regional economic impacts of events: A comparison of methods. *South African Journal of Economic and Management Sciences, 18*(2), 155–176.

Viol, M., Todd, L., Theodoraki, E., & Anastasiadou, C. (2018). The role of iconic-historic commemorative events in event tourism: Insights from the 20th and 25th anniversaries of the fall of the Berlin Wall. *Tourism Management, 69*, 246–262.

Zhong, D., & Luo, Q. (2018). Knowledge diffusion at business events: The mechanism. *International Journal of Hospitality Management, 71*, 111–119.

Entrepreneurship Education in Portuguese Tourism: Fostering Ideation, Innovation and Initiative

Francisco Banha, André Rui Graça, and Francisco Miguel Banha

Abstract At a time when it is important to start preparing the future and devise new ways for economic and financial recovery, we look back to the some of the factors that helped Portugal take a significant leap in the quantity and quality of services related to tourism in the last decade. This chapter will address how state intervention, through an approach based on the triple-helix model, has helped new projects emerge. We analyse Turismo de Portugal's strategy and the Fostering Innovation in Tourism (FIT) programme and how it has been stimulating the economy around tourism through development of entrepreneurial and business skills. In order to do so, we will resort the Tourism Creative Factory programme case study to see not only the methodologies used to foster and promote an entrepreneurial culture throughout the different regions of the country, but also to assess the role that this programme had in attaining the strategic goals set by Turismo de Portugal. The analysis of this case study will add to the literature on entrepreneurship education, as well as to research related to the history and evolution of tourism in Portugal.

1 Introduction

Due to its geographical position, good weather, moderate cost of living, low criminality rates and the quality of its endogenous resources and products, Portugal has been riding the tide and making the most of its touristic potential—especially in

F. Banha (✉)
Universidade do Algarve (FEUALG), CEFAGE-UÉvora and (CinTurs)-UALG, Faro, Portugal
e-mail: fmbanha@ualg.pt

A. R. Graça
Universidade da Beira Interior/Centro de Estudos Interdisciplinares - Universidade de Coimbra, Coimbra, Portugal
e-mail: andre.graca@ubi.pt

F. M. Banha
Universidade Católica Portuguesa, Lisbon, Portugal
e-mail: fmbanha@ucp.pt

J. Leitão et al. (eds.), *Tourism Entrepreneurship in Portugal and Spain*, Tourism, Hospitality & Event Management, https://doi.org/10.1007/978-3-030-89232-6_13

the last 10 years. The big qualitative and quantitative leap, as we will see, was made possible not only by big investments from already-established players, but also by a plethora of enterprises carried out by small companies, led by a new generation of entrepreneurs. Effectively, while the former investors have focused on a more traditional approach to hospitality, the latter have created (mainly) small, boutique projects that have been the disruptive driving force of this new chapter in the history of tourism in Portugal.

This myriad of initiatives and businesses centred around tourism has given the country new economic breadth and provided plenty of case studies in terms of creativity, attention to detail and innovation in the fields of hospitality and tourist experience. In the following pages we will analyse how this significant increase in entrepreneurial activity came about, i.e. how it was propelled and how it took shape through a series of initiatives meant to enable, train and empower people to ideate and create their projects. We will use Turismo de Portugal's FIT programme and its impact as case study. With this analysis, we aim to add to the literature on how strategic state intervention and coordination with people and businesses can help entrepreneurship and innovation in tourism. Moreover, this research matters given the scarcity when it comes to studies on entrepreneurship education[1] (and its implementation) in Portugal.[2]

Tourism (and its related activities) has become one of the largest and fastest-growing economic industries in the world. Currently, tourism directly contributes to about 4.4% to GDP, 6.9% of employment and 21.5% of service exports of the OECD economies. According to recent data, the tourism sector was bound to continue to grow. It was expected to present a weight of 17% of the Portuguese GDP and 20.9% of employment by 2025 (WTTC, 2016).

However, this was all before the outbreak of the Covid-19 pandemic (an era where tourism was in clear expansion in Portugal). There are no studies, as of yet, that demonstrate the impact of the ongoing pandemic[3] (only projections). Nevertheless, Portuguese tourism players have reported average losses of 60% on their revenues (AHRESP, 2021) and dozens of small companies either apply for state support or file for bankruptcy by the hour. If and when the pandemic is over, ceteris paribus, it is predictable that this economic sector will somehow recover and continue to grow, but we cannot predict the direction of its growth or how things will play out in the near future. For that reason, we shall speak in general terms when considering the future and will mention mainly past events that led us thus far. In

---

[1] We use this term to refer to the transfer and transmission of entrepreneurial skills. It is indeed the established term in the academic field and literature, and it is also the concept most closely associated with the approach of the case study analysed, according to the OECD (Lackéus, 2015).

[2] In this particular, we ought to highlight the work carried out by Daniel et al. (2017), entitled *Tourism Education: What About Entrepreneurial Skills?* This paper provides a very interesting and pioneer overview on the transmission of entrepreneurial skills in tourism courses. In a way, we intend to take that study further, provide an up-to-date state of the art, as well as the analysis of a case study concerning the way Turismo de Portugal articulated its strategy on entrepreneurship.

[3] Although the negative impact is rather clear (especially from March 2020 to March 2021).

fact, the size of the damage provoked by the pandemic and the governmental action enforced to prevent its spread reflect the size of the industry itself and an indicator of a mature sector.

The continuing growth of this sector has a significant impact on labour markets, since tourism is a significant source of employment, due to its high labour-intensive nature, mainly in a market like the Portuguese which is strongly influenced by tourism in its territory. The tourism sector has been the largest economic export activity in Portugal. In 2019, it was responsible for 52.3% of services exports and 19.7% of total exports (Turismo de Portugal, 2019).

This economic sector (which is now struggling to reinvent itself within the measure of the possible to fight back the impact of lockdowns and survive economic crises) has seen considerable development in the last decades. Its outlook as potential income source has also changed, as it has also been increasingly seen as attractive to young people and investors wanting to create value. Thus, there has been a rather considerable increase in the number of educational programmes offered by schools and higher education institutions in this field (Airey, 2005, 2016; Wang et al., 2013). The registered activity rate in Portugal, six months after concluding a tourism-related degree, is 90.0%. Of these students, 90.0% found a placement in less than three months. Moreover, 19.4% were integrated in the hotel or restaurant units where they did their internship (Turismo de Portugal, 2021a).

Nevertheless, due to the highly competitive environment of the tourism industry, entrepreneurship and innovation are crucial for dealing with current consumer, technological and environmental trends (Daniel et al., 2017). Innovation in tourism may contribute to the competitiveness of a state and lead to national prosperity and wealth. A specialised network focused on new ideas and business models associated with tourism, with the ability to revitalise and enhance innovation and the competitive growth of the sector (Hall & Williams, 2008). The Portuguese public body for tourism, Turismo de Portugal, has a network of schools that is also associated with the innovation initiative since decision-makers intend that students, when they finish their courses, can participate with their ideas in these ideation, acceleration and open innovation programmes. During their studies, they will have been stimulated to develop these ideas in the entrepreneurship syllabus (Turismo de Portugal, 2021b).

Students entering the tourism sector, with its high levels of volatility and rapid globalisation, need different skills and abilities to achieve meaningful and successful professional lives (Dredge et al., 2015). Thus, governments worldwide, such as the Portuguese and Spanish governments, adopt programmes aimed at taking action in supporting innovation to enhance national innovativeness (Hall, 2008). The FIT programme—Fostering Innovation in Tourism—by Turismo de Portugal is one of these programmes and it will be at the centre of our analysis. Indeed, several studies have elaborated on governmental interference in the area of innovation, government involvement on a national scale and the contribution that this involvement has brought into national systems of innovation (Edquist & Chaminade, 2006; Rodrigues et al., 2015).

In the context of tourism, various types of innovations that exist have been examined (Budiman et al., 2020; Rodrigues et al., 2015; Volo, 2005). Yet, there is a lack of focus on the factors that contribute to the innovation factor and the role played by government (Batra, 2016; Edwards et al., 2008; Jugmohan & Giampiccoli, 2017). In fact, the definition of innovation is not yet stabilised. Therefore, it is widely discussed by several scholars the features and nature of innovation in an ever-changing, ever-evolving environment—in this case, tourism. Thus, a successful working definition has yet to be developed (Hjalager, 2010; Sardak et al., 2016).

In the tourism industry, innovation is often divided into five categories: Product or service innovations; Process innovations; Managerial or organisational innovations; Marketing innovations and Institutional innovations (Hall, 2009; OECD, 2018, 2020; OECD & European Union, 2005). Moreover, innovation can be influenced by the government's decisions, because it controls several policy instruments that can be used to influence and foster its various categories (Michael & Pearce, 2009). Typical government actions to promote innovation may include engaging in activities like R&D, funding the education sector, tax breaks, promoting the creation of innovation clusters and deployment of new technology (De la Vega & Stankosky, 2006). In fact, governmental influence is so great that it basically shapes the rules of the game through regulation, taxation and legal processes (Basheer et al., 2019; Mei et al., 2013).

Despite the many arguments for government involvement, others also stress the negative sides of public sector interest (Bryant et al., 2017; Hartley, 2005). Some claim that innovation policies have difficulty in helping up and coming economic activities that require long-term development (Barbero et al., 2021; Boschma, 2005). Although government involvement is inevitable due to its role as the ultimate guarantor of social rules and minimal conditions, an ultra-liberal approach to the subject might find that regulations limit the ability of individuals and groups to freely organise, debate and take initiatives by themselves to enhance allegedly, attain maximum efficiency—especially in highly-regulated societies (Etzkowitz, 2003). Nevertheless, it seems clear that tourism innovation at a national level is not effectively possible without government intervention, as it is crucial in the general economic development.

Considering, for instance, the example of a researched conducted in Norway regarding the role of governments in tourism innovation (Mei et al., 2013), the results confirmed that public investment through government does have a key facilitating role to play in terms of providing the framework in which the private sector can innovate—although true innovation and successful outcomes will depend on the involvement and drive from actors in both the private and public sectors. These results, despite being collected from Norway (which is a wealthy and well-developed country, and therefore with its own national, geographical and regional idiosyncrasies) may be applicable to other countries with similar stages of development in the field of tourism.

Another approach that worked well in regions with non-competing destinations was the Canadian Product Club Programme.[4] Created in 1996, it promoted product development for 45 product categories through research and development facilitation (Weiermair, 2006). Two hundred and one co-ordinating partners and more than 5000 businesses cooperated through this product club programme. The author concluded that:

> this poses the well-known policy question as to how much governments ought to spend on innovation projects in tourism on account of externalities and prohibitive risk and cost of process and product innovation and the indivisibility of research and development activities vis-à-vis a multitude of small user firms. (Weiermair, 2006)

Secondly, if it is clear that governments should get involved, the question is how money should be spent. A mention that deserves to be pointed out is the provision of a well-functioning, free market. In fact, the difference between North American versus European tourism companies' or entrepreneurial firms' mandatory regulation and point of view is substantial when innovation barriers are considered—with one key difference being the amount of bureaucracy existing in European cases. In the context of the above-mentioned literature we will provide insights into the Portuguese case—especially when it comes to programmes aimed at educating entrepreneurs and guide people through the process of setting up a business in the context of tourism.

2 The Portuguese Case

Tourism in Portugal is regulated and represented by a secretary of state, incorporated within the ministry of economics and digital transition. The institution that puts in practice the ministerial guidelines and serves as link between the government and the business community is Turismo de Portugal (TdP). This entity, which is a public institute and the national authority in all things regarding tourism, operates in two fronts, inwards and outwards; its mission rests on four main pillars: promote Portugal as travel destination; support the development of infrastructures and investment in the sector; develop and train human resources and regulate and oversee gambling activities. In order to carry out its purpose, TdP is divided into three sections: the planning section, which is in charge of devising strategies and carry out quality control; the business section, responsible for the development and maintenance of the business activity related to tourism in the country; and the support section, focused on activities such as communication, finance, HR and legal matters.

In order to materialise its mission with regard to the training of skilled professionals for the tourism and hospitality sector, TdP runs a network of 12 vocational

---

[4]For more on the discussion about innovation and sustainable tourism is Canada, see Thout et al. (2010).

schools, spread throughout the country. These schools, which have been inspired by the best international institutions (such as *EHL Lausanne Tourism School*), have been providing the different Portuguese regions with thousands of young graduates, that have received intensive training, looking to enter the labour force. Thus, TdP schools have been pivotal in setting higher standards and uplifting the overall quality of services in tourism and hospitality in the country.

In addition to the core training programmes, which are mainly job-oriented and focus on the basic specificities and know-how of hospitality, TdP has been expanding the scope of training through various initiatives. In 2017, TdP released an official document containing a global, 10-year strategy, for tourism in Portugal. Among the different goals and topics in the *Estratégia Turismo 2027*,[5] one stands out: the importance of investing more in innovation and entrepreneurship.[6] Indeed, the stimulation of innovation and entrepreneurship was signalled as one of the 10 overall goals of this strategy and much of its operationalisation is meant to be carried out through the fostering of an entrepreneurial culture, in which different stakeholders and players can become connected and promote the growth of the economy centred around tourism in particular and the overall economic growth and prosperity of the country in general.

In line with the guiding principles of its mission and the objectives enshrined in this document, the business section of TdP created the Fostering Innovation in Tourism (FIT)[7] programme. This initiative, which we found to be aligned with the triple-helix model (Etzkowitz & Leydesdorff, 1995), linked the TdP[8] with a network of business incubators and education programmes in the realms of business ideation and business acceleration. The goal is to empower the community[9] with the necessary knowledge and motivation to develop and shape business models related to tourism and, ultimately, enable entrepreneurs to engage with the entrepreneurial ecosystem[10] in order to start new, viable businesses. According to TdP the FIT programme has two strands (and therefore looks for two types of training in companies related to entrepreneurship education): one related to ideation, which is

---

[5] https://estrategia.turismodeportugal.pt/sites/default/files/Estrategia_Turismo_Portugal_ET27.pdf

[6] It should be noted that, before the final version of this document there was a document put to public scrutiny and discussion, in 2016. This was the year of the first edition of the Tourism Creative Factory. The success of this programme was acknowledged in the 2017 document by the then minister of economics, Manuel Caldeira Cabral in his opening note and the 2017 document underlines the importance of entrepreneurship and innovation more emphatically probably due to the globally positive impact of their 2016 initiatives around those topics.

[7] http://business.turismodeportugal.pt/pt/Conhecer/Inovacao/programa-fit/Paginas/default.aspx

[8] Especially through its schools. However, the programme is open to the broader community of entrepreneurs-to-be.

[9] We would like to underline that, the Tourism Training Talent programme (included in the TdP 2027 strategy) is aimed at providing students from TdP vocational schools with soft skills—many of them related to entrepreneurship. As we will see, FIT and its associated programmes fall into this scope.

[10] For a detailed account of the Portuguese entrepreneurial ecosystem and the main barriers to its development, see Banha et al. (2017).

carried out in partnership with the TdP schools; and another dedicated to acceleration and open innovation—which is promoted by the business incubators network. TdP also supports the presence of start-up companies at international tourism fairs.

The importance of the FIT programme has been paramount in changing the paradigm in the country when it comes to the scope of entrepreneurship education and its impact.[11] For instance, in 2020, 16 programmes were selected,[12] involving 327 startups and projects. With regard to the two strands, several companies have collaborated regularly with TdP to promote the entrepreneurship education initiatives that are inherent to FIT. However, two stand out: GesEntrepreneur (ideation) and Fábrica de Startups (acceleration and open innovation). The former through the Tourism Creative Factory programme (TCF) and the latter through the Tourism Explorers and Discoveries programmes. We will now look into the TCF case study in order to provide an account of the methodologies used and how they created an impact in the entrepreneurial ecosystem throughout the years.

3 Case Study: Tourism Creative Factory

As mentioned, the TCF is an entrepreneurship education programme, provided to TdP by the company GesEntrepreneur, focused on business ideation. Starting in 2016, TCF had, so far, five editions. According to reports given by GesEntrepreneur to TdP (and validated by this institute), the hard numbers boil down to: involvement with eight schools (seven in Portugal's mainland and one in the Azores), 316 project applications, 189 projects (approved after the first stage) and 281 entrepreneurs. However, even with such loquacious numbers, it is important to tell the story behind them and how TCF, through its methodologies, has successfully worked to maintain its partnership with TdP, encompass its demands and challenges and be able to attain the above-mentioned numbers.

Founded in 2006, GesEntrepreneur has been working with (and within) the Portuguese entrepreneurial ecosystem. Resorting to a "learning-by-doing" approach, the company has consistently worked with several public and private institutions, such as intermunicipal communities, Government of Azores or EDP—Energias de Portugal (the county's main electric utilities company). With regard to the

[11] Ideation programmes, which eventually lead to venture creation, according to the OECD (Lackéus, 2015): "allows for the development of entrepreneurial passion and perhaps even entrepreneurial identity in some learners. The value created as formal part of curriculum can be so significant that it sometimes leads to real-life economic growth for the collaboration partners outside the educational institution. The final output of the embedded approach is more entrepreneurial people creating new kinds of value in all domains of society and all walks of life".

[12] Four in the realm of ideation, five acceleration programmes and seven open innovation programmes. It should be noted that all entities that promote these programmes answer a public call released by TdP. The programmes are selected and contracted based on the merit of the proposal.

involvement in the FIT programme, GesEntrepreneur designed TCF to help improve the output (both in quantitative and qualitative terms) of viable business ideas. In other words, to promote the development of entrepreneurial and business skills, through training, in those in charge of their projects. TCF also allows its participants to engage with regional entrepreneurial ecosystems, the possibility to integrate circular economy initiatives and connect with qualified mentors/businesspeople that will enrich their experience.

In practical terms, the TCF programme consists of a six-step course with a multidisciplinary approach to business (product design, marketing, financial education, business management and tourism). The courses usually run for 10 or 14 weeks. During this timeframe there is a dimension of continuous self-improvement.[13] Also, given that the FIT programme is decentralised, so has been the TCF.[14] This can be better understood by explaining the process; the six steps are the following[15]:

1. Open call

   This is the first stage of the TCF programme. As the name implies, this a period when a call is released and interested applicants submit their proposals online. A preliminary screening occurs at this moment, i.e., only projects that fall within the scope of, at least, one of the eligible business areas (a factor that is pre-determined by TdP) are accepted. Since the TCF programme is sponsored by TdP, the TCF experience is free for applicant who sees their projects selected. The projects with most potential (around 40) are selected and continue to the next stage of the process. The selection criteria are: the degree of innovation; viability; overall potential impact for the tourism sector; maturity of the project and merit of the team.
2. Bootcamps

   Bootcamps (applied to the TCF context) are intensive training moments in which there is a strong interaction between the participants and the training staff. They occur in different cities and usually last for three days. On the one hand, bootcamps allow participants to be fully immersed in a setting where they must

[13]Entrepreneurship skills presuppose that they are known and understood (JO EU, 2018: C 189/11) "(...) project planning and management methods (...) strategic thinking, critical and constructive reflection on the evolving creative processes and about innovation (...) ability to work individually and as a team, to mobilize resources (people and things) and to develop a sustainable activity (...) ability to make financial decisions related to cost and value (...) communicate and to negotiate effectively with others and to deal with uncertainty, ambiguity and risk as elements of the informed decision-making process".

[14]One of the TCF's goals is to bring to people outside the Lisbon area specialised training and entrepreneurial culture. During this process, people will be exposed to a course based on the best international practices and promote networking activities.

[15]To some extent, TCF mirrors the broader methodology used in other programmes and initiatives promoted by GesEntrepreneur—which are based on international best practiced and adapted to the Portuguese context (especially its entrepreneurial ecosystem and geography). For a detailed overview and explanation of these programmes, see Banha et al. (2021).

develop their communication skills, take part in training activities and focus on improving their projects upon receiving feedback and inputs from different coaches. On the other hand, bootcamps aim to assess the specific needs of each team, as well as the strengths and weaknesses of each project, in order to address those issues through a custom-made plan.

3. Regional Demonstration Day (whenever possible)

   To conclude the bootcamps, the communication skills and the maturity of each project and team are put to the test. In a public session, where the entrepreneurial community or that specific region, coaches, staff and TCF ambassadors are present, the participants must pitch their ideas. The interventions are recorded for future reference and analysed by the TCF staff in order to evaluate what should be addressed first in the following step and track the evolution of the project and the team throughout the programme. This step is part of the overall methodology. However, due to practical and specific reasons, sometimes, the public session does not occur. All of the other elements are still carried out.

4. Business modelling

   This stage focuses on the development of the projects. To that end, contact with several business modelling tools and concepts is provided to participants who take classes on: pitching ideas; minimum viable product; pivoting; branding; business model/lean canvas; branding; business model; legal and compliance basics; fundraising and strategic communication. Since the TCF has the ambition of being a far-reaching programme, this stage also occurs in different areas of Portugal, in partnership with the nearer TdP schools.

   Throughout the years, point of contact in these schools has been designated as ambassadors, who work closely together with the TCF staff. The programme endeavours to provide an experience as homogeneous as possible, regardless of the area, i.e., put participant in contact with same number of mentors and tutors, and allow them to participate in side-events as well. Given that there is a great concentration of ideation programmes in Lisbon for obvious reasons, the TCF assumes the social commitment of reaching different areas and provide a broad territorial coverage. This stage usually lasts for about 2 months.

5. Mentoring

   During this stage the participants are put in contact with experts in the field. They are given access to classes and workshops on specific aspects of tourism and engage in networking activities. The mentors, who are already-established personalities in their fields of business, interact formally with participants (in one-on-one sessions or in classes) and informally (such as in the casual setting of a business lunch or dinner). Some of these mentors are also investors or business angels and may provide interesting advice in various aspects of business or network tips.

6. National Demonstration Day

   This is the zenith of the TCF programme. After successfully completing the business modelling stage, participants will have the chance to publicly present their improved projects. This event will take place in a high-profile venue in a Portuguese city. Tourism experts, investors and representatives of potentially interested companies will be among the audience. In the end, a certificate of conclusion is issued, along with a letter of recommendation stating the development stage and potential of each project.

The TCF programme has had the same structure since its inception and continues to improve, following a *kaizen* philosophy based on the feedback provided by previous participants, the profile they present and the needs they show in general. Recently, due to the covid-19 pandemic, some changes had to be put in place to accommodate the new reality of education methods and the hospitality sector in the country. Based on the best possible knowledge, i.e., studies and projections available throughout the world, GesEntrepreneur identified one particular field of business (within the broader scope of TdP) that show resilience, still has room to grow during the pandemic situation and will, hopefully, be more economically relevant in the future: gastronomy. This adaptation is part of the sustainable entrepreneurship *ethos* of the TCF. Resorting to a similar structure, the 2021 edition of the TCF programme had the subtitle "food edition",[16] since it focuses specifically on gastronomy, endogenous products and initiatives around wine tourism.

4 Outcomes

Having presented here the history and functioning of the TCF programmes as part of the broader strategy by TdP to boost entrepreneurship in Portugal in the field of tourism, we should highlight some of the results of the FIT Programme.

With regard to the acceleration and innovation strand, a report from Fábrica de Startups,[17] which traces the evolution of their two programmes (Discoveries and Tourism Explorers) from 2015 to 2020, states that, in eight editions (five from the Discoveries Programme and the remaining three pertaining to the Tourism Explorers programme), 1124 applications were received, 380 startups were created and 932 entrepreneurs were reached. Of those 380 startups, 60% were still active in 2020, 30% received external investment to stimulate their growth and 39% of those startups had managed to internationalise their business/service. If we add the TCF programme results to these numbers,[18] from 2015 to 2020, 1443 projects submitted their application to these three programmes and a total of 1213 entrepreneurs actively participated in them (an average of 243 per year).

Moreover, the TCF programme has been the genesis of some successful businesses (across the whole country) that have contributed to change the landscape of

---

[16] https://tourismcreativefactory.com

[17] https://www.fabricadestartups.com/blog/25/2020/relatorio-de-impacto-startups-de-turismo

[18] It should be noted, once more, the difference in the nature of the initiatives on ideation and those on acceleration. The TCF, by its own nature and size of its structure, supervises fewer projects than acceleration projects.

travel and accommodation services. We shall mention a few examples. Amicis Gin[19] was created in 2016. After a partnership with a business angel, the brand was created. The product utilises endogenous resources from the Beiras regions in its formula and has already a solid implementation in the Portuguese market. Coimbra Airport Shuttle[20] is also another good example. Given the fact that the Portuguese mainland is served, mostly by three airports (in Porto, Lisbon and Faro—which serves specifically the Algarve), this service allows an unprecedented access to people from the Centro region to both Porto and Lisbon airports. Another example is Eco Soul Guest House Ericeira,[21] a small and cosy Guest House, located between the beach and the valley of Sao Juliao, inserted in the World Surfing Reserve of Ericeira. The name Eco Soul refers to the decisions that the founders made when they started this project on what concerns the sustainability and environment issues. Abrigo Queijo Serra da Estrela DOP[22] is one of the latest investments made by a public venture capital firm and consists on a Museum that will gather and show all the material and immaterial heritage on the ancestral manufacture of Serra da Estrela Cheese. Finally, TryPortugal,[23] which has attracted funding from a venture capital firm, has also built a reputation in being a provider of customised experiences in the context of active tourism, using as background the cultural and natural heritage of the country.

Overall, the FIT programme and its associated initiatives have actively contributed not only to boost the quantity and quality of services in Portugal but has also become an example of continued commitment and investment in entrepreneurship education programmes.

5 Final Remarks

Entrepreneurship education is a field in expansion in Portugal, with much room for further implementation. The FIT programme and the results we have presented are important evidence of the impact that this type of education can produce in a very unequal territory, marked by regions of very high or very low density in terms of population and wealth. The TCF programme has strived to reach various zones of Portugal and help entrepreneurs make the most of the characteristics and resources of their regions. Hopefully, this text will shed new light on the state of the art concerning entrepreneurship education in Portugal (especially when it comes to

[19] https://amicisgin.com/en/

[20] https://www.airportshuttle.pt

[21] https://www.ecosoulericeira.com/pt-pt/?fbclid=IwAR1FvuHLeJ7Ggw8GCW-bCyiMBLzX9POVEMQlNWMOg8XI-X-3ycBgRz15yT0

[22] https://abrigoqueijoserradaestrela.pt/en/

[23] https://tryportugal.pt

the tourism and hospitality sector) and help explain the qualitative and quantitative leap that the country has taken in the last decade.

It should be noted that only an efficient operationalisation of the strategic guidelines of TdP can guarantee the fostering of innovation in tourism. The concrete data presented here should not only contribute to the academic discussion around the innovation in tourism in Portugal and Spain but may also contribute to the wider research about the relevance of entrepreneurship education. Indeed, the success of the FIT programme[24] seem to bring to the fore proof that these initiatives do have a significant reach and fulfil their purpose. Perhaps this example may be able to shape the creation of future evidence-based policymaking around entrepreneurship education or encourage further research on the topic. Ultimately, all of these initiatives are bigger than the sum of the parts. This is a mosaic, consisting of hundreds of pieces (some larger, some smaller) that coexist within the same frame and are, somehow, interconnected.

Acknowledgments An especial acknowledgement is due to the Turismo de Portugal Public Institute. Namely its board of administrators and its staff for providing crucial information regarding the implementation of the TCF project in Portugal.

Special acknowledgement is also due to the whole TCF project staff. They made the development of such a great experience possible. Special thanks to all teachers, students, mentors, entrepreneurs and TCF technicians involved.

Very special thanks to António Francisco Mourato for assisting with proof-reading, an effort that contributed to the improvement of this paper.

Declaration of Conflicting Interests The authors declare no potential conflicts of interest with respect to the research, authorship and/or publication of this text. This paper did not receive any funds from third parties. The first author intervenes in this paper in his academic capacity (and not corporate), therefore implying he strictly abode to academic ethics and methodologies and practiced consistently a self-reflexive oversight regarding his contribution.

References

AHRESP. (2021). Retrieved from https://ahresp.com/2020/12/inquerito-impacto-covid-novembro-2/

Airey, D. (2005). Growth and development. In D. Airey & J. Tribe (Eds.), *An international handbook of tourism education* (pp. 13–24). Elsevier Ltd..

Airey, D. (2016). Tourism education: Past, present and future. *Turisticko poslovanje, 2016*, 9–12. https://doi.org/10.5937/TurPos1617009A

Banha, F., Almeida, H., Rebelo, E., & Orgambídez-Ramos, A. (2017). The main barriers of Portuguese entrepreneurship ecosystem: Interpretive Structural Modeling (ISM) approach. *Tourism & Management Studies, 13*(2), 60–70.

[24] In 2018, the UNWTO awarded the TdP schools and their Tourism Training Talent initiative with the first place in the 14th edition of the UNWTO prizes. It was considered an example of training of future generations of human resources in the sector (https://escolas.turismodeportugal.pt/en/p/about-us/). It should be noted, that TCF had already been involved with TdP and its schools since 2016.

Banha, F., Saúde, S., & Flores, A. (2021). Entrepreneurship education: A dimension of citizenship education. In S. Saúde, M. A. Raposo, N. Pereira, & A. I. Rodrigues (Eds.), *Teaching and learning practices that promote sustainable development and active citizenship* (pp. 107–132). IGI Global.

Barbero, J., Mandras, G., Rodríguez-Crespo, E., & Rodríguez-Pose, A. (2021). Quality of government and regional trade: Evidence from European Union regions. *Regional Studies, 55*(7), 1240–1251. https://doi.org/10.1080/00343404.2021.1873934

Basheer, M., Ahmad, A., & Hassan, S. (2019). Impact of economic and financial factors on tax revenue: Evidence from the Middle East countries. *Accounting, 5*, 53–60. https://doi.org/10.5267/j.ac.2018.8.001

Batra, A. (2016). Bridging the gap between tourism education, tourism industry and graduate employability: intricacies and emerging issues in Thailand. *ABAC Journal, 36*, 78–89.

Boschma, R. A. (2005). Rethinking regional innovation policy: The making and breaking of regional history. In G. Fuchs & P. Shapira (Eds.), *Rethinking regional innovation and change: Path dependency or regional breakthrough?* (pp. 249–271). Springer.

Bryant, C., Akkari, C., Bousbaine, A., Delusca, K., Daouda, O., Sarr, M., & Azzeddine, M. (2017). The unintended negative consequences of government actions and initiatives in selected environmental, social and economic domains: Opportunities for co-construction approaches. *Journal of Settlements and Spatial Planning, 8*, 79–88. https://doi.org/10.24193/JSSP.2017.2.01

Budiman, A., Nugraheni, T., & Purnomo, P. (2020). The effect of architecture of arts education tourism towards interest in learning arts for high school students. *Harmonia: Journal of Arts Research and Education, 20*, 117–125. https://doi.org/10.15294/harmonia.v20i2.24845

Daniel, A. D., Costa, R. A., Pita, M., & Costa, C. (2017). Tourism education: What about entrepreneurial skills? *Journal of Hospitality and Tourism Management, 30*, 65–72.

De la Vega, A. F. R., & Stankosky, M. (2006). Knowledge management and innovation: What must governments do to increase innovation? *Journal of Knowledge Management Practice, 7*(4), N/A.

Dredge, D., Schott, C., Daniele, R., Caton, K., Edelheim, J., & Munar, A. M. (2015). The tourism education futures initiative. *Anatolia, 26*(2), 340–346.

Edquist, C., & Chaminade, C. (2006). *Industrial policy from a systems-of-innovation perspective*. Centre for Innovation, Research and Competence in the Learning Economy (CIRCLE).

Edwards, D., Martinac, I., & Miller, G. (2008). Research agenda for innovation in sustainable tourism. *Tourism and Hospitality Research, 8*(1), 56–61.

Etzkowitz, H. (2003). Innovation in innovation: The triple helix of university-industry-government relations. *Social Science Information, 42*(3), 293–337.

Etzkowitz, H., & Leydesdorff, L. (1995). The triple helix - university-industry-government relations: A laboratory for knowledge based economic development. *EASST Review, 14*(1), 14–19.

Hall, C. M. (2008). *Tourism planning: Policies, processes and relationships* (2nd ed.). Prentice Hall.

Hall, C. M. (2009). Innovation and tourism policy in Australia and New Zealand: Never the twain shall meet. *Journal of Policy Research in Tourism, Leisure and Events, 1*(1), 2–18.

Hall, C. M., & Williams, A. M. (2008). *Tourism and innovation*. Routledge.

Hartley, J. (2005). Innovation in governance and public services: Past and present. *Public Money & Management, 25*(1), 27–34.

Hjalager, A.-M. (2010). A review of innovation research in tourism. *Tourism Management, 31*(1), 1–12.

Jugmohan, S., & Giampiccoli, A. (2017). Community-based tourism development: A possible educational gap. *Anthropologist, 30*, 52–60. https://doi.org/10.1080/09720073.2017.1353667

Lackéus, M. (2015). *Entrepreneurship in education: What, why, when, how*. OECD.

Mei, X. Y., Arcodia, C., & Ruhanen, L. (2013). The national government as the facilitator of tourism innovation: Evidence from Norway. *Current Issues in Tourism, 18*(12), 1172–1191.

Michael, S. C., & Pearce, J. A. (2009). The need for innovation as a rationale for government involvement in entrepreneurship. *Entrepreneurship & Regional Development, 21*(3), 285–302.

OECD. (2018). *Oslo manual: Guidelines for collecting and interpreting innovation data* (4th ed.). OECD.
OECD. (2020). *OECD tourism trends and policies 2020*. OECD.
OECD & Statistical Office of the European Communities. (2005).
Rodrigues, I., Williams, A., & Hall, C. (2015). Tourism innovation policy: Implementation and outcomes. *Annals of Tourism Research, 49*, 76–93.
Sardak, S., Dzhyndzhoian, V., & Samoilenko, A. (2016). Global innovations in tourism. *Innovative Marketing, 12*, 45–50. https://doi.org/10.21511/im.12(3).2016.04
Thuot, L., Vaugeois, N., & Maher, P. (2010). Fostering innovation in sustainable tourism. *Journal of Rural and Community Development, 5*(1), 76–89.
Turismo de Portugal. (2019). Retrieved from http://www.turismodeportugal.pt/pt/Turismo_Portugal/visao_geral/Paginas/default.aspx
Turismo de Portugal. (2021a). Retrieved from https://escolas.turismodeportugal.pt/p/6-motivos-para-nos-escolher/
Turismo de Portugal. (2021b). Retrieved from https://escolas.turismodeportugal.pt/p/tourism-creative-factory/
Volo, S. (2005). A consumer-based measurement of tourism innovation. In M. Peters & B. Pikkemaa (Eds.), *Innovation in hospitality and tourism* (pp. 73–87). Haworth Hospitality Press.
Wang, J., Ayres, H., & Huyton, J. (2013). Is tourism education meeting the needs of the tourism industry? An Australian case study. *Journal of Hospitality & Tourism Education, 22*, 8–14. https://doi.org/10.1080/10963758.2010.10696964
Weiermair, K. (2006). Product improvement or innovation: What is the key to success in tourism? In *Innovation and growth in tourism*. OECD Publishing.
WTTC. (2016). *Travel & tourism economic impact 2015 - Portugal*. World Travel & Tourism Council.

Francisco Banha holds a Ph.D. in Economics and Management Sciences from the University of Algarve, where he is a guest assistant professor. He is also a guest assistant professor at ISEG—Lisbon School of Economics and Management. He is the CEO and chairman of the Gesbanha Group and founder of GesEntrepreneur.

André Rui Graça holds an MBA from the University of Coimbra and a Ph.D. from University College London. He is a guest assistant professor at the University of Beira Interior, a researcher at CEIS-20/University of Coimbra and a freelance consultant.

Francisco Miguel Banha holds an MSc in Management from the Portuguese Catholic University, where he is a teaching assistant. He has worked as a consultant at KPMG and Deloitte and is currently the CFO of Gesbanha Group.

Factors Affecting Portuguese and Spanish Hotel Guests' Safety Perception Under COVID-19: Insights for the Development of Personalized Hotel Offers

Cristina Estevão, Paulo Duarte, Adília Cabral, Ana María Campón-Cerro, and Uci Yuliati

Abstract Portugal and Spain have both cultural differences and similarities. The current study evaluates the hotel safety perception by Portuguese and Spanish tourists during the COVID-19 pandemic. It discusses the differentiated changes needed to be implemented by the hotel industry to address each country's tourists' safety concerns. A sample of 533 Portuguese and Spanish hotel guests was inquired using an online questionnaire to assess safety perception and service needs under the COVID-19 situation by exploring seven critical dimensions: financial risk, front-desk digitalization, entertainment, front-desk service, hygiene and sanitation, hotel characteristics, and space management, and food and beverage service.

The results show that there are differences regarding safety perception and preferences between Spanish and Portuguese hotel guests. Findings indicated that the differences spread across all categories with special incidence in hygiene-related factors. The differences vary with gender, and Women seem to be more sensitive to safety as they present more differences than men. The level of education and having children also affects the perception of hotel safety and the requisites of guests. The need to prove hygiene care and have the hotel certificate for COVID-19 emerged as key factors in restoring the client's trust.

The research on the effects of pandemics on hotel guests' safety perception is scarce. Therefore, the proposed framework for assessing hotel clients' perceptions and the findings are particularly important for Iberian hotel managers to adopt an innovative approach and shape their offer to better address specific needs and

C. Estevão (✉) · P. Duarte
NECE-Research Center in Business Sciences, Universidade da Beira Interior, Covilhã, Portugal
e-mail: cristina.estevao@ubi.pt

A. Cabral
Polytechnic Institute of Coimbra, Coimbra, Portugal

A. M. Campón-Cerro
Universidad de Extremadura, Cáceres, Spain

U. Yuliati
Universitas Muhammadiyah Malang, Malang, Indonesia

J. Leitão et al. (eds.), *Tourism Entrepreneurship in Portugal and Spain*, Tourism, Hospitality & Event Management, https://doi.org/10.1007/978-3-030-89232-6_14

preferences of Portuguese and Spanish clients in the face of the COVID-19 pandemic.

1 Introduction

Peace, safety, and security are essential conditions for tourism development in any country, region, or destination (Cavlek, 2002). However, over the past decades, the tourism industry has been severely hampered by the increasing lack of security caused by crime, terrorism, food security, natural disasters, and health problems (Breda & Costa, 2006).

The tourism industry is most vulnerable to disasters and usually experiences the effects of a disaster across organizations and geographic boundaries (Jiang & Ritchie, 2017). According to Sydnor et al. (2011), any crisis or disaster ends up making a certain region/country lose some tourists and damaging its image. Even a small disaster may have huge impacts on tourism and cause severe consequences to the hotel industry (Sydnor et al., 2011).

The hospitality sector represents an essential piece for tourists' satisfaction and is extremely important for the tourism industry since it is responsible for many jobs (McManus et al., 2008). However, to remain competitive, the hotel industry must preserve its current customers and attract new ones (Nunkoo et al., 2019). Innovation is considered a critical source of competitive advantage (Dess & Picken, 2000; Tushman & O'Reilly, 1996) and the most important determinant of organizational performance (Mone et al., 1998). Since the hotel industry operates in a very competitive environment, having the best marketing strategies by strengthening the offer in vital dimensions, like the cleanliness, diversity, and quality of services and safety, is crucial to ensure profitability and long-term survival (Ching, 2007). Hence, hotel managers must emphasize innovation to increase service quality and improve customer satisfaction as a key strategy. However, to achieve these objectives, the hotel industry needs to know customers' concerns in difficult times and use innovation to provide a personalized answer to guests' needs and worries (Mártinez-López & Vargas-Sánchez, 2013).

Prideaux et al. (2003) suggest that one thing that could be very demanding for the hotel industry's innovation capacity could be disasters, especially pandemics caused by a new strain of influenza or other unknown diseases. Their concern turned out to be right, as confirmed by the arrival of COVID-19, which was officially considered a pandemic on March 11, 2020, by the World Health Organization (2020).

To assist the hotel industry in understanding where and how to innovate under a pandemic crisis, the current study focuses on identifying what dimensions hotel industry are more critical for Portuguese and Spanish hotel guests, as Portugal and Spain are two of the most visited countries in the world. These study outcomes are useful for hotel managers, who by better understanding the safety concerns of tourists from each country can innovate and customize the services provided to better address and resolve the guests' anxieties and fears. Hence, this study discusses the differentiated changes that should be implemented to better address the concern of each type of customer according to seven critical dimensions: financial risk, digitalization of reception, entertainment, front desk service, hygiene and sanitation, hotel characteristics, and space management, and food and beverage service. The research findings provide a theoretical contribution for the profiling of the two countries tourists aims but also a managerial contribution by guiding the innovative management of hotel–client relationship by Iberian hotel managers to shape their offer to better respond to the specific needs and preferences of Portuguese and Spanish customers in the face of the COVID-19 pandemic.

2 Literature Review

The economic disruption resulting from the COVID-19 epidemic outbreak is having a major impact in tourism markets (González-Torres et al., 2021). Travel restrictions around the globe have led to staggeringly low occupancy levels in hotels (Pillai et al., 2021).

Different types of crises and disasters have had different effects on the hotel industry. The COVID-19 pandemic, in particular, has forced the hospitality industry to take unprecedented measures to respond to unanticipated challenges and negative consequences affecting the DNA of hotels at their core (Rivera, 2020). In this uncommon scenario, the hotel industry must identify the critical factor under pandemic and segment the customers according to how they decide when choosing a hotel (Choorichom, 2011). The emergence of a major pandemic, such as COVID-19, has reinforced the need to update the existing knowledge as travelers' expectations have changed (Hu et al., 2021). A shift in consumers' expectations is likely to impact their perceptions of hotel performance and requisites (Hu et al., 2021), thus demanding novel approaches to the services provided.

Companies must realize the best ways to create value for their target markets and develop strong, profitable long-term relationships with customers (Lee & Hwang, 2011). To achieve these objectives, the hospitality industry should implement demographic profiling and enhance the knowledge and understanding of customer demographic differences concerning expectation, perception, and quality of service (Tsiotsou & Vasioti, 2006).

According to Lee and Hwang (2011), demographic variables are especially useful sources of information to understand customers' needs and buying habits. Thus, hotel managers may innovate and differentiate their strategies to reach distinct market segments. Humans behave differently based on their sociodemographic individualities, which play an essential role in dealing with and responding to their daily behaviors and health threats (Foroudi et al., 2021). However, for Tefera and Migiro (2017), segmenting customers based solely on their age, gender, or marital status may not provide a substantial advantage in how they are approached as hotel guests. Instead of classifying customers using demographic attributes, hotel managers must understand the relationship between sociodemographic characteristics and their buying behavior to attend to their individual needs and desires to meet or even exceed their expectations.

Palan (2001) suggests that the choices and behavior may differ between men and women in their choices in general and the emphasis each places on the attributes of choice. For this reason, gender has been the subject of study as a key demographic factor in influencing individual behaviors in the hospitality industry for some time (Seo & Hwang, 2014).

Similarly, Maoz (2007) considers that the tourists' country of origin plays an important role in shaping traveler's behavior and motivations due to distinctive cultural characteristics. Early research by Tsiotsou and Vasioti (2006) suggested that employment and family status did not affect customers' expectations, perception, and satisfaction of services, while education and age significantly predicted customer satisfaction. In contrast, the family composition seems to play a notable role in vacation decisions. Families with children generally have a different choice process and criteria for vacation and accommodation selection (Bronner & de Hoog, 2008; Gram, 2005). Children can influence the early stages of decision-making regarding the vacation period, choice of destination, hotel, and activities to be developed (Jenkins, 1979).

Acting in a globalized environment, hotel companies are also advised to identify differences in behavior associated with nationality to improve the prediction of customer behavior and decision-making and adjust their strategies accordingly (Andreu et al., 2017; Tran et al., 2019). According to Foroudi et al. (2021), the COVID-19 outbreak has resulted in psychological, economic, and sociocultural changes that will impact hospitality demand for an uncertain period. Hotels need to be prepared to change their value proposition to match the customers changing beliefs and perceptions.

Based on the evidence in the existent literature, the next section of this research explores the impact of the aforementioned variables on hotel guest perceptions and behavior to aid hotel managers in designing innovative plans to cope with the changes in customers' demands.

3 Methods

To fulfill the objective of evaluating Portuguese and Spanish clients' perceptions and expectations regarding hotel services safety during the COVID-19 pandemic, an online questionnaire was designed to collect hotel guests' data.

Considering the study's setting, it was problematic to find studies addressing safety and health risk perception in hotels explicitly during pandemic crises. Though many studies address the topic of risk perception in travel and tourism connected to disasters, no studies were found addressing specific and objective measures for the impact of pandemics on hotel clients' perception and confidence. To overcome this problem, a panel of tourism experts was assembled. The researchers and four hotel managers used the Delphi method to define the relevant dimensions and measures to evaluate them. A list of items covering seven critical categories: (1) Financial risk; (2) Front-desk digitalization and automation; (3) Entertainment; (4) Front-desk service; (5) Hygiene, sanitation, and cleanness; (6) Hotel characteristics and space management; and (7) Food and beverage service; was prepared based on information collected hospitality professionals. The list was distributed among the panel members, who were instructed to evaluate each item for later debate in an online meeting. Using this interactive process, a compromise on the final list was attained after five online meetings.

After defining the items, an online questionnaire was prepared to include the 52 items measured using five-point Likert-type scales for agreement and importance, two questions about travel habits, one ranking question to evaluate respondents' priorities on selected hotel attributes, and a group of questions aimed at characterizing the respondents. The questionnaire was made available in Portuguese and Spanish in May 2020. The survey was publicized through multiple channels, namely social media pages and groups, mailing lists, and personal contacts from the researchers. A final sample comprising 401 Portuguese and 132 Spanish hotel guests was obtained. Data analysis included descriptive statistics and ANOVA with post hoc tests. These statistical methods have been extensively used in hospitality research (Rivera & Pizam, 2015) Microsoft Excel and SPSS 27 software were used to produce the results.

4 Results

4.1 Sample Characterization

The final sample was composed of 67.5% (360 out of 533) females and 32.5% males, aged between 18 and 90 years old, almost evenly distributed by the four age categories defined. Married people represent almost half of the sample (48.2%). Globally the sample was well educated, as 78.5% holding a college degree. Concerning occupation, 59.3% are employed in public and private sectors, and only 18.6% were students. Globally, the samples from Portugal and Spain display similar characteristics. The complete information on sample characteristics is presented in Table 1.

4.2 Priorities for Hotel Choice

Respondents were also asked to order the importance of a set of factors for choosing a hotel vis-à-vis COVID-19. The results in Fig. 1 indicate that the most important factor is the perception of hotel health safety, which was mostly ranked first, followed by the limitation of people in public spaces. The size of the hotel and the availability of alternative food services are regarded as not important since they were

Table 1 Characteristics of respondents (N = 533)

	Nationality				Total	
	Portuguese		Spanish			
	N	%	N	%	N	%
Gender						
Female	282	70.3	78	59.1	360	67.5
Male	119	29.7	54	40.9	173	32.5
Total	401	100.0	132	100.0	533	100.0
Age						
18–25	78	19.5	37	28.0	115	21.6
26–40	106	26.4	62	47.0	168	31.5
41–50	101	25.2	20	15.2	121	22.7
>50	116	28.9	13	9.8	129	24.2
Total	401	100.0	132	100.0	533	100.0
Marriage status						
Single (never married)	144	35.9	62	47.0	206	38.6
Married or cohabitating	192	47.9	65	49.2	257	48.2
Divorced	65	16.2	5	3.8	70	13.1
Total	401	100.0	132	100.0	533	100.0
Education						
Basic education	4	1.0%	1	0.8%	5	0.9%
High school	50	12.5	5	3.8	55	10.3
Professional degree	26	6.5	23	17.4	49	9.2
Bachelor's	165	41.1	64	48.5	229	43.0
Master degree	92	22.9	21	15.9	113	21.2
Doctoral degree	58	14.5	18	13.6	76	14.3
Other	6	1.5	0	0.0	6	1.1
Total	401	100.0	132	100.0	533	100.0
Employment status						
Unemployed	26	6.5	12	9.1	38	7.1
Employee (private sector)	126	31.4	40	30.3	166	31.2
Employee (public sector)	115	28.6	35	26.5	150	28.1
Entrepreneur	18	4.5	3	2.3	21	3.9
Liberal professional	23	5.7	2	1.5	25	4.7
Student	66	16.5	33	25.0	99	18.6
Retired or pre-retired	16	4.0	4	3.0	20	3.8
Other	11	2.6	3	2.3	14	2.7
Total	401	100.0	132	100.0	533	100.0

the factors ranked most times in the last positions. Broadly, the pattern of the curves shows that the priorities for hotel choice are very similar between the two samples.

The results from the ranking of priorities for hotel choice suggest that the preferences are similar for both samples. Therefore, hotel managers do not need to

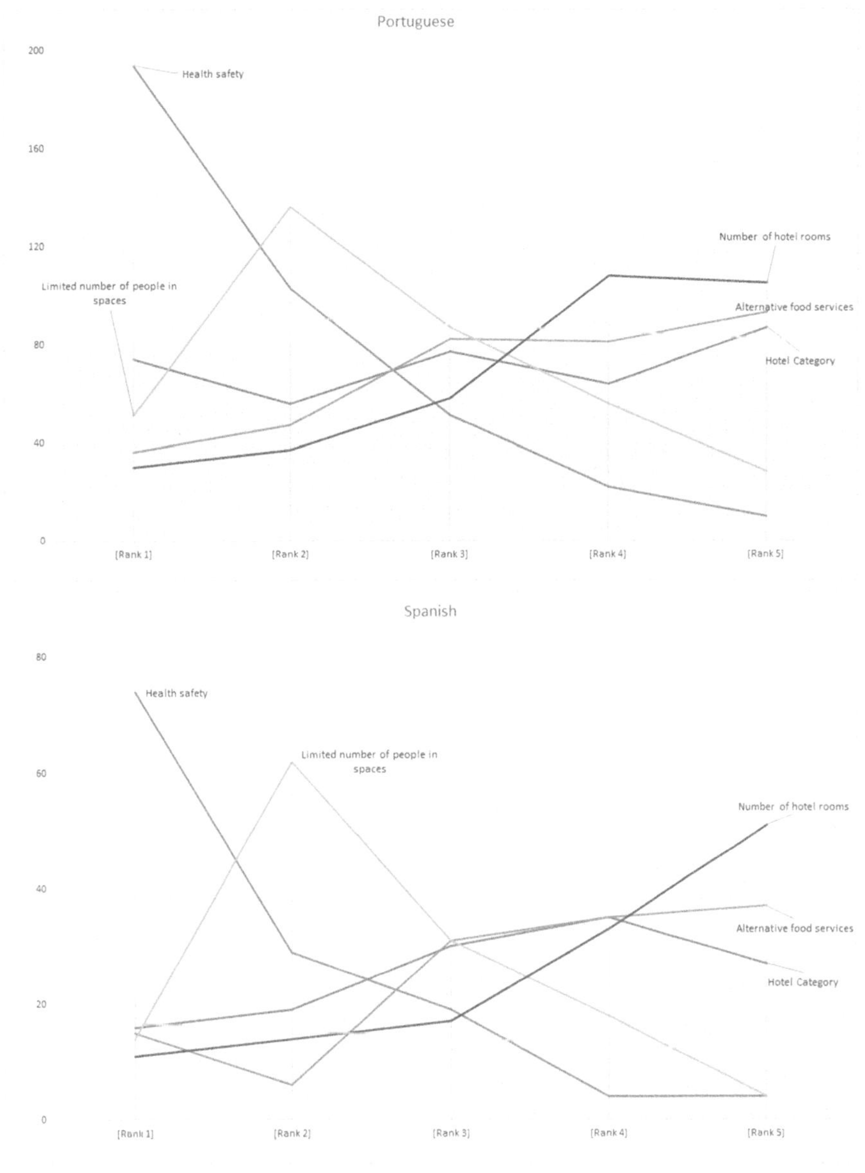

Fig. 1 Ranking of hotel choice priorities for Portuguese and Spanish

put especially emphasis on any of these attributes when marketing the hotel to potential guests from one specific country.

4.3 Analysis of Importance and Differences in Critical Factors

To decide which areas to successfully innovate, hotel managers need to know what factors the clients value the most in a pandemic context. Since a five-point Likert-type scale was used, we consider as most important the factors presenting a mean greater than four. According to Table 2, this resulted in 17 factors distributed across five critical dimensions for Portuguese tourists and 17 factors spread across six dimensions for Spanish tourists.

An important difference is linked to the preference for hotels with entertainment in open spaces, without agglomerations which scored higher among Spanish. The Spanish also seem to be more responsive to hotels having special agreements with local healthcare units in case of getting sick during the stay. In general, the results in Table 1 indicate that despite little quantitative differences between the means, Portuguese and Spanish hotel guest concerns seem to coincide. Without surprise, hotel guests are especially concerned with factors related to hygiene and cleaning as this dimension is vital to ensure safety regarding contagion. Hotel hygiene and cleanliness were already extremely important to customers before COVID-19 (Pillai et al., 2021), but it has become the focal point for any recovery plan during the pandemic. Like the current results, Park et al. (2019) found that customers rank hotel hygiene and cleanliness as the most important factor when booking a hotel.

Other dimensions demanding attention from hotel managers are those related to space organization and management, food and beverage services, and assuring appropriate front desk service. Hu et al. (2021) reinforce this idea by stating that hygiene is relevant in common areas (meal service and reception) as a priority during the pandemic and adds that some core hotel attributes (room, bed, and price) are perceived as less important during the pandemic.

The financial risk associate with the reservation seems to more important for Spanish guests. However, the difference in the mean is not statistically significant, preventing us from proposing measures for a specific market.

To provide meaningful indications for hotel managers on how to innovate by customizing the offer for the Portuguese and Spanish markets, the responses from the two groups were compared using the *t*-test for comparing independent samples. The results are presented in Table 3. The results show that Portuguese and Spanish hotel guests have a different opinions in 17 from the total 52 factors.

These differences constitute an opportunity for hotel managers to innovate in their approach to each market. Interestingly the Portuguese hotel guests seem to be more worried about the hotel characteristics and space management, and food and beverage services, besides hygiene e-sanitation. In contrast, the Spanish are more concerned with financial risk, front-desk service, and entertainment. Regarding hygiene and sanitation, they seem to value different things. Portuguese are more aware of the general risks involved in any hotel stay and are more concerned at a general level with hotel practices. For its side, the Spanish seem to value specific

Table 2 Descriptive statistics

Dim.		Factor	Portuguese (N=401)		Spanish (N=132)	
			Mean	Std. Deviation	Mean	Std. Deviation
FR	FR01	I only book if I am assured that I can change the reservation at any time	4,02	0,978	4,21	1,063
	FR02	I do not mind paying extra for pandemic insurance when I am booking.	3,36	1,145	3,59	1,084
	FR03	I attach more importance to ensuring Covid-19's prevention conditions than to the price	3,74	0,945	3,84	1,025
	FR04	For me, it is very important that hotels offer free health and personal accident insurance	3,42	1,120	3,51	1,037
	FR05	I would give preference to hotels that allow the extension of the stay at reduced prices if it is impossible to continue the trip.	4,07	0,829	4,12	0,856
	FR06	A higher price corresponds to higher health quality	2,48	1,098	2,51	1,037
FDA	FD01	New technological ways of customer service will negatively affect service personalization	2,86	1,132	2,96	1,155
	FD02	New technological ways of customer service can positively increase service satisfaction	3,35	0,921	3,36	1,013
	FD03	Be attended by digital assistants will compromise my satisfaction with the service	2,93	1,041	3,13	1,059
	FD04	I only choose hotels with qualified staff capable of responding to situations of anxiety in case of crisis	3,30	1,037	3,20	1,115
	FD05	Technologies that minimize direct contact will decrease the trust	2,61	1,063	2,85	1,115
	FD06	Digital technologies in hotels will prevent the creation of empathy with the client	3,33	1,154	3,30	1,159
ENT	ET01	I prefer hotels that offer personalized counseling in choice of experiences at the destination	3,61	0,947	3,83	0,821
	ET02	I prefer hotels with entertainment in open spaces, without agglomerations	3,94	0,851	4,11	0,902
FDS	FDS01	At reception, it is important to ensure health safety conditions	4,68	0,533	4,60	0,770
	FDS02	The temperature measurement at the time of arrival will make me feel uncomfortable	2,01	1,033	2,20	1,196
	FDS03	I prefer to have a personalized service to minimize the risk of contagion from Covid-19	3,84	0,902	4,00	0,933
	FDS04	I get upset if attended by employees with mask/visor and gloves	1,68	0,839	1,58	0,901
	FDS05	The digitization of the hotel service is vital to ensure the customer's confidence regarding	3,39	0,967	3,39	1,103
	FDS06	Be able to check in and out by appointment	3,38	1,205	3,69	0,982
	FDS07	Having differentiated entrances and exits for customers	3,56	1,169	3,83	0,985
	FDS08	Being serviced by appropriately uniformed employees with mask/visor and gloves	4,18	0,939	4,46	0,725
	FDS09	Be informed daily of the occupation rate, clients with Covid-19 and hygiene and disinfection practices	3,93	1,128	3,95	1,104
HGS	HG01	Hotels should only reopen if they meet all hygienic conditions to avoid contamination	4,73	0,544	4,64	0,568
	HG02	I trust hotel safety against the risk of contagion	3,49	0,889	3,75	0,859
	HG03	The Covid-19 certification ensures that there is no risk of contagion.	2,59	1,033	2,91	1,059
	HG04	Health entity certifications reinforce the degree of trust when selecting the hotel	3,85	0,810	3,86	0,923
	HG05	In hotels there are always risks resulting from deficiencies in the hygiene of the rooms	3,77	0,841	3,48	0,969
	HG06	In hotel stays there is always a health risk associated with the poor hygiene of the common and service areas	3,76	0,877	3,45	0,968
	HG07	Hotel chains have stricter health safety policies	3,23	0,943	3,37	0,936
	HG08	Hotel chains have greater ability to react to dangerous situations such as pandemics	3,13	0,970	3,25	1,022
	HG09	Hotels adopt a responsible behavior towards the risk of contagion	4,66	0,574	4,69	0,526
	HG10	Hotels periodically close for general sanitization and disinfection	4,10	0,986	4,15	0,815
	HG11	The Safety and health practices in hotel choice	4,50	0,682	4,34	0,750
	HG12	Promote Online Check-In and Check-out to avoid agglomeration	4,11	0,927	4,12	0,856
	HG13	Offer masks, disinfectant gel and gloves in rooms	4,29	0,901	4,39	0,768
	HG14	Favoring hotels that have special agreements with local healthcare units	3,75	1,111	4,13	0,992
HSM	HC01	Staying in a hotel with a small number of rooms is safer	3,24	1,023	2,96	0,960
	HC02	Rural accommodations and isolated units will be preferred over large hotels	3,79	0,905	3,74	1,024
	HC03	I prefer a large hotel	2,60	0,912	2,60	0,964
	HC04	I prefer a higher class hotel	3,36	0,949	3,10	1,057
	HC05	It bothers me that some services are not available to minimize the risk of contagion.	3,25	1,307	2,97	1,330
	HC06	Hotels have infrastructure to ensure physical and psychological comfort in case of	4,05	0,959	4,14	0,789
	HC07	Hotels ensure safety distance in common spaces	4,45	0,789	4,48	0,659
	HC08	Hotels with limited number of clients in common spaces	4,39	0,812	4,49	0,599
FBS	FBS01	Because of the pandemic, I have become more sensitive to food security issues	3,88	1,043	3,58	1,099
	FBS02	The meals should not be served in buffet	4,15	1,047	4,04	1,007
	FBS03	I feel more confident in hotels that have kitchnet/kitchen where I can prepare/take meals	3,47	1,051	3,42	1,019
	FBS04	Because of the Covid-19 the room service is safer.	3,37	0,940	3,49	0,860
	FBS05	Meals being served in reduced capacity locations	4,21	0,862	3,88	0,865
	FBS06	The meals be served by appointment	4,06	0,956	3,86	0,966
	FBS07	Having an option for take-away meals provided by the Hotel kitchen	4,11	0,959	4,25	0,785

FR Financial Risk, *FDA* Front-Desk Automation, *ENT* Entertainment, *HSC* Hygiene, sanitation, and cleanness, *HSM* Hotel characteristics and space management, *FBS* Food and beverage service

Table 3 *t*-Test results for the factors mean comparisons

Dim.		Factor	Mean Difference pt-sp	Sig. (*p*-value) (2-tailed)
FR	FR02	I do not mind paying extra for pandemic insurance when I am booking.	-0,23	0,041
FDA	FD05	Technologies that minimize direct contact will decrease the trust	-0,24	0,025
ENT	ET01	I prefer hotels that offer personalized counseling in choice of experiences at the destination	-0,22	0,017
FDS	FDS06	Be able to check in and out by appointment	-0,31	0,008
	FDS07	Having differentiated entrances and exits for customers	-0,26	0,020
	FDS08	Being serviced by appropriately uniformed employees with mask/visor and gloves	-0,28	0,002
HG	HG02	I trust hotel safety against the risk of contagion	-0,26	0,003
	HG03	The Covid-19 certification ensures that there is no risk of contagion.	-0,32	0,002
	HG05	In hotels there are always risks resulting from deficiencies in the hygiene of the rooms	0,29	0,001
	HG06	In hotel stays there is always a health risk associated with the poor hygiene of the common and service areas	0,31	0,001
	HG11	The Safety and health practices in hotel choice	0,16	0,021
	HG14	Favoring hotels that have special agreements with local healthcare units	-0,38	0,000
HSM	HC01	Staying in a hotel with a small number of rooms is safer	0,27	0,007
	HC05	It bothers me that some services are not available to minimize the risk of contagion.	0,28	0,034
FBS	FBS01	Because of the pandemic, I have become more sensitive to food security issues	0,31	0,004
	FBS05	Meals being served in reduced capacity locations	0,33	0,000
	FBS06	The meals be served by appointment	0,20	0,037

things like the hotel have been awarding with a COVID-19 safety certification by health authorities.

A particular relevant finding is that the differences are mainly associated with factors presenting an average rating below four. Factors FDS08 and HG11 are the only ones with mean ratings above four, presenting statistically significant differences between Portuguese and Spanish Hotel guests. This result suggests that hotel managers do not need to innovate in differentiating their offer related to the factors considered most important by the respondents, suggesting the possibility of a common approach with small adaptations for the Iberian market.

4.4 Analysis of the Moderating Impact of Demographic Characteristics

To understand the impact of respondents' characteristics on the evaluation of the critical factors when booking a hotel under a pandemic situation, the differences were controlled by gender, education, and family type (e.g., having children).

The results indicate that the behavior of women is more heterogeneous than that of men. It was possible to identify 15 factors (FR02, FD03, FD05, ET01, FDS02, FDS06, FDS07, FDS08, HG01, HG03, HG05, HG06, HG14, HC01, HC05, and FBS05) where the opinion between Portuguese and Spanish women differ to a significance level of 0.05, against only 8 factors (FD02, FDS08, HG02, HG03, HG06, HG14, HC08, and FBS05) for men.

This finding supports the previous indication by Shemwell and Cronin (1995) that women and men differ in their approaches to assess services. Wolf and Zhang (2016) provide further details by reporting that women place more value on their uncertainties, time and money constraints, and decision consequences, while men place more value on information, goals, and motivation. Given that different genders have different perspectives on decision-making (Khare, 2011), providing different appeals based on the gender of hotel guests may be a significant and novel improvement to the current selling approach (Chen et al., 2017).

Comparing the results for similar genders, it stands out that Portuguese and Spanish women's opinions differ mainly in the factors related to front-desk service and hygiene, sanitation, and cleanliness. Men's opinions also differ mostly in the hygiene, sanitation, and cleanliness critical dimension. Interestingly, FDS08, HG03, HG06, HG14, and FBS05 are the factors where both women and men have in common regarding different opinions. Based on this finding, it can be considered that these factors are the ones requiring more attention from hotel managers because they are differently evaluated by both genders regardless of their nationality.

The analysis of differences in the perception of Iberian hotel guests regarding safety under COVID-19 revealed mixed results. Respondents with basic and high school (secondary) education have a relatively similar opinion on the critical factors affecting hotel safety under pandemic states. For hotel guests holding basic education, the only difference is related to factor FSB02 regarding meals being served in the buffet. Those who hold a high school degree display differences in four factors (FD02, HG01, HG09, and HG14). Surprisingly, three of them are related to hygiene, sanitation, and cleanliness.

The results on the differences among the respondents holding higher education degree revels a distinctive pattern. Among respondents holding a bachelor's degree, it is possible to identify 16 factors (FD05, ET01, ET02, FDS06, FDS07, FDS08, FDS09, HG02, HG03, HG05, HG07, HG10, HG13, HG14, HC01, HC08, and FBS07) with significant differences between Portuguese and Spanish hotel guests based on the *t*-test for a level of significance of 0.05. Holding a master's degree seems to influence the perception of 13 factors (FDS01, FDS09, HG01, HG02, HG05, HG09, HG10, HG11, HC07, FSB01, FSB02, FBS05, and FSB06). However, many of them are distinct from those identified for the previous segment, which is somewhat puzzling. In particular, holding a master's degree seems to largely affect the perception of food and beverage-related factors, contrary to the results presented by the sample holding a bachelor's degree.

Finally, Portuguese and Spanish guests holding a doctorate differ in evaluating seven factors (HG01, HG05, HG06, HC02, HC04, FBS04, and FBS05) related to hotel safety under pandemic. The difference is related to only three critical dimensions: hygiene, sanitation, and cleanliness; hotel characteristics and space management; and food and beverages service.

In summary, the findings suggest that Portuguese and Spanish hotel guests with the lowest level of education and highly educated present fewer differences in how they evaluate hotel safety.

These findings suggest a meaningful association between the level of education and the perception, demands, and preferences of hotel guests. Hotel managers who are willing to provide innovative and differentiated offers to each market must account for education. The findings suggest that it may be relevant to assess guests' profiles and design offers targeted at specific segments to improve guests' safety perception, preference, satisfaction, and loyalty. The current results also reinforce the indication by Tsiotsou and Vasioti (2006) that education can be used to rank consumers in terms of satisfaction, with less educated people displaying lower levels of satisfaction with tourism services, while more educated people tended to be more satisfied.

Finally, as children are always a matter of concern for parents, namely on what is related to safety, we proceed to assess if having children could influence the evaluation of the factors associated with safety perception. The greater the number of children and adolescents in the family, the greater their influence in decisions about the destination, activities, and spending (Jenkins, 1979). Younger children may even amplify the influence in the choices simply because of the extra need for child care and parents' concerns with meals (Thornton et al., 1997).

The results in Table 4 indicate the differences in safety factors perception between Portuguese and Spanish guests according to the number of children and their age range. Findings indicate that clients without children present a higher number of differences regarding the perception of safety factors. Having children seems to have a conciliatory effect regarding guests' opinions as the number of differences in the perception decreases for those having children, and there seems to be an inverse association between children's age and the number of significant differences in the perception. Therefore, we are convinced that having children and their age category influence the perception of the safety factors.

The current findings suggest that hotel managers must consider the family profile when advertising the hotel safety measures and services.

To provide a global overview of the study findings, Table 5 presents the average of the Portuguese and Spanish scores on the factors affecting hotel guests' safety perception and the statistically significant differences associated with the demographic characteristics analyzed.

5 Conclusions

The hospitality sector is considered one of the most affected by the consequences of the pandemic COVID-19. The indefinite persistence of the pandemic duration increases anxiety about the ability to recover from this dramatic situation (Wieczorek-Kosmala, 2021). Portugal and Spain have worldwide recognition as tourism destinations by excellence. However, Portugal being one of the world's most tourism-dependent economies, with tourism accounting for over 16% of the gross domestic product in 2018, is particularly vulnerable to crises and disasters affecting the tourism industry, like COVID-19. Considering this economic exposure

Table 4 Factors presenting statistically significant differences at 0.05 level

	Factors' means and statistical differences significance		
	No children ($N_{pt} = 323$; $N_{sp} = 88$)	Children 0–3 years ($N_{pt} = 45$; $N_{sp} = 22$)	Children 4–11 years ($N_{pt} = 10$; $N_{sp} = 11$)
FD01	pt = 2.81; Sp = 3.09; $p = 0.042$	pt = 3.32; Sp = 2.42; $p = 0.020$	
FD03	pt = 2.88; Sp = 3.17; $p = 0.020$	pt = 3.36; Sp = 2.58; $p = 0.017$	
FD04		pt = 3.64; Sp = 2.58; $p = 0.008$	
FD05		pt = 3.18; Sp = 2.25; $p = 0.018$	
FD06		pt = 3.86; Sp = 2.75; $p = 0.003$	
ET01	Pt = 3.62; Sp = 3.85; $p = 0.022$		Pt = 3.36; Sp = 3.91; $p = 0.017$
ET02			Pt = 3.93; Sp = 4.59; $p = 0.001$
FDS02	Pt = 1.96;2.28 Sp=; $p = 0.021$		
FDS06	Pt = 3.37; Sp = 3.74; $p = 0.003$		
FDS08	Pt = 4.18; Sp = 4.50; $p = 0.000$		
HG01		Pt = 4.82; Sp = 4.25; $p = 0.028$	
HG02	Pt = 3.48; Sp = 3.75; $p = 0.012$		
HG03	Pt = 2.58; Sp = 2.97; $p = 0.002$		
HG05	Pt = 3.76; Sp = 3.50; $p = 0.022$		Pt = 3.80; Sp = 3.32; $p = 0.046$
HG06	Pt = 3.76; Sp = 3.50; $p = 0.018$		Pt = 3.76; Sp = 3.27; $p = 0.047$
HG07	Pt = 3.27; Sp = 3.49; $p = 0.049$		
HG11	Pt = 4.51; Sp = 4.34; $p = 0.041$		
HG14	Pt = 3.77; Sp = 4.11; $p = 0.005$	Pt = 3.55; Sp = 4.33; $p = 0.031$	
HC01	Pt = 3.20; Sp = 2.98; $p = 0.048$		
HC04	Pt = 3.38; Sp = 3.13; $p = 0.027$		
HC05	Pt = 3.22; Sp = 2.89; $p = 0.037$		Pt = 4.09; Sp = 3.90; $p = 0.006$
FSB01	Pt = 3.87; Sp = 3.57; $p = 0.019$	Pt = 3.91; Sp = 3.25; $p = 0.014$	

(continued)

Table 4 (continued)

	Factors' means and statistical differences significance		
	No children ($N_{pt} = 323$; $N_{sp} = 88$)	Children 0–3 years ($N_{pt} = 45$; $N_{sp} = 22$)	Children 4–11 years ($N_{pt} = 10$; $N_{sp} = 11$)
FSB05	Pt = 4.21; Sp = 3.89; $p = 0.002$		
FSB06	Pt = 4.03; Sp = 3.81; $p = 0.049$		
FSB07		Pt = 3.77; Sp = 4.50; $p = 0.027$	

of Portugal to the tourism activity, the main purpose of the current article is to provide usable knowledge about tourism consumers' risk concerns, service expectations, and demands regarding the hospitality sector under pandemics. These insights on tourists' behavior are vital for hotel managers to develop customized offers crated to low tourists' risk perception. Additionally, it is being expected that customers' perception of risk decreases over time as vaccination progress, and the governments implement policies aimed at assuring travel safety and stimulating the tourism industry (Foroudi et al., 2021).

In the meantime, Iberian hotel managers must adopt action to weaken tourists' fears to sustain and stimulate the demand for hotel services. Hotels safety image is a remarkably important valuable intangible asset for the hospitality industry (Bor et al., 2018). It is a determinant factor influencing tourists' destination choice and hotel booking, according to O'Neill and Belfrage (2005).

From the current results, it can be concluded that assuring hotel's safety and communicate the safety practices must be the primary objective of Iberian hotel managers to retain Portuguese and Spanish demand. In general, the concerns of tourists from both countries are much comparable despite some differences. This suggests that hotel managers may use a similar approach for the two segments in several critical areas, with clear benefits in terms of efficiency. This approach is especially viable for the financial risk, front-desk digitalization and automation, front-desk services, and space management dimensions. In each dimension, single factors emerge as more important, and these must be comprehended and addressed by hotel managers. For example, ensuring health conditions at reception, namely by being serviced by employees using masks and gloves, are considered remarkably more important than other factors (see Table 2).

Limiting the number of people in common space also emerged as a primary measure to be implemented, especially for Spanish tourists. Studying innovative ways of managing guests' circulation in common areas could significantly positively impact safety perception. Likewise, the development of alternative food services alternatives is highly recommendable since this service is considered of high risk for potential contagious. In general Spanish tourists appear to be more responsive to hotels having exclusive agreements with local healthcare units.

Factors associated with hygiene and sanitation are obviously at the top of both countries' tourists' priority lists. Even so, the Spanish hotel guests seem more

Table 5 Summary of the findings

Dim.		Average		Statisticaly Significant Differences (0.05 level)										
		Portuguese	Spanish	Global	Gender		Education					Children		
					Female	Male	Basic	Secondary	Bachelor's	Master's	Doctorate	No children	0-3 years	4-11 years
FR	FR01	4,02	4,21											
	FR02	3,36	3,59											
	FR03	3,74	3,84											
	FR04	3,42	3,51											
	FR05	4,07	4,12											
	FR06	2,48	2,51											
FDA	FD01	2,86	2,96											
	FD02	3,35	3,36											
	FD03	2,93	3,13											
	FD04	3,30	3,20											
	FD05	2,61	2,85											
	FD06	3,33	3,30											
ENT	ET01	3,61	3,83											
	ET02	3,94	4,11											
FDS	FDS01	4,68	4,60											
	FDS02	2,01	2,20											
	FDS03	3,84	4,00											
	FDS04	1,68	1,58											
	FDS05	3,39	3,39											
	FDS06	3,38	3,69											
	FDS07	3,56	3,83											
	FDS08	4,18	4,46											
	FDS09	3,93	3,95											
HGS	HG01	4,73	4,64											
	HG02	3,49	3,75											
	HG03	2,59	2,91											
	HG04	3,85	3,86											
	HG05	3,77	3,48											
	HG06	3,76	3,45											
	HG07	3,23	3,37											
	HG08	3,13	3,25											
	HG09	4,66	4,69											
	HG10	4,10	4,15											
	HG11	4,50	4,34											
	HG12	4,11	4,12											
	HG13	4,29	4,39											
	HG14	3,75	4,13											
HSM	HC01	3,24	2,96											
	HC02	3,79	3,74											
	HC03	2,60	2,60											
	HC04	3,36	3,19											
	HC05	3,25	2,97											
	HC06	4,05	4,14											
	HC07	4,45	4,48											
	HC08	4,39	4,49											
FBS	FBS01	3,88	3,58											
	FBS02	4,15	4,04											
	FBS03	3,47	3,42											
	FBS04	3,37	3,49											
	FBS05	4,21	3,88											
	FBS06	4,06	3,86											
	FBS07	4,11	4,25											

concerned with the hotel's capacity to ensure safety and respond to deal with consequences of undesired situations (e.g., infection). In contrast, the Portuguese seem more aware and willing to accept the risks. Consequently, COVID-19 safety certification by health authorities is more important for hotels targeting Spanish tourists.

As several differences are associated with gender, however, the findings suggest that communication targeted at men can be more homogeneous than the one targeted at women from both countries. Therefore, hotel managers can innovate by modeling communication targeted at the specific gender of the tourist from each country, with a special focus on women. The current findings support the findings by Aziz et al. (2018), portraying women's power as decisionmakers inside the family. They usually feel responsible for deciding what will be best for the family, ensuring that all parties are satisfied during their stay at the destination. The current findings point to front-desk activities and hygiene and sanitation as the most important dimensions and also the ones where Portuguese and Spanish women tend to disagree the most. Even though previous studies have shown that children are important members of family decision-making (Lee & Collins, 2000), the mother has more influence on the final decision-making based on her authority and motivation levels.

Nevertheless, the importance of having children to consider must not be disregarded. The needs of families with young children on vacation change the request for new services that hotels can provide to improve guests' experience (Khoo-Lattimore et al., 2015). Interestingly, based on the current results, the differences are lower in families having children. Moreover, the differences tend to blur with the increase in the age of the children. In summary, having children seems to increase the consensus in evaluating hotel services' critical factors under pandemics, suggesting that a common approach is more suitable for the segment of families with children, especially with older ones.

Huang and Xiao (2000) suggest that income affects leisure-based tourist behavior, especially regarding the length of vacation and accommodation services. Since income is often associated with the education level, this variable may also have an important role in hotel assessment. In fact, the results show that as the number of differences between Portuguese and Spanish guests increases along with the increase in education, at least until the bachelor's degree. Surprisingly Spanish holding a bachelor's degree presents a higher average in all dimensions except for HC01. However, the pattern reverses for the respondents holding a master's and a PhD degree, with Portuguese respondents presenting higher averages than Spanish respondents. This relationship between education and the evaluation of hotel services, particularly the discrepancies, deserve more attention and future research to determine how this information could shape current offer and stimulate novel service offers.

Compared to the results by Bor et al. (2018), which suggest that the most important attributes when choosing a hotel were perceived value for money, followed by the condition of the hotel facilities, service delivery, accessibility of hotel and services, perceived service quality, perceived attractiveness, and employee provision, the current findings show that under crisis, like a pandemic, new critical attributes emerge and need to be accounted for. Furthermore, the demographic profile of customers could be useful to shape the offer to attend to specific demands related to the unique context. When choosing the type of accommodation, tourists often can make alternative decisions. To remain competitive despite the context, hotels must continuously analyze the profile of their customers and be prepared to quickly adapt their offer to respond to changes in their target public caused by an unexpected crisis (Hedlund et al., 2012).

The current study shows significant differences in how Portuguese and Spanish tourists perceive service hotels under pandemic crises. Globally, it can be concluded that to cope with the consequences of the COVID-19 pandemic effectively. The Iberian hospitality industry would benefit from developing profiles based on the demographic characteristics of potential guests to innovate in terms of services and communication. Customizing the communication of hotel services and add-ins according to the target public allows a better fit with customers' particular needs, wants, and concerns. The present study provides the base for future research devoted to improving theoretical knowledge on the determinant dimensions and characteristics more suitable for implementing customized answers to different segments of hotel guests. At a short-term managerial level, the current findings help hotel managers get a clearer idea of what they need to do to be able to comply with customers' expectations under the current pandemic and future pandemic contexts.

6 Limitations and Future Research

The current research has some limitations that deserve to be noticed. Despite the sample size, the total sample is unbalanced regarding the origin of responses and some demographic characteristics. This limitation should be considered when evaluating the conclusions. Future research should try to get a more balanced sample. Additionally, further research should be considered outside the context of pandemics to expand the current findings to normal market conditions. Other factors and characteristics should also be considered to provide a broader vision of the differences in hotel guests' behavior and allow alternative segmentation criteria. Future research should also explore reasons behind the change of behavior, namely the inversion in the averages scores between Portuguese and Spanish hotel guests holding bachelor's, master's, and PhD degrees.

Acknowledgment The authors would like to thank NECE—Research Center in Business Sciences funded by the Multiannual Funding Programme of R&D Centres of FCT—Fundação para a Ciência e a Tecnologia, under the project UIDB/04630/2020.

References

Andreu, R., Claver, E., & Quer, D. (2017). Foreign market entry mode choice of hotel companies: Determining factors. *International Journal of Hospitality Management, 62*, 111–119. https://doi.org/10.1016/j.ijhm.2016.12.008

Aziz, Y., Hussin, S. R., Nezakati, H., Nerina, R., & Hashim, H. (2018). The effect of socio-demographic variables and travel characteristics on motivation of Muslim family tourists in Malaysia. *Journal of Islamic Marketing, 9*, 00–00. https://doi.org/10.1108/JIMA-03-2016-0016

Bor, T., Kieti, D., & Rotich, D. (2018). Hotel choice attributes and demographic characteristics of tourists in North Rift Region, Kenya. *African Journal of Hospitality, Tourism and Leisure, 7*, 1–12.

Breda, Z., & Costa, C. (2006). Safety and security issues affecting in bound tourism in the People's Republic of China. In Y. Mansfeld & A. Pizam (Eds.), *Tourism, security, and safety: From theory to practice* (pp. 187–208). Elsevier Butterworth-Heinemann.

Bronner, F., & de Hoog, R. (2008). Agreement and disagreement in family vacation decision-making. *Tourism Management, 29*(5), 967–979. https://doi.org/10.1016/j.tourman.2007.12.001

Cavlek, N. (2002). Tour operators and destination safety. *Annals of Tourism Research, 29*(2), 478–496. https://doi.org/10.1016/S0160-7383(01)00067-6

Chen, A., Lu, Y., & Wang, B. (2017). Customers' purchase decision-making process in social commerce: A social learning perspective. *International Journal of Information Management, 37* (6), 627–638. https://doi.org/10.1016/j.ijinfomgt.2017.05.001

Ching, C. (2007). Applying the stochastic frontier approach to measure hotel managerial efficiency in Taiwan. *Tourism Management Perspectives, 28*(3), 696–702. https://doi.org/10.1016/j.tourman.2006.04.023

Choorichom, J. (2011). Factors influencing the selection of hotels/resorts in Lanta Yai Island, Krabi, Thailand by international travelers. *Humanities and Social Sciences, 4*, 125–148.

Dess, G. G., & Picken, J. C. (2000). Changing roles: Leadership in the 21st century. *Organizational Dynamics, 28*(3), 18–34. https://doi.org/10.1016/S0090-2616(00)88447-8

Foroudi, P., Tabaghdehi, S. A. H., & Marvi, R. (2021). The gloom of the COVID-19 shock in the hospitality industry: A study of consumer risk perception and adaptive belief in the dark cloud of a pandemic. *International Journal of Hospitality Management, 92*, 102717. https://doi.org/10.1016/j.ijhm.2020.102717

González-Torres, T., Rodríguez-Sánchez, J.-L., & Pelechano-Barahona, E. (2021). Managing relationships in the tourism supply chain to overcome epidemic outbreaks: The case of COVID-19 and the hospitality industry in Spain. *International Journal of Hospitality Management, 92*, 102733. https://doi.org/10.1016/j.ijhm.2020.102733

Gram, M. (2005). Family holidays. A qualitative analysis of family holiday experiences. *Scandinavian Journal of Hospitality and Tourism, 5*, 2–22.

Hedlund, T., Marell, A., & Gärling, T. (2012). The mediating effect of value orientation on the relationship between socio-demographic factors and environmental concern in Swedish tourists' vacation choices. *Journal of Ecotourism, 11*, 16–33. https://doi.org/10.1080/14724049.2011.626859

Hu, F., Teichert, T., Deng, S., Liu, Y., & Zhou, G. (2021). Dealing with pandemics: An investigation of the effects of COVID-19 on customers' evaluations of hospitality services. *Tourism Management, 85*, 104320. https://doi.org/10.1016/j.tourman.2021.104320

Huang, A., & Xiao, H. (2000). Leisure-based tourist behavior: A case study of Changchun. *International Journal of Contemporary Hospitality Management, 12*, 210–214. https://doi.org/10.1108/09596110010320850

Jenkins, R. (1979). The influence of children in family decision-making: Parents' perceptions. *Association for Consumer Research, 6*, 413–418.

Jiang, Y., & Ritchie, B. W. (2017). Disaster collaboration in tourism: Motives, impediments and success factors. *Journal of Hospitality and Tourism Management, 31*, 70–82. https://doi.org/10.1016/j.jhtm.2016.09.004

Khare, A. (2011). Mall shopping behaviour of Indian small town consumers. *Journal of Retailing and Consumer Services, 18*(1), 110–118. https://doi.org/10.1016/j.jretconser.2010.10.005

Khoo-Lattimore, C., Prayag, G., & Cheah, B. L. (2015). Kids on board: Exploring the choice process and vacation needs of Asian parents with young children in resort hotels. *Journal of Hospitality Marketing & Management, 24*(5), 511–531. https://doi.org/10.1080/19368623.2014.914862

Lee, C., & Collins, B. (2000). Family decision making and coalition patterns. *European Journal of Marketing, 34*, 1181–1198. https://doi.org/10.1108/03090560010342584

Lee, J. H., & Hwang, J. (2011). Luxury marketing: The influences of psychological and demographic characteristics on attitudes toward luxury restaurants. *International Journal of Hospitality Management, 30*(3), 658–669. https://doi.org/10.1016/j.ijhm.2010.12.001

Maoz, D. (2007). Backpackers' motivations the role of culture and nationality. *Annals of Tourism Research, 34*(1), 122–140. https://doi.org/10.1016/j.annals.2006.07.008

Mártinez-López, A., & Vargas-Sánchez, A. (2013). Factores com un Especial Impacto em el Nível de Innovacción del Sector Hotelero Español. *Tourism & Management Studies, 9*(2), 7–12.

McManus, S., Seville, E., Vargo, J., & Brunsdon, D. (2008). Facilitated process for improving organizational resilience. *Natural Hazards Review, 9*, 81–90. https://doi.org/10.1061/(ASCE)1527-6988(2008)9:2(81)

Mone, M. A., McKinley, W., & Barker, V. L. (1998). Organizational decline and innovation: A contingency framework. *The Academy of Management Review, 23*(1), 115–132. https://doi.org/10.2307/259102

Nunkoo, R., Teeroovengadum, V., Ringle, C. M., & Sunnassee, V. (2019). Service quality and customer satisfaction: The moderating effects of hotel star rating. *International Journal of Hospitality Management, 91*, 102414. https://doi.org/10.1016/j.ijhm.2019.102414

O'Neill, J., & Belfrage, E. E. (2005). A strategy for estimating identified intangible asset value: Hotel affiliation contribution. *The Appraisal Journal, 73*(1), 78–86.

Palan, K. (2001). Gender identity in consumer behavior research: A literature review and research agenda. *Academy of Marketing Science Review, 10*, 1–24.

Park, H., Kline, S. F., Kim, J., Almanza, B., & Ma, J. (2019). Does hotel cleanliness correlate with surfaces guests contact? *International Journal of Contemporary Hospitality Management, 31*(7), 2933–2950. https://doi.org/10.1108/IJCHM-02-2018-0105

Pillai, S. G., Haldorai, K., Seo, W. S., & Kim, W. G. (2021). COVID-19 and hospitality 5.0: Redefining hospitality operations. *International Journal of Hospitality Management, 94*, 102869. https://doi.org/10.1016/j.ijhm.2021.102869

Prideaux, B., Laws, E., & Faulkner, B. (2003). Events in Indonesia: Exploring the limits to formal tourism trends forecasting methods in complex crisis situations. *Tourism Management, 24*(4), 475–487. https://doi.org/10.1016/S0261-5177(02)00115-2

Rivera, M., & Pizam, A. (2015). Advances in hospitality research: "from Rodney Dangerfield to Aretha Franklin". *International Journal of Contemporary Hospitality Management, 27*(3), 362–378. https://doi.org/10.1108/ijchm-03-2014-0146

Rivera, M. A. (2020). Hitting the reset button for hospitality research in times of crisis: Covid19 and beyond. *International Journal of Hospitality Management, 87*, 102528. https://doi.org/10.1016/j.ijhm.2020.102528

Seo, S., & Hwang, J. (2014). Does gender matter? Examining gender composition's relationships with meal duration and spending in restaurants. *International Journal of Hospitality Management, 42*, 61–70. https://doi.org/10.1016/j.ijhm.2014.06.010

Shemwell, D. J., & Cronin, J. J. (1995). Trust and commitment in customer/service-provider relationships. *Journal of Customer Service in Marketing & Management, 1*(2), 65–75. https://doi.org/10.1300/J127v01n02_07

Sydnor, S., Stafford, K., Tews, M., & Adler, H. (2011). Toward a resilience model for the hospitality & tourism industry. *Journal of Human Resources in Hospitality & Tourism, 10*, 195–217. https://doi.org/10.1080/15332845.2011.536942

Tefera, O., & Migiro, S. (2017). The relationship between demographic and tripographic factors and customers' expectation, perception and service quality: A case of star rated hotel guests in Ethiopia. *African Journal of Hospitality, Tourism and Leisure, 6*, 1–20.

Thornton, P. R., Shaw, G., & Williams, A. M. (1997). Tourist group holiday decision-making and behaviour: the influence of children. *Tourism Management, 18*(5), 287–297. https://doi.org/10.1016/S0261-5177(97)00017-4

Tran, L. T. T., Ly, P. T. M., & Le, L. T. (2019). Hotel choice: A closer look at demographics and online ratings. *International Journal of Hospitality Management, 82*, 13–21. https://doi.org/10.1016/j.ijhm.2019.03.015

Tsiotsou, R., & Vasioti, E. (2006). Using demographics and leisure activities to predict satisfaction with tourism services in Greece. *Journal of Hospitality & Leisure Marketing, 14*(2), 69–82. https://doi.org/10.1300/J150v14n02_05

Tushman, M. L., & O'Reilly, C. A. (1996). Ambidextrous organizations: managing evolutionary and revolutionary change. *California Management Review, 38*(4), 8–29. https://doi.org/10.2307/41165852

Wieczorek-Kosmala, M. (2021). COVID-19 impact on the hospitality industry: Exploratory study of financial-slack-driven risk preparedness. *International Journal of Hospitality Management, 94*, 102799. https://doi.org/10.1016/j.ijhm.2020.102799

Wolf, A., & Zhang, L. (2016). The effect of customization and gender on customers' attitude. *International Journal of Hospitality Management, 56*, 28–32. https://doi.org/10.1016/j.ijhm.2016.04.001

World Health Organization. (2020). WHO Director-General's opening remarks at the media briefing on COVID-19. Retrieved from https://www.who.int/director-general/speeches/detail/who-director-general-s-opening-remarks-at-the-media-briefing-on-covid-19%2D%2D-11-march-2020

The manufacturer's authorised representative in the EU is Springer Nature Customer Service Centre GmbH, Europaplatz 3, 69115 Heidelberg, Germany. If you have any concerns regarding our products, please contact ProductSafety@springernature.com

Printed and bound by CPI Group (UK) Ltd, Croydon, CR0 4YY
30/09/2026
02225706-0005